Science & Technology of COMPUTER MUSIC

Victor A. Stanionis

Iona College

Kendall Hunt
publishing company
4050 Westmark Drive • P O Box 1840 • Dubuque IA 52004-1840

ISBN 978-0-7575-6228-0

Printed in the United States of America

Brief Contents

Contents

Preface

Many scientists over the centuries have been fascinated with and have studied the periodic relationships inherent in the musical scale. From the time of Pythagoras, research in the field of acoustics by scientists such as Kepler, Descartes, Galileo, Newton, Euler, Bernoulli and Helmholtz testify to this fact. Using the mathematical methods developed by Fourier, an understanding of how sounds are generated and differ from one another became possible. With such knowledge, devices that synthesize and simulate sounds produced by musical instruments became a reality.

Computer music is an application of science, computer technology and electrical technology in which music, historically produced by acoustic wind, string, and percussion instruments, is synthesized electronically, orchestrated using computer software, and played through an audio reproduction system. It allows one to simulate the sounds and duplicate the orchestration of past musical compositions and performances. In addition, new orchestrations are simply accomplished using computer software and the creation of new instrumental sounds and structures becomes possible. A whole new art form for expressing musical creativity and imagination becomes possible.

To introduce people to the art form, its scientific and technological underpinnings, and its impact on society and consumers, we have created a course called Computer Music. Computer Music is a course that involves a "hands-on" approach so as to increase appreciation, knowledge, and skills in music and, at the same time, develop an understanding of science, technology and elementary quantitative methods that are intrinsic to computer music. Consumer concerns are also addressed. Elementary quantitative methods including algebra, trigonometry, and graphical analysis are treated as needed. Binary notation and coding are introduced when computer and MIDI messages are discussed. Two hour supervised laboratory or activity sessions each week, coupled with additional lab availability, provide the "hands-on" experience for student fulfillment, development and mastery of skills, reinforcement of class lectures and fun. Appendix A contains a list of laboratory activities used in the course along with a short summary of each one.

The key to successfully developing scientific and technological literacy in this course is the integration of the science topics on a "need to know" basis within the course. We have developed a strategy around the MIDI (Musical Instrument Digital Interface) technology. First a simple MIDI application is considered in order to motivate the students to develop the requisite scientific understanding and build further skills which allow for more complicated MIDI applications.

Interspersed with the applications of the MIDI software, fundamentals of science dealing with waves, sound, electricity, amplifiers, receivers, electromagnetism, speakers, recording media, computers, MIDI interface, synthesizers, coding and the associated quantitative methods are developed in an integrated fashion.

This textbook evolved from a monograph entitled *Computer Music: Science and Technology of a New Art*, co-written by the author as part of the monograph series of the New Liberal Arts Program (1980-1992), a project of the Alfred P. Sloan Foundation. The NLA Program had the goal of assisting in the introduction of quantitative reasoning and concepts of modern technology within liberal education. The Program was based on the conviction that college graduates should have been introduced to both areas if they are to live in the social mainstream and participate in the resolution of policy issues.

Acknowledgments

THe author is most grateful for the persistence and encouragement received from Sue Saad, associate editor at Kendall-Hunt and also Shauna O'Garro, Colleen Zelinsky, and Kimberly Terry, editors at Kendall-Hunt for their guidance and suggestions..

The author would like to thank the publisher Kendall-Hunt for obtaining from the authors of *The Science of Hi-Fidelity*, Kenneth Johnson, Willard Walker and John Cutnell, and from the author of *Physics of Hi-Fi,* Orest Symko. permission to use the many line drawings and diagrams from their texts. Also the author thanks the Research Foundation of the State University of New York at Stony Brook for permission to use material from the monograph *Computer Music: Science and Technology of a New Art* written by Victor A. Stanionis and Hugh Berberich. Carmine Bonanno and Fred Romano of Voyetra Turtle Beach Inc. are also to be thanked for allowing use of illustrations from their sequencer software *Record Producer Deluxe.*

Special thanks are given to Hugh Berberich whose lecture and demonstrations on synthesizers inspired the initiation and design of the course on Computer Music and for his co-authorship of the monograph, *Computer Music: Science and Technology of a New Art.* We are also grateful for his assistance and advice in the early development of the course, Computer Music.

Victor A. Stanionis, Ph.D.
Professor of Physics

CHAPTER 1

The MIDI System

"Would those of you in the cheaper seats clap your hands. All the rest of you, if you'll just rattle your jewelry."

John Lennon, Royal Variety Performance, 1963

English singer and songwriter of the "Beatles"

A system is a collection of parts that work together to achieve a common goal. A typical MIDI music system consists of a computer, a computer interface (MIDI), a synthesizer, a musical input device (such as a keyboard), and a hi-fidelity audio reproduction system. This constitutes the anatomy of the MIDI Music System.

Goals

After studying this chapter, you should be able to:

- Identify the components of a MIDI music system.
- Define the parts of a computer system.
- Distinguish between hardware and software in a computer system.
- List some of the main hardware components of a computer system and their functions.
- Distinguish between systems software and applications software.
- Define MIDI and what it does.
- Define a synthesizer and describe what it does.
- Name the building blocks of the audio reproduction system and describe their individual functions.

The MIDI Computer Music System

A MIDI system, as illustrated in Figure 1.1, can replicate the sounds of acoustical instruments such as violins, flutes, pianos, or drums, generate a host of new sounds, and play them simultaneously as would a band or orchestra at a concert. It is analogous to having a genie locked in your computer who will produce a band or orchestra to play your music exactly as you want it played and at your bidding.

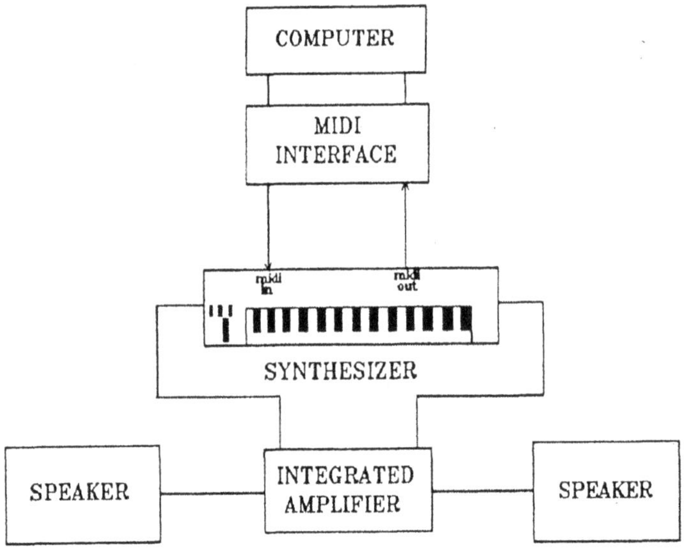

Figure 1.1 MIDI system

From *Computer Music: Science and Technology of a New Art*, Stanionis, Berberich. Copyright © 1989 by Research Foundation of State University of New York. Reprinted by permission.

What Is a MIDI Music System?

With an internally or externally connected Musical Instrument Digital Interface (MIDI) and the appropriate software, a computer provides the means for recording existing pieces, such as compositions by Bach, entered directly from the computer keyboard, by using a mouse, a MIDI equipped piano keyboard or a MIDI controller. The computer also allows one to edit all the messages recorded so that incorrect notes may be corrected and rhythms adjusted. In laboratory activities we shall focus on three methods for entering musical information into the computer: using the mouse, step-entry with a piano keyboard, and real-time entry using a piano keyboard.

Using the mouse allows us to enter each note of the musical piece with the correct duration and pitch. With step-entry, we use a piano keyboard to enter each correct note one at a time and adjust its duration using the software. In real-time entry, we set the software for recording and play the piece using the piano keyboard directly into the computer. A musical piece can be recorded by playing it very slowly and then using the computer to replay it at normal speed, thus compensating for any playing deficiencies one may have.

The computer can be used to lay down separate tracks one at a time (e.g. bass, tenor, alto, etc.) for a complicated piece. The same computer can then be called on to play them all back simultaneously through the MIDI system. Once the piece has been edited, it can be printed on paper, in whole or in terms of individual parts. Sound samples, collected from outside sources on a tape recorder, can be transferred, digitized, edited and stored in the computer library and called upon when needed.

In this book we shall be studying how complex forms of sound waves can be analyzed by feeding the output from sampling devices or tape recordings into a computer equipped with a software program that allows for the Fourier decomposition of the sound into its component frequencies. To enter a sound into a sampling synthesizer, or any electronic sound processing system, the sound waves must be converted from a series of pressure waves into a fluctuating voltage by a device such as a microphone. The fluctuating voltage is then converted to a digital signal by an analog-to-digital converter and stored in digital form. The sampler can store the waveforms in a table in memory. The sampling synthesizer can use the waveform table to create different sounds and notes.

Using the new sound, scales and chords can be played through the synthesizer giving the piece a unique quality. In order to be heard, the signal from the synthesizer is converted to a fluctuating voltage by a digital-to-analog converter and played through a conventional hi-fidelity system. Some older synthesizers use analog techniques for creating sound, using electrical circuits, while digital synthesizers, using mathematical equations, produce sound from electrical circuits built into the synthesizer. In theory, an entire rock concert could be generated by one person with a computer and the appropriate MIDI hardware and software. Many television productions use MIDI systems to generate their sound tracks.

The Computer

The computer in the MIDI music system acts as a master controller. Using the appropriate software the computer can record music on separate tracks, and you can assign a different instrument to each track. During playback the computer acts like a multi-track recording studio, which can playback a tenor track, bass track, alto line and a soprano line all at once. In effect you have an orchestra of instruments, each with their own part. The computer, which records events (not actual sounds), such as keystrokes, notes and control settings and stores this information in memory, allows you to play back all the MIDI messages later by sending them to the synthesizer.

Using a computer allows you to change or edit wrong notes, remove them at will, and adjust rhythms and timing, as well as the speed of playback, without changing the pitch of the music being played. Most software will allow you to print the musical score in standard musical notation.

A brief review of the terminology used in computing might be useful. The computer system can be divided into two subgroups: hardware and software. When you look at a computer, what do you see? In profile, we notice the hardware evidently consisting of a display unit or monitor, an enhanced keyboard, mouse, and a rectangular box that sits on the desktop or floor. The box which has many wires connecting it to the other peripherals is called the system unit (Fig. 1.2).

Figure 1.2 Computer system

It is the system unit that contains the motherboard with CPU (central processing unit) whose function is to enact and process arithmetic and logic (ALU) functions and exercise control functions. We find here volatile memory units, random access memory (RAM) chips, and read only memory chips (ROM), which retain their information even when the power is turned off. The system unit also contains all sorts of circuit cards that allow the computer to generate sound and video, and connect to networks through a NIC (network interface card), a wireless communication card, or telephone modem.

Today's computers contain all sorts of mass memory storage devices such as floppy disk drives, CD-ROM drives that can read and write, DVD players and writers, and hard disk drives. In addition, the system unit also provides ports for connecting peripheral devices. By adding a MIDI connector to the game port, which is usually part of the sound card in a ordinary computer, we can provide the MIDI interface which is needed for the computer music system. We can also use a USB port. Usually there is a parallel port for the printer connection. USB ports allow for the connection of a host of input and output devices such as: MIDI devices, external digital video and still cameras, DVD readers and writers, and external hard disc drives. "FireWire" or IEEE 1394 ports also provide a similar accessibility.

The software that is used on the computer is usually divided into systems software and applications software. The systems software provides the interface between the user and the applications software the user wants to run on the computer. For example, the Windows operating system controls the functions of the computer. It permits you to run a word-processing program. It also allows you to run sequencer software, so that you are able to record a musical piece by several different methods. It allows you to edit the recorded information, print it in musical notation, and control its playback. The software provides the capability for orchestration and enhancement with effects.

MIDI Interface

Central to the entire modern computer music system is the exploitation of the MIDI interface. It provides both the novice and experienced player opportunities in expanding the sound capabilities of instruments and enhancing their own playing abilities.

What Is MIDI?

MIDI is an acronym for Musical Instrument Digital Interface, a standard established by manufacturers of electronic musical instruments that provides for compatibility among instruments of different manufacturers. It is characterized by a software and hardware set of specifications. In particular, the standard specifies the format of information transmitted through it, the type of physical connection, MIDI port, that is built into the instrument, and the cable used to connect ports. It also specifies the voltage and transmission rate of the signals sent over the cables so compatibility will exist between instruments. Instruments that can receive and send MIDI codes are called MIDI devices. MIDI information is digitally coded in binary form. Any musical instrument can be a MIDI device provided it satisfies the MIDI specifications. However, the common characteristic of all MIDI devices is the presence of a microprocessor which is used to read, control, and send signals.

Electronic instruments that do not have microprocessors, like the electric guitar, can be equipped with microprocessors so that MIDI messages can be processed. In addition, light displays synchronized to musical instruments and sound amplifiers can be equipped with MIDI interfaces and controlled. Computers are ideal MIDI devices provided they are equipped with the appropriate MIDI hardware and software.

How Is It Used?

The MIDI device such as a MIDI keyboard, can transmit messages indicating when a key on a keyboard is first pressed, when it is released, and the nuances of the note played on the keyboard. Its messages can describe the pitch and loudness of the note and whether any of the electronic controls on an electronic instrument are being used. Prerecorded scores or songs can be played together from several MIDI devices in synchronization. It allows the insertion of samples that have been previously stored to be used when desired. By connecting several MIDI musical instruments in series or through multiple ports on a computer, the quality or fullness of the sound may be enhanced as each instrument contributes to the signal. A computer equipped with a MIDI interface and appropriate sequencer software allows a musical composition to be recorded in digital form, allowing for replay at a later time to recreate the original.

Most computers do not come with a built-in MIDI port. In the past, a serial port could be converted to a MIDI port with the addition of a MIDI interface. The MIDI interface usually has a single plug that goes into the serial port on the back of the computer. In addition, it will have a "MIDI In" port and a "MIDI Out" port and sometimes a MIDI Thru port (Fig. 1.3). The interface converts the signal from the computer to the correct voltage to send to devices connected to the MIDI interface and will also convert signals or voltages coming from various MIDI devices so that the computer recognizes them. MIDI events or messages are sent in serial form (i.e. one after the other in binary coded form) over the MIDI cables. Today, it is possible to connect a PC adapter cable to a Windows-compatible sound card's joystick port, converting it to a MIDI In/Out port. Alternatively, a USB MIDI adapter cable can be connected to the computer's USB port to convert it to a MIDI In/Out port.

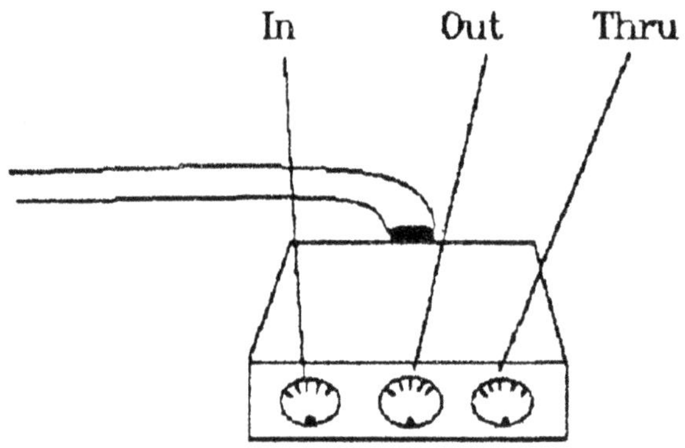

Figure 1.3 MIDI interface

From *Computer Music: Science and Technology of a New Art*, Stanionis, Berberich. Copyright © 1989 by Research Foundation of State University of New York. Reprinted by permission.

When the MIDI standard was set by the musical instrument manufacturers, they chose serial transmission primarily to avoid electrical interference generated by parallel transmission. The MIDI cable was specified to use a five-pin plug called the DIN plug, a standard connector in common use. It consists of five pins (Fig. 1.4).

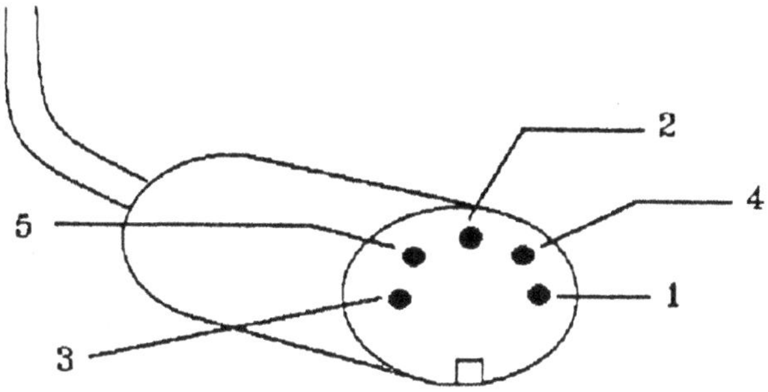

Figure 1.4 DIN plug
From *Computer Music: Science and Technology of a New Art*, Stanionis, Berberich. Copyright © 1989 by Research Foundation of State University of New York. Reprinted by permission.

Pins one and three are not used. Pins four and five transmit the MIDI signal. Pin 2 is connected to a wire that is wrapped around the others which shields them from electrical interference and is connected to ground.

Because DIN connectors have been found to have durability problems associated with their fragile pins, the three conductor XLR connectors with stronger pins (Fig. 1.5) may be substituted.

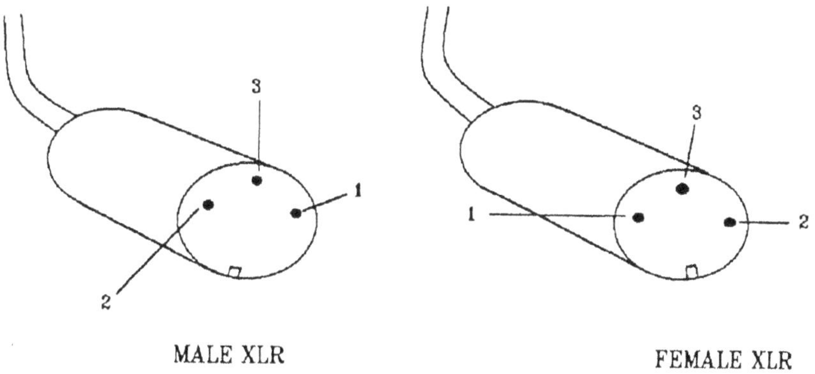

Figure 1.5 XLR connector
From *Computer Music: Science and Technology of a New Art*, Stanionis, Berberich. Copyright © 1989 by Research Foundation of State University of New York. Reprinted by permission.

Synthesizers

Electronic synthesizers which create musical tones electronically have been around since the early part of this century. They were rarely used outside of university and industrial laboratories. "Switched on Bach", recorded by Walter Carlos (now Wendy Carlos) in 1968 and played on the MOOG synthesizer, was the first time that the general public became acquainted with electronic music. CDs of "Switched on Bach" are still available and worth listening to in order to get some idea of what excited the general public at the time. The synthesizer enabled rock bands to generate their own "sound." You will find more about the development and history of the synthesizer in the next chapter on "The History of Computer Music."

Musicians in subsequent years tried to incorporate the sounds of the synthesizer in their own concerts. Two major drawbacks became apparent. The synthesizers produced a rather simple sound compared to the complex sounds produced by traditional musical instruments. It was "thin", meaning it lacked the many overtones or harmonics necessary to generate a complex "fat" sound. The second deficiency was the fact that the synthesizers could play only one note at a time as opposed to many notes at once, which is called a chord. To solve these problems two synthesizers could be played at once or connected electronically. Incompatibility between synthesizers of different manufacturers made such complicated hook-ups almost impossible. One had to be an electrical engineer to wire up several synthesizers to be played in unison.

In 1982, at the June show of the National Association of Music Merchants (NAMM), the MIDI standard or set of specifications was agreed upon by the manufacturers and formalized in 1983 as MIDI 1.0. This allowed for compatibility among instruments from different companies. Synthesizers were among the instruments most affected.

A synthesizer is essentially a system for producing an audio signal which will generate sound after passing through a sound reproduction system. It ordinarily consists of a set of sound generators, or oscillators, a keyboard, control panel, microprocessor, memory, auxiliary controllers, and a MIDI interface (Fig. 1.6).

Additive synthesis

Subtractive synthesis

Figure 1.6 Synthesizer

From *Science of Hi-Fidelity* by Johnson, Walker, Cutnell. Reprinted by permission of the authors

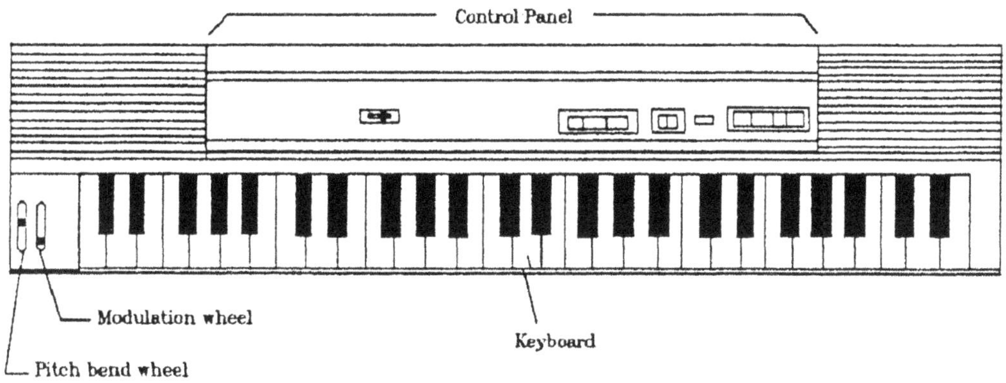

Fig. 1.7 Synthesizer keyboard

From *Computer Music: Science and Technology of a New Art*, Stanionis, Berberich. Copyright © 1989 by Research Foundation of State University of New York. Reprinted by permission.

A synthesizer keyboard (Fig. 1.7) is different from a standard piano keyboard which uses a felt-lined hammer to strike several strings tuned to the same pitch. The MIDI keyboard usually has fewer keys than a piano keyboard but can be transposed over any octave range so as to be able to play notes from 0 to 127 as illustrated in Figure 1.8.

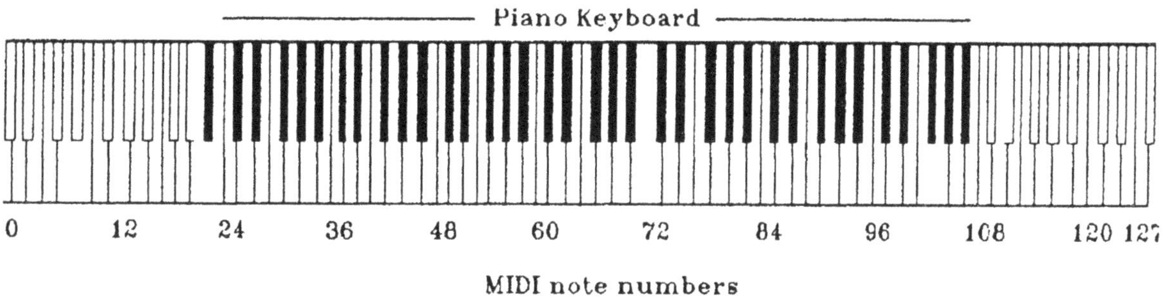

Figure 1.8 Piano keyboard and MIDI range

From *Computer Music: Science and Technology of a New Art*, Stanionis, Berberich. Copyright © 1989 by Research Foundation of State University of New York. Reprinted by permission.

The synthesizer keyboard uses electronics to modify the sound produced by sound generators. A keyboard player can impact the quality of the music by the way in which he/she strikes the keys. The nuances of how the key is pressed and released are detected by electronic sensors monitoring the key. Some less expensive keyboards detect only note-on and note-off. Velocity sensitive keyboards are capable of detecting how fast the key moves and converts this motion into a voltage that modifies the sound coming from the sound generators.

MIDI System Limitations

At the time the MIDI standard was established certain compromises were made between cost and performance. As a result there are inherent limitations in what can be done using a MIDI system. Computer and audio reproduction systems also have limitations resulting from compromises and technology limits. Despite these limitations, the MIDI based computer music system is astonishingly versatile and competent.

MIDI messages are transmitted in a **serial mode**, one after another. This means that a C major chord made up of the three notes C-E-G played simultaneously would be sent in serial fashion as G-E-C over the MIDI cable. Since the transmission rate for MIDI information is high, the delay time between notes would be unnoticeable and it would sound as if all three were played at the same time. However, when a large amount of data is sent over the MIDI line, time delays may be observed. Rhythm or drum tracks out of synchronization would be especially noticeable.

Timing problems may also occur when many synthesizers and other MIDI devices are networked. Too much data passing through the MIDI output can lead to timing problems resulting in the lack of synchronization of instruments. A few MIDI systems use USB ports that operate at 12,000,000 bps and don't exhibit the delays associated with serial port transmission.

"MIDI choke" can occur when an instrument's buffer is filled and overflows. Any additional information arriving at the MIDI port is then ignored and lost. There is also a phenomenon known as **"computer choke"** which occurs when the computer's microprocessor can not handle the information it receives fast enough, resulting in lost information.

Interestingly, not all MIDI devices understand all **MIDI messages**. There are messages that apply to synchronization of MIDI recording and playback equipment. If an instrument does not have those functions, it may ignore messages coming from such devices.

The MIDI standard also limits the **number of channels** that are available to a maximum of sixteen different musical lines per output port. If more than sixteen are needed, problems arise. The MIDI synthesizer is limited by the number of notes it can sound simultaneously. The early MIDI synthesizers were eight or sixteen note polyphonic. Today we have synthesizers which provide a polyphony of thirty-two, sixty-four or even one hundred and twenty-eight notes.

Another inherent limitation is that most MIDI equipped instruments are set for the twelve note **Western scale**. For musical scales from other cultures, the MIDI instruments must be designed to replicate those scales.

The quality of sound reproduction depends directly on the **quality of the synthesizer** and its capability to reproduce accurately the sound of acoustical instruments or generate new sounds. It also depends on the **quality of the sound reproduction system itself**.

Another limitation has been the problem of incorporating into a MIDI performance the **interpretation** that a real musician gives to music in a live performance with acoustical instruments. The creative performance details of a live musician are hard to replicate and the MIDI performance may sound machine-like, mechanical, and unnatural.

General MIDI

After musicians started to use the MIDI systems specified by the MIDI 1.0 standard, problems arose. The MIDI 1.0 specification did not require that particular instrumental sounds be assigned to particular patch numbers. Manufacturers arbitrarily assigned instrumental sounds to different patch numbers. So when using one synthesizer keyboard to control another, the request that a flute patch play on one synthesizer, could be the patch number on the second synthesizer for a tuba.

In late 1991, the MIDI Manufacturers Association adopted General MIDI (GM1) System Level 1 which added some new requirements to the MIDI 1.0 specification, which is still the basic standard for all MIDI devices. Generally speaking, General MIDI increased the compatibility of devices of different manufacturers and MIDI composers. A standard was set for the 128 patches which must appear in a specific order, and this standard is called General MIDI (GM1). Sound modules and sound cards now adhere to this standard as they all assign the same instrument to a standard patch number. For example, all General MIDI devices would have an acoustic grand piano assigned to patch number 1.

Another requirement for a General MIDI device is that they all be multitimbral, that is be able to play MIDI messages on all 16 channels, with a different GM patch or instrumental sound for each channel. In addition, piano key numbers were standardized. All the keys on a piano keyboard are assigned a number but different manufacturers used different numbers for different keys. On all General MIDI devices note number 60 was assigned to middle C and the other piano keys were assigned numbers accordingly. Drum machines were assigned to channel 10. As far as polyphony was concerned, the minimum number of notes a module could play simultaneously had to be 24 or greater. The GM Standard also set other minimum requirements for a General MIDI device, such as being able to respond to pitch and modulation wheels.

In 1999, a number of additions were made to General MIDI 1 creating General MIDI 2 which increased the number of sounds available to be played and amount of control for editing and musical performance. It also required all General MIDI 2 devices to be compatible with the General MIDI 1 standard.

Audio Reproduction System

To hear the sounds created by a MIDI music system, a hi-fidelity audio system is essential. After all, the sounds you hear emanate from speakers powered by an electrical amplification system. The faithful generation of sound from electrical impulses requires a high quality audio system if the sounds are to simulate their real musical instrument counterparts.

An audio system is also a system made up of parts working together to achieve some common objective. It has four major parts: a signal or sound source, a preamplifier, a power amplifier, and speakers, as shown in Figure 1.9.

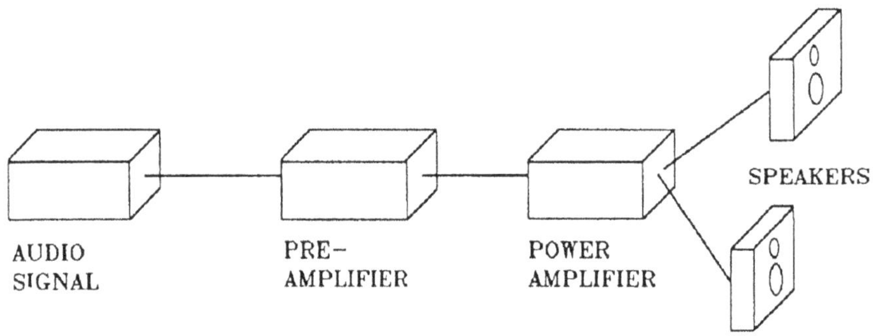

Figure 1.9 Hi-fidelity audio system

From *Computer Music: Science and Technology of a New Art*, Stanionis, Berberich. Copyright © 1989 by Research Foundation of State University of New York. Reprinted by permission.

The signal or sound source can be the output from a MIDI synthesizer, or more conventionally, a turntable, CD player, tuner, tape deck, television, VCR, or DVD system. The preamplifier, or pre-amp for short, can be thought of as the control unit for the system where volume, bass, treble, and balance can be adjusted. The amplifier produces the equivalent of a photographic enlargement of a photograph by increasing the size of the electrical signals so that they will be large enough to drive the speakers, which will produce sound.

Just as the piano revolutionized music in the first half of the nineteenth-century, the computer is doing so today. The computer, through digital sound processing. represents sounds by means of numbers which can then be manipulated. The composer can now "create" his own instruments with a microcomputer and have it play the score he has composed. New musical structures are now possible, as well as performance capabilities that would have been impossible earlier because of human physical limitations.

In this book, we are going to study the computer music system by focusing on each part, starting first with the audio reproduction system and the underlying scientific principles and technology associated with it. This will allows us to then move on to an analysis of the synthesizer and the MIDI interface. It is assumed that you already have a good working knowledge of the operation of the computer hardware and software.

Review Questions

1. List the four major components of the MIDI computer music system.
2. Describe the role played by each of the components in the MIDI computer music system.
3. List the main hardware components in a computer system and describe their functions.
4. What is systems software? What is applications software?
5. Define MIDI.
6. What does the MIDI protocol specify?
7. What are some of the limitations associated with the MIDI standard?
8. What is General MIDI level 1?
9. What is General MIDI level 2?
10. Name the building blocks of a hi-fidelity audio system and describe the function of each.

Multiple Choice Questions

Choose the best answer.

1. MIDI is an acronym which stands for ______.
a. musical interface for digital instruments
b. musical instrument digital interface
c. Moog instrument digitizing interface
d. monitoring internal data instrument

2. The computer in a MIDI music system acts as a master ____.
a. controller
b. source of sound
c. synthesizer
d. generator

3. The computer in a MIDI system records _____.

a. sounds
b. instruments
c. events
d. sequels

4. The computer system can be divided into two subgroups: _____.
a. CPU and monitor
b. program and system unit
c. hardware and software
d. input and output

5. The box that contains the motherboard, disc drives, memory, etc., is called the _____.
a. computer
b. CPU
c. control unit
d. system unit

6. Software used by the computer is usually classified as ______ software.
a. utilities and operating
b. word-processing, spreadsheets, database
c. arithmetic and logic
d. system and applications

7. MIDI is a software and hardware set of _____.
a. specifications
b. interfaces
c. digitized instruments
d. digitized code

8. In a MIDI DIN connector there are ____ pins.
a. 2
b. 3
c. 4
d. 5

9. In a DIN connector, pins _____ are not used.
a. 1 and 2
b. 4 and 5
c. 1 and 3
d. 2, 4, and 5

10. Because the pins in a DIN connector are fragile, they may be replaced with an XLR connector which has _____ pins.
a. 2
b. 3
c. 4
d. 5

11. Synthesizers create _____ electronically.
a. audio signals
b. songs
c. incompatibility
d. MIDI choke

12. One limitation of the MIDI standard is called _____.
a. MIDI Moog
b. MIDI parallel mode
c. MIDI choke
d. MIDI compression

13. The four basic building blocks in an audio reproduction system are the sound sources, _____, power amplifier, and speakers.
a. MIDI
b. preamplifier
c. synthesizer
d. sonic interface

14. Systems software in a computer system includes programs such as _____.
a. word processing
b. database
c. spreadsheets
d. utilities

15. Entering music into the computer using _____ allows us to enter each note with the correct pitch and duration one note at a time using a piano keyboard.
a. step entry
b. timed entry
c. computer DIN keys
d. the MIDI

16. The MOOG was a(n) _____.
a. 1950s rock band
b. analog synthesizer
c. organ used by Bach
d. the precursor of "mosh"

17. NAMM was _____.
a. an analog synthesizer
b. a MIDI standard
c. a professional association
d. a type of jam session

18. Theoretically a piano keyboard can have _____ keys.
a. 61 keys
b. 88 keys
c. 96 keys
d. 128 keys

19. When the buffer in an instrument is filled and overflows, this is called _____.
a. overflow
b. MIDI choke
c. computer choke
d. Heimlich choking

20. MIDI messages are transmitted in _____.
a. serial mode
b. parallel mode
c. simultaneous mode
d. synchronized mode

21. What does a patch number designate?
a. error correction code
b. polyphonic sequence
c. short recorded piece
d. instrumental sound

22. In General MIDI 1 the _____ numbers were standardized.
a. patch
b. channel
c. song
d. XLR

23. A device is multitimbral if it can play ______.
a. many instrumental sounds at once
b. many notes at once
c. a computer MIDI program
d. several synthesizers at once

24. The volume, bass, and treble controls are found in the _____.
a. power amplifier
b. preamplifier
c. sound source
d. ALU

25. The audio component that acts like a photographic enlarger is the ______.
a. power amplifier
b. preamplifier
c. sound source
d. ALU

26. A _____ is made up of parts that work together to achieve a common goal.
a. system
b. group
c. unit
d. transistor

27. General MIDI 1 assigned a specific _____ to each patch.
a. tone
b. channel
c. number
d. mode

28. General MIDI 2 was issued in _____.
a. 1996
b. 1997
c. 1998
d. 1999

29. A typical signal or sound source in a hi-fi system would be the _____.
a. preamplifier
b. power amplifier
c. speaker
d. tuner

30. The main accomplishment of the MIDI 1.0 standard was that it_____.
a. standardized connectors
b. standardized voltage transmission rates
c. standardized patch numbers
d. provided for compatibility among manufacturers of synthesizers

Bibliography and Web Resources

Anderson, Craig. *MIDI for Musicians*, Amsco Publications, New York, NY, 1986.

Baird, Joel, editor. "*Understanding MIDI*", *Musician*, special edition, Amordian Press, Inc, Gloucester, MA, 1986.

Bassett, Rick and Jonathan Matykin. "*A MIDI Musical Offering*", *PC Magazine*, V7, No.20, pp. 229-272, November 1988.

Boom, Michael. *Music Through MIDI*. Microsoft Press, Redmond, WA, 1987.

De Furia Steve and Joe Scacciaferro. *The MIDI Resources Book*. Third Earth Publishing Inc., Pompton Lakes, NJ, 1987.

De Furia Steve and Joe Scacciaferro. *The MIDI Book*, Third Earth Publishing Inc., Pompton Lakes, NJ 1988.

Milano, Dominico, editor. *Mind over MIDI*. Hal Leonard Books, Milwaukee, WI, 1987.

Stanionis, Victor and Hugh Berberich. *Computer Music: Science and Technology of a New Art*. Research Foundation of State University of New York, Stony Brook, NY, 1989.

History of MIDI - **http://www.physics.smu.edu/~olness/www/02fall1320/TrevorChapman/MIDI%20Web%20Page2/midi/history.htm**

How does MIDI work? - **http://www.physics.smu.edu/~olness/www/02fall1320/TrevorChapman/MIDI%20Web%20Page2/midi/how.htm**

MIDI note number for different octaves - **http://www.harmony-central.com/MIDI/Doc/table2.html**

Short MIDI Overview - **http://www.rbjones.com/rbjpub/music/mus007.htm#what**

General MIDI - **http://www.borg.com/~jglatt/tutr/gm.htm**

General MIDI 2 - **http://www.midi.org/about-midi/gm/gm2_spec.shtml**

Chapter 2

History of Computer Music

"Behold the marvelous art of Archimedes, I allude to the Hydraulic Organ; so many members, so many parts, so many joints, so many sound conduits, so much tonal effect, so many combinations, so many pipes, and all at one touch."

Quintus Septimius Florens Tertullian, 200 AD

Roman theologian and Christian apologist

Goals

After studying this chapter, you should be able to:

- Trace the history of mechanical synthesizers like the organ.
- Outline the development of the computer from ancient times.
- Know the role of punch cards in computing.
- Understand the evolution of magnetic tape in recording media.
- Know the history of disc recording.
- Understand the origins of computer music with its roots in science, computer technology and electrical technology.

What is music? The definitions of music have changed dramatically over time. Webster defines music as "the art of giving structural form and rhythmic pattern to combinations of sounds produced instrumentally or vocally." In this chapter we shall consider the history of the computer music system which has its roots spread out in mechanical and then electronic synthesizers, computer evolution, and recording media.

Mechanical Synthesizers

The Hydraulis

It is perhaps in the ancient hydraulis (Fig. 2.1) that we find the origins of synthesized music. It consisted of one or more sets of metal pan pipes of different sizes. A small cistern, called the pnigeus, was turned upside down and placed inside a barrel of water. A set of pumps forced air into the pnigeus, forming an air reservoir. A large pipe leads from the pipe chest to the top of a dome immersed in water. The water rising in the dome, after being pressed down by the air pumped in, compresses it and supplies a steady flow to the pipes, generating sound.

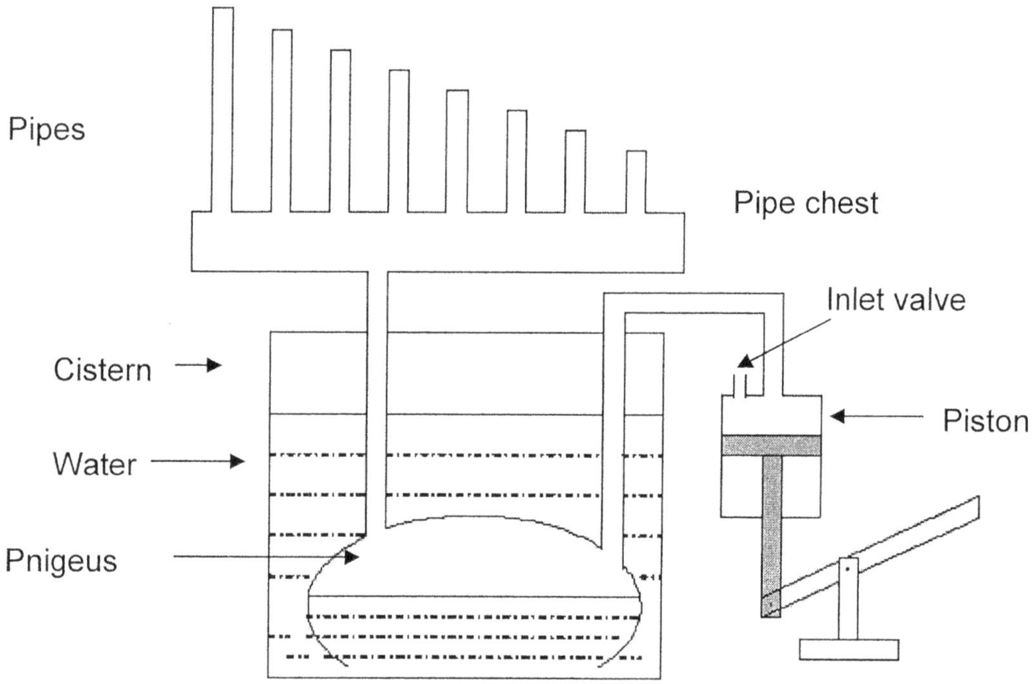

Figure 2.1 Hydraulis Model

Above the pipe chest are three channels through which wind can flow controlled by taps. Three sets of pipes are placed above the three channels sounding the unison, octave and super octave. Air is admitted to the pipes from the appropriate channel by pressing on the keys, A schematic drawing of the operation of the hydraulis is shown in Figure 2.1

The hydraulis was probably the first keyboard instrument and the ancestor of the organ. It is believed that the hydraulis was invented in the third-century B.C. by the engineer Ctesibius in Alexandria. It was used in Rome in theaters and for all sorts of festivals, banquets, games and circuses. It is said that Nero played the hydraulis as Rome burned. Remember the violin hadn't been invented yet. In the third-century A.D., the water mechanism was replaced by bellows.

It disappeared in the West with the collapse of the Roman Empire, but it is said that Constantine II sent an organ, descended from the hydraulis as a gift in 257 A.D. to Pepin the Short, King of the Franks and father of Charlemagne. It was copied and spread throughout Europe. It was used widely both in the Western and Eastern Empires, making its way into churches around 900 A. D. By the thirteenth-century it was accepted by the Catholic Church and used widely in the churches.

Up until about 1500 A. D., each organ pipe could sound only one tone. It was then discovered that stops could be used which allowed separate groups of pipes to sound alone. Within ten to twenty years the organ developed all the sound and effects we associate with it today, mimicking the sound of various instruments and becoming the first synthesizer. Mozart and Beethoven were court organists along with Handel, Mendelssohn, Liszt, Franck, and Dvorak, amongst others.

Parallel Development of the Computer

Most important in the evolution of computer music was the invention of the electronic computer itself. Its history begins with the abacus (Fig. 2.3), which has its origin in about 500 B.C. Abacus is a Latin word which has its root in the Greek word "abax" or "abacon", tablet or table. The abacus in ancient times was essentially a mechanical aid for counting that kept track of sums and carries for the person doing the calculations mentally.

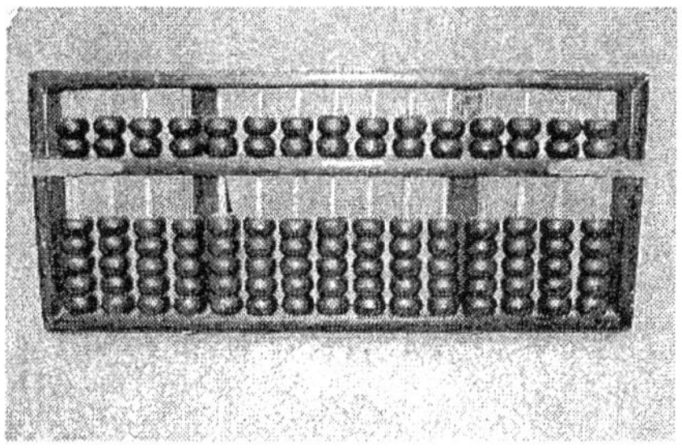

Figure 2.3 Abacus

Each bead in the upper section has a value of **5** and each bead in the lower section has a value of **1**. Beads are considered counted, when moved towards the beam that separates the two decks.

The right-most column is the ones column; the next adjacent to the left is the tens column; the next adjacent to the left is the hundreds column, and so on. As arithmetic, which consisted of counting numbers that have been written, gained in popularity in the Middle Ages the abacus fell into disuse in Europe. It can be thought as a precursor of the computers we have today. A calculating machine called the the Pascaline, invented by Blaise Pascal in 1641, was similar to a design of Leonado DaVinci's. It was also a precursor of the modern computer and the digital synthesizer.

One of the problems scientists and mathematicians faced in the 1800s were the mistakes mathematicians made in producing mathematical tables and the errors printers made in printing mathematical tables, such as logarithmic and trigonometric tables. The mathematicians who produced the tables were called "computers". In 1823, Charles Babbage received a British government grant to build a machine called the Difference Engine, that could compute mathematical tables of values and print them. However, he ran into difficulties and before he could complete it he came up with a better idea, the Analytical Engine, which was inspired by Jacquard's loom. This machine was capable of doing any calculation just as Jacquard's loom could weave any pattern.

Babbage's assistant, Augusta Ada Byron, daughter of the English poet Lord Byron, showed how the machine could be programmed using punch cards to perform calculations. She is now acknowledged as the first computer programmer and in 1979 a modern computer language was named ADA in her honor. Babbage applied to the British government for financial support for his project. However, the British government was not impressed, since Babbage had not yet completed building the Difference Engine. The Analytical Engine led to the development of the modern computer.

In his book, **Engines of the Mind**, Joel Shurkin stated:

> *"One of Babbage's most serious flaws was his inability to stop tinkering. No sooner would he send a drawing to the machine shop than he would find a better way to perform the task and would order work stopped until he had finished pursuing the new line. By and large this flaw kept Babbage from ever finishing anything."*[1]

A hundred years passed before someone built a machine like Babbage's Analytical Engine. In 1944, Howard Aiken completed the Harvard Mark I Automatic Sequence Controlled Calculator at Harvard University. In many ways it was the embodiment of Babbage's design. The Mark I was said to be the closest thing to the Analytical Engine ever built. It used punched paper tape much in the same way Babbage envisioned using punched cards. It used electricity rather than steam for power and for transfering information. As a computer, it was not fast enough, being only five or six times faster than a human using a mechanical calculator.

The ENIAC

J. Presper Eckert and John William Mauchly of the University of Pennsylvania conceived and designed the all electronic ENIAC (Electronic Numerical Integrator and Computer) and finished it in 1946. It was first used in solving atomic energy problems. It had approximately 18,000 vacuum tubes, weighed 30 tons, and consumed 200 kilowatts of electrical power. Some say that the lights in Philadelphia dimmed when it was turned on. It eventually became known as the Univac when acquired by the Remington Rand Corporation. It was improved on and succeeded by such computers as the EDVAC, noted for using binary notation and the stored program concept (when a program is stored in memory in the same coded form as data).

These first generation computers developed during the 1940s and 1950s used bulky, energy consuming and heat generating vacuum tubes for their operation. By the late 1950s, the transistor invented at Bell Labs in 1947 began to replace the vacuum tube. Noted for their small size, low heat generation, and long lives, they powered computers in the second generation through the early sixties. Integrated circuits were introduced in the early 1960s leading to the third generation of computers. The integrated circuits incorporated hundreds of transistors on a single silicon chip. Now thousands upon thousands of transistors can be placed on single chips to power the computers of today. See Figure 2.4.

Vacuum tube Transistor Integrated circuit

Figure 2.4 Vaccum tube, transistor, integrated circuit

One of the first applications of the integrated circuit in a consumer product was the hand-held electronic calculator. Texas Instruments was the first to mass produce silicon transistors in 1954 and invented the integrated circuit -- the micro-chip -- in 1958. The electronic hand-held calculator followed in 1967. Shipments of calculators bearing the TI brand began five years later. In the Texas Instrument SR-10 (Fig. 2.5), the SR stood for slide rule and the 10 for the 10 digit display. It used rechargeable nickel cadmium batteries and had an LED display. It could add, subtract, multiply, divide and take the square root of positive numbers, and in 1972 it sold for $169.

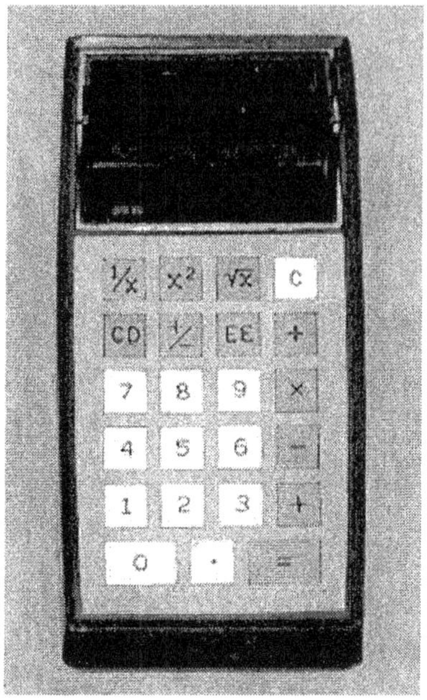

Figure 2.5 Slide-rule calculator

By 1977 the first microcomputers to gain the attention of the general public were put on sale by Apple Computer (the Apple II) and the Commodore Company (the Commodore PET). With the development of microcomputers from the 1980s to the present time, we have the essential tool needed to operate a computer music system as we know it today.

The Dawn of Electricity

Electricity was first used in the design of a musical instrument in 1759. Throughout the history of music made by machines new ways were devised to make musical sounds and play music. The "electric harpsichord", clavecin electrique, was invented by Jean-Baptiste de Laborde in 1759 and built in 1761, by Abbe de Laborde, in Paris. It consisted of a short harpsichord-like keyboard that activated clappers charged with static electricity to ring bells. These creative inventions eventually evolved into the computer music systems we have today.

Keyboard

We see the seed of the electronic keyboard illustrated in the electromechanical piano developed by Hipps who used a keyboard to activate electromagnets which turned on small electric generators to produce sound. In 1887, the player piano was introduced. It used rolls of paper with small holes cut into them to reproduce a performance. The holes were cut by using a recording player piano. People could purchase these rolls and play them at home. much as MIDI files are traded and used today.

Telharmonium

One of the first precursors of the modern synthesizer was the Telharmonium (or dynamophone) invented by Thaddeus Cahill, who patented it in 1897. The Telharmonium was essentially a collection of 145 dynamos that produced alternating currents of different audio frequencies, controlled by a set of polyphonic, velocity-sensitive keyboards. Acoustic horns were used to amplify the sound (twenty years before electronic amplifiers were built). The music generated by two musicians was sent out to telephone receivers, where it could be heard without lifting the receiver.

The Telharmonium weighed about 200 tons and was 60 feet long. The cost was $200,000. The instrument occupied the entire floor of "Telharmonic Hall" on 39th Street and Broadway, in New York City, for 20 years; It wasn't exactly a portable machine. Cahill planned to transmit music to hotels, restaurants, theaters and homes over telephone lines. Its prohibitive cost and the fact it interfered with local telephone calls led to its demise. In 1950 Arthur.T. Cahill, brother of Thaddeus, tried to find a home for the prototype instrument. However, his search was futile and the machine was sold for scrap. No recordings made by the Telharmonium have been found. However, the principles underlying the Telharmonium were used in the Hammond organ designed in the early 1930s.

Theremin

During their experiments with vacuum tubes in the early twentieth century electronic engineers found the principle of beat frequency, or heterodyning oscillators. This occurs when two high frequency sound waves of different frequencies combine to create a third lower frequency that is equal to the difference between them. However, the frequency varied when a human came near the vacuum tubes because of the body's capacitance. Leon Termen used the body's capacitance as a control mechanism for the instrument. He devised a production model in 1920 which was played by moving one's hands near and around a metal loop for volume, and around an antenna for pitch. The sound generated resembled the sound from a violin string. This model was exhibited at the Moscow Industrial Fair of 1920 and Lenin commissioned 600 models of the Theremin (Fig. 2.6) to be built.

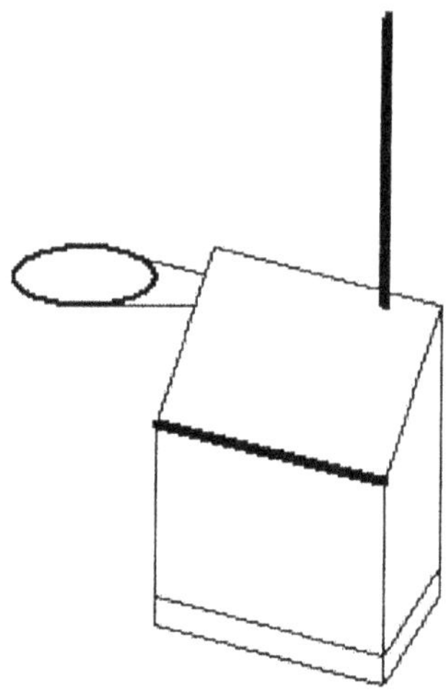

Figure 2.6 Theremin

In 1927 Termen left the Soviet Union for the United States where he was granted a patent for the Theremin in 1928. The Theremin was marketed and distributed in the USA by RCA during the 1930s and continues, in a transistorized form, to be manufactured by Robert Moog's 'Big Briar' company. It also appeared in several pop records of the 1960's (the Beach Boys "Good Vibrations") and was used for effect rather than as an instrument. Bands such as Led Zeppelin brought the Theremin into the public eye for a short time. The sound of the Theremin was used in several movie soundtracks during the 1950s and 1960s.

Hammond Organ

In April 1935 Laurens Hammond, an ex-watchmaker, built the Hammond Organ His design was closely related to Cahill's Telharmonium, but it was much smaller. It generated sounds in the same way. In the Hammond organ additive synthesis was used (as in the Telharmonium) and the fundamental and up to eight harmonics were available controlled by means of drawbars and preset keys. It became popular with all sorts of Jazz, Blues, and Rock musicians as well as with the general public for home use.

Jacquard's Loom

The history of computer music is intimately related to the development of the digital computer itself. In 1803, a French silk weaver, Joseph-Marie Jacquard, revolutionized patterned textile weaving by controlling the weaving of threads on a silk loom by using patterns of holes punched into a set of cards. Fancy designs could be woven automatically by one man and a machine. Charles Babbage and Augusta Lovelace used this technique in devising a program for Babbage's Analytical Engine. (Rumor has it that the hand weavers of the day put out a "contract" on Jacquard.)

In 1890, the U.S Census began the job of counting the 62 million people estimated to be living in the U.S. Based on the usual technique of placing tally marks on a roll of paper next to each name and then adding them by hand for each column, it was projected that the census would not be completed in the mandated ten year period. The Census Bureau in the 1880s invited inventors to submit their designs for counting the population. Herman Hollerith, a lecturer at MIT, responded by using Jacquard's punched cards along with an electronic sorting machine to not only count the number of people in the U.S., but also to sort them by occupation, state and other variables.

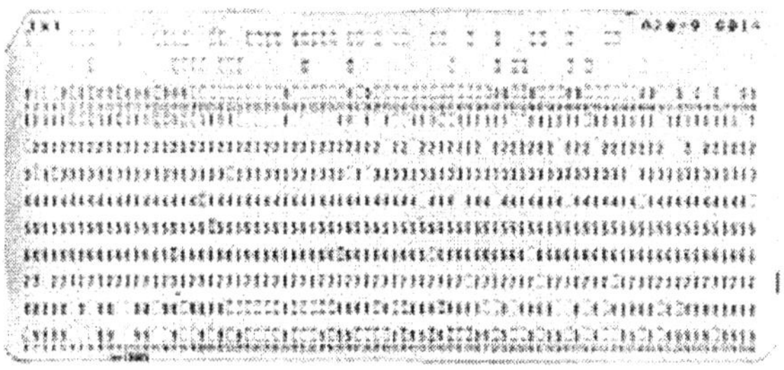

Figure 2.7 IBM 80 column punched card

Herman Hollerith is thought of as being the father of modern automatic computation. The punched card (Fig. 2.7) was the basis for storing and processing information. He was able to build the first punched-card tabulating and sorting machines as well as the first key punch. His machines were used for censuses in Russia, Austria, Canada, France, Norway, and several other countries. In 1911 Hollerith's company merged with two others to form the Computing-Tabulating-Recording Company (CTR) and, after being bought by James Watson, Sr. in 1913, changed its name to the International Business Machines Corporation (IBM) in 1924.

Computers and Electronics

In 1877, Thomas Edison developed one of his first great inventions. He noticed that the tape of his telegraph transmitter emitted a noise similar to spoken human words. This led him to develop the tin foil phonograph. He made the first recording of a human voice ("Mary had a little lamb.") on the first tinfoil cylinder phonograph on December 6, 1877.

Joseph Begun graduated in 1929 from the Institute of Technology in Berlin, Germany. It was there that he wrote his book entitled *Magnetic Recording*. In 1934, he built the world's first tape recorder used for broadcasting and he later created the first consumer tape recorder called the Sound Mirror.

Hi-fi or (hi-fidelity) recording had its beginnings in Germany in 1941 when German Radio broadcast an experimental stereo FM music program in Berlin. They used a customized magnetophon, a magnetic tape recorder made by AEG, a German electronics company. Jack

Mullin was sent to the newly liberated Paris by the Allied Military in 1944 to set up an electronics lab. His mission was to examine captured German electronics and to submit reports to the Signal Corps and to Allied intelligence. Don Rushin in *The Magic of Magnetic Tape* states:

> On May 16, 1946, Mullin was scheduled as the speaker at the regular meeting of the San Francisco chapter of the Institute of Radio Engineers, held at the studios of radio station KFRC. A demonstration of the German tape recording equipment had been promised, and the room was packed. Mullin played recordings of an orchestra, vocalists and a pipe organ that he had made on some of the tapes he brought back with him. The reaction was little short of a sensation.[2]

Mullin, using two of the German tape recorders, introduced the American audience to high-fidelity. He also improved upon the design of the magnetophon. Mullin was famous for his early broadcast of a Bing Crosby performance. Radio networks appreciated the fidelity and convenience of tape and in the late 1940's switched to tape from magnetic disks for recording purposes. ABC bought the 12 audio recorders Ampex Corporation was rushing to complete. Crosby had been instrumental in persuading the network to buy the machines, copied from Mullin's German originals.

Cassette Tape

The first compact audio-cassette was invented and released by the Philips Company from the Netherlands in 1962. They used a polyester 1/8 inch tape produced by BASF. Soon after, Norelco began to sell its Carry-Corder dictation machine that used the new cassette tape. The demand for the tape was increased by its ability to record music for personal use.

In reviewing the history of computer music we learned that the Germans were broadcasting hi-fidelity FM programs using tape recorded music in 1941. Hi-fidelity is usually defined as **the faithful reproduction of original sound**. When we study wave motion and sound later on, we shall find that the human ear is sensitive to sound vibrations between 20 and 20,000 cycles per second. FM broadcasts generate vibrations between 30 - 15,000 cycles/s (1 cycle/s = 1 Hz) which is suitable for hi-fidelity reproduction, while AM broadcasts broadcast sound vibrations between 200 - 5,000 Hz, which is not hi-fidelity reproduction and is suitable for voice reproduction but not music.

The Germans were already using a form of magnetic tape in their recorders in 1944. The American 3M Company almost simultaneously developed a coated tape with a smooth surface and uniform dispersion of ferromagnetic powder that could be used to record electromagnetic signals with better sound quality than phonograph discs. The development of magnetic tape led to a new art form in 1948 called "musique concrete." It is created by recording various sounds on tape and modifying them in some way by playing the tape backwards, stretching the tape, splicing short segments, making a tape loop, etc. Using these techniques and overlaying the sounds allowed one to create a song. The process required a large amount of time to be effective.

Synthesizer

The creation of the synthesizer seemed to solve the "time" problem. The RCA Mark II created in 1957 for RCA by Harry Olson and Hebert Belar was basically a vacuum tube analog synthesizer controlled by paper tape. It used a multi-track tape machine to record sounds.

Robert Moog came up with the idea of using electronic circuits to create sound in 1963. He built the first playable synthesizer with a keyboard in 1963 and demonstrated it in 1964 at the Audio Engineering Society Convention. Wendy Carlos (then Walter) released "Switched-on Bach" in 1968 which rose to the top of the musical charts and was awarded three Grammies. The Beatles bought one as well as Mick Jagger. The significance of Moog's voltage controlled synthesizer was that it was essentially an analog computer. Moog's synthesizers fell into disuse with the advent of the digital computer in the early 1980s and the MIDI 1.0 standard in 1983.

The problem initially faced by using digital computers for sound reproduction was that the information they used was discrete and essentially expressed in terms of numbers. It wasn't until the 1960s when Max Mathews at Bell Labs wrote programs such as Music I that made it possible for the digital computer to produce sounds. Data was entered into the computer in terms of mathematical functions that were then used to calculate the form of the waveforms for musical sounds. These were stored in digital form and could be recalled at will at the command of the user.

The conversion from analog to digital accelerated as digital computers became faster and could sample sound more quickly. Speed was important, as the computer had to specify thousands of numbers to produce a single second of sound.

Recording on Disc

Thomas Edison's tin foil phonograph (Fig. 2.8) used tin foil cylinders and each recording was good for one play. Alexander Graham Bell followed up with the graphophone that used a wax cylinder that could play many times but each cylinder had to be made individually, making mass reproduction impossible.

Figure 2.8 Tin foil phonograph
http://www.nps.gov/edis/edisonia/tae_bio.html

In 1887, German immigrant Emile Berliner, invented a system of recording on flat discs which could be played over and over and from which many copies could be made. Disc players, notably Victrolas introduced by the Victor Talking Machine Company in 1906, became popular around 1912.

In 1912 Edison's scientists developed a disc that would outperform their competitors' shellac-made records and used a vertical-cut method for the grooves. It was made of molded phenol and formaldehyde mixed with wood-flour and a solvent which was then shaped into a heat-resistant disc. The stylus would then bob up and down in the groove instead of side to side. A ten-inch record would run for 5 minutes per side at approximately 80 rpm. By 1920 Edison was the only disc company not using steel needles or the lateral method of grooves. In 1929, Edison closed his disc business.

Shellac was found to be the best material for producing records. It is a resin made from the secretions of the lac insect. Shellac is a natural thermoplastic. It is soft and flowing when heated, but rigid and hard at room temperature. In the 1930s shellac was replaced by synthetic resins. The 78 rpm record became the standard recording medium for music for many years. It had a tendency to break easily and could only play for five minutes before a new disc had to be used to continue listening to a longer musical work.

The rotational speed of most 10 inch records varied somewhat from one manufacturer to another, but most turned at between 75 and 80 revolutions per minute and most 'Gramophone' machines were capable of some adjustment. Eventually 78 rpm became the common standard.

In 1948, a new format was introduced by the Columbia Company in which a 12 inch vinyl disc spinning at 33 1/3 rpm could play up to 25 minutes per side. In 1949 RCA Victor introduced a seven inch disc that played at 45 rpm and could hold as much sound as the older 78 rpm. Its size made it attractive as well as the automatic record changer developed by RCA.

However, the new formats required new equipment to play them and as a result the 78 rpm records remained on the market for many years. See Figure 2.9.

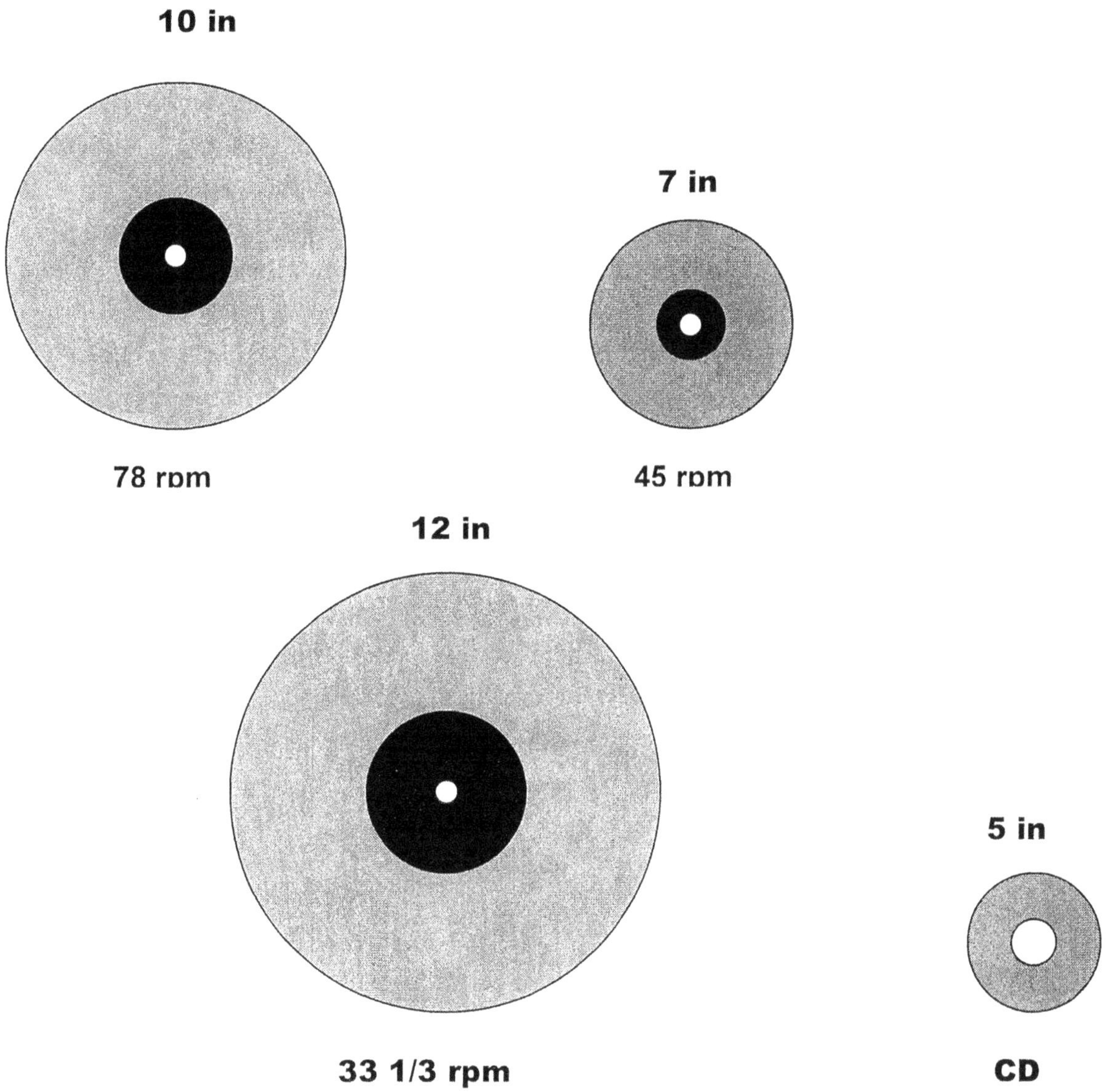

Figure 2.9 Recording discs

With the advent of rock and roll in the 1950s the new 7 inch 45 rpm discs captured the market and effectively drove the 78 rpm records into oblivion. Bill Haley and the Comets, and Elvis Presley led the parade of rock and roll artists and was followed by the Beatles in the 1960s. By the end of the 1960s the 12 inch LP was replacing the 45 rpm record.

In the 1950s, records and phonographs were monaural but their fidelity was improved using the new record formats and materials. By 1958, in the interests of producing music at hi-fidelity, stereophonic phonographs and discs appeared and by the 1960s almost all new records and phonographs were stereophonic and hi-fidelity. Finally, the 78 rpm records disappeared from the marketplace. Some claim that the last one issued in the U.S. was Chuck Berry's "Too Pooped to Pop", released in February 1960.

In 1982, compacts discs (CDs) were introduced in Japan and Europe and arrived in the United States in 1983. Their hi-fidelity sound and durability allowed them to capture the market.

The development of the CD players resulted in portable and automobile CD players in 1984, and CD-ROM drives in computers in 1985. In 1988, we see the development and use of the CD-Recordable/Recorder technology. In 1996 the DVD (digital versatile disk or digital video disk) is introduced and mass-marketed. Today, we find DVD players in most homes and computers. The next development lies in the next generation of DVDs, with either blue laser technology or HD-DVD. In 208, the blue laser technology triumphed and production of the HD-DVD player was discontinued.

Early in 2001, Apple Computer introduced a new program called iTunes for the Macintosh that converted audio CDs into compressed digital audiofiles, played Internet radio, and allowed for the organization of audio files. Later that year, Apple Computer released a small device it called the iPod. It was a digital music player and held about 1,000 songs on a miniature hard disk.

MIDI Is Born

As previously discussed, in 1982, a group of synthesizer manufacturers met at the NAMM (National Association of Music Merchants) Convention and discussed a standard for digital transmission and reception between all types of electronic synthesizers and instruments. The name MIDI (Musical Instrument Digital Interface) was chosen for this standard and the first MIDI product that was released was the Sequential Prophet-600. In 1983, the MIDI 1.0 standard was issued and published in October by the manufacturers. At this time the MMA (MIDI Manufacturers Association) was set up with the responsibility of maintaining and developing the standard.

The standard assured compatibility between MIDI instruments produced by different manufacturers and the sales of synthesizers escalated tremendously. Problems arose, leading to an extension of the MIDI 1.0 protocol which evolved into General MIDI.

General MIDI 1

In 1991, the MIDI Manufacturers Association adopted General MIDI 1 (GM1) or General MIDI System Level 1 which added some new requirements to the MIDI 1.0 specification, which is still the basic standard for all MIDI devices. A standard was set for numbering instrumental patches, so that all MIDI modules would play the same instrument for the same patch number. Without this requirement musical arrangements would have different instruments play depending on the MIDI synthesizer which was being used.

Another requirement of General MIDI 1 was that the MIDI module be fully multitimbral and be able to play all 16 channels simultaneously with a different patch on each channel. It was also decided that all patches sound an A440 pitch upon receiving note number 69. Middle C would then be note number 60 in all compliant modules. Some MIDI modules had limited polyphony. To address this discrepancy GM1 required that a MIDI module should have a minimum of 24 note polyphony. A number of other requirements such as requiring each MIDI module to be velocity sensitive, and each pitch bend wheel to default to plus or minus two semitones.

A "drum map" was specified in which drum sounds were assigned to specific note numbers and a drum machine part had to be recorded on channel 10. A GM1 Module was also required to respond to a certain controller messages.

General MIDI 2

In 1999 a number of additions were made to General MIDI 1 creating General MIDI 2 (GM2), or General MIDI System Level 2, which increased the number of sounds available to be played and amount of control for editing and musical performance. It required MIDI modules to have 32 note polyphony, two simultaneous drum channels, 10 and 11, a number of control change assignments and additional Universal System Exclusive messages. It also required all General MIDI 2 devices to be compatible with the General MIDI 1 standard.

Review Questions

1. What is music?
2. What was the hydraulis?
3. What was the significance of the hydraulis?
4. What is an organ?
5. What is an abacus and how is it used?
6. What is the Difference Engine and what was it supposed to do?
7. What is the Analytical Engine and what was it supposed to do?
8. What is the transistor and how was it used?
9. What is an integrated circuit and what was it used for?
10. What was the Telharmonium?
11. What was the Theremin?
12. What was Joseph-Marie Jacquard famous for?
13. Who was Herman Hollerith?
14. Who were the first to broadcast over the radio in high fidelity?
15. What role did magnetic tape play in early broadcasting?
16. What was "musique concrete"?
17. What was the major problem with "musique concrete" and how was it solved?
18. Describe the evolution of disc recording from Edison to the CD.
19. What is MIDI and when was it born?
20. What is an iPod?

Multiple Choice Questions

1. A hydraulis was a _____.
a. type of jack
b. water powered organ
c. fiddle used by Nero
d. type of Roman fountain

2. The abacus was _____.
a. the first computer
b. a counting aid
c. a type of Difference Engine
d. an early synthesizer

3. The first computer programmer was _____.
a. Charles Babbage
b. Robert Moog
c. Augusta Byron
d. Joseph-Marie Jacquard

4. Punch cards were used first by ______ in weaving machines.
a. Herman Hollerith
b. Charles Babbage
c. Joseph-Marie Jacquard
d. Thomas Watson

5. The first electronic computer to use binary notation and the stored program concept was the _____.
a. UNIVAC
b. EDVAC
c. ENIAC
d. IBM 360

6. The _______ was a precursor of the modern synthesizer.
a. abacus
b. Theremin
c. Telharmonium
d. harmonica

7. The Hammond organ generated sound using _____.
a. additive synthesis
b. subtractive synthesis
c. the Telharmonium
d. using punched cards

8. Herman Hollerith used _____ for the 1890 census.
a. an abacus
b. an Analytical Engine
c. Jacquard's loom
d. punched cards

9. The first recording of a human voice saying "Mary had a little lamb." was made on the tin-foil phonograph invented by _____.
a. Herman Hollerith
b. Thaddeus Cahill
c. John Hammond
d. Thomas Edison

10. In 1941, the first stereo FM broadcast was made by _____.
a. RCA
b. German Radio
c. Ampex
d. Bell Labs

11. The faithful reproduction of musical sound is called _____.
a. real-time recording
b. high fidelity
c. stereo sound reproduction
d. "musique concrete"

12. Splicing pieces of magnetic tape together to create a musical presentations was called _____.
a. high fidelity
b. "musique concrete"
c. tape editing
d. magnetic recording

13. The RCA Mark II was essentially a(n) ________.
a. analog computer
b. analog synthesizer
c. transistorized synthesizer
d. digital synthesizer

14. The Victrola used flat discs or records for playing sound and were usually made out of _____.
a. shellac
b. vinyl
c. polyvinyl chloride
d. Corian

15. Shellac is a resin made from the secretions of the _____ insect.
a. mosquito
b. fly
c. lac
d. shel

16. The standard speed of the first records was set at _____rpm.
a. 33 1/3
b. 45
c. 78
d. 98

17. The advent of rock and roll in the 1950s flourished using ______ rpm discs.
a. 33 1/3
b. 45
c. 78
d. 98

18. Hi-fidelity and stereophonic sound recordings appeared on the market in _____.
a. 1941
b. 1947
c. 1958
d. 1972

19. CDs were introduced in the United States in _____.
a. 1958
b. 1972
c. 1977
d. 1983

20. DVD discs were introduced on the market in _____.
a. 1977
b. 1983
c. 1996
d. 2001

21. The letters DVD stand for _____.
a. digital variable disc
b. digital versatile disc
c. data video disc
d. digitized video data

22. The MIDI 1.0 standard provided for ____ between MIDI instruments of different manufacturers.
a. connections
b. compatibility
c. incompatibility
d. interchangeability

23. The General MIDI 1 standard is _____.
a. an extension of the MIDI 1.0 protocol
b. the predecessor of MIDI 1.0
c. another name for MIDI 1.0
d. none of the above

24. When we say a synthesizer is multitimbral, we mean it can_______.
a. play several notes at the same time
b. play several instrumental sounds simultaneously
c. control several synthesizers simultaneously
d. play polyphonic sounds

25. The term polyphonic is used to indicate that the instrument can _____.
a. play several notes at the same time
b. play several instrumental sounds simultaneously
c. control several synthesizers simultaneously
d. play multitimbral sounds

26. A 78 rpm vinyl record could play music for about _____ minutes.
a. 5
b. 10
c. 15
d. 30

27. The _____ made the 7 in diameter 45 rpm record popular.
a. synthesizer
b. record changer
c. turntable
d. laser

28. In an iPod music is stored on a _____.
a. floppy disk
b. hard disk
c. sound card
d. laser disk

29. Max Mathews at Bell Labs wrote the program _____ that made it possible for the computer to produce music.
a. Sequencer Plus Mark III
b. Music I
c. Voyetra Gold
d. Digital Orchestration Pro

30. The MOOG was _____.
a. a type of organ
b. a dance craze
c. Bach's piano
d. an analog synthesizer

Bibliography and Web Resources

[1] Sherkin, Joel. *Engines of the Mind.* WW. Norton Co., NY 1984, 1996.

Computers

Abacus -
http://www.ee.ryerson.ca:8080/~elf/abacus/history.html

Eniac –
ENIAC: The Triumphs and Tragedies of the World's First Computer. Walker & Co. ISBN 0802713483

Analytical Engine -
http://www.maxmon.com/1830ad.htm - Analytical engine

Calculators -
http://www.thocp.net/hardware/ti_calculators.htm

Loom - Jacquard -
http://www.maxmon.com/1800ad.htm

Organs

Hydraulis operation -
http://www.water.eat-online.net/english/arts/hydraulis.htm

Story of the Hydraulis -
http://users.ipa.net/~tanker/organs.htm

Hydraulis - Ernest. D, *The Evolution of Electronic Music*, Schirmer Books, 1977

Hydraulis and organs -
http://romanliturgy.net/organ_history.html

Hammond and organ -
http://theatreorgans.com/hammond/

Telharmonium -
http://www.obsolete.com/120_years/machines/telharmonium/

Phonos

Edison's discs -
http://inventors.about.com/library/inventors/bledisondiscphpgraph2.htm

Phonograph chronology -
http://www.recording-history.org/HTML/phonochrono.htm

Tin foil Phonograph -
http://www.recording-history.org/HTML/phono_technology2.htm

Victrola logo -
http://www.victor-victrola.com/

Victrola -
http://alephnull.net/20s/victrola/victrola.html

Synthesizers

Theremin - -
http://www.thereminworld.com/learn.asp

RCA Mark II -
http://emfinstitute.emf.org/exhibits/rca.html

Moog -
http://www.obsolete.com/120_years/machines/moog/

Tape Machines

Hollerith -
http://www.columbia.edu/acis/history/hollerith.html

Musique concrete -
http://www.musespace.com/writings/essays/musique.html

Begun -
http://inventors.about.com/od/bstartinventors/a/Joseph_Begun.htm

[2]Mullin broadcast
http://www.tvhandbook.com/History/History_tape.htm)

Chapter 3

Audio Systems

"Mary had a little lamb."

Thomas Edison, 1877

Inventor of light bulb and phonograph
and holder of 1093 patents

In this chapter we shall study the basic building blocks or components used in assembling a hi-fidelity audio reproduction system. The music we hear from the computer music system depends on the quality of the audio reproduction system. A hi-fidelity audio system can be considered as being made-up of four major parts: the sound sources, preamplifier, power amplifier and the speakers (Fig. 3.1).

Goals

After studying this chapter, you should be able to:

- Identify the components of a hi-fidelity audio system.
- Identify various sound sources.
- Know what surround sound is.
- Understand the function of a preamplifier.
- Understand the function of an amplifier.
- Understand the role of the speakers.
- Know the function of an integrated amplifier.
- Understand what a receiver is.
- Be able to identify and distinguish between a compact system and a console system.
- Know what a boom box is.

Hi-Fidelity

Hi-fidelity is defined as a total faithfulness to the original sound. It was German Radio in Berlin that used hi-fidelity sound for the first time in an experimental FM broadcast in 1941. With the advent of rock and roll in the early 1950s recordings were not only hi-fidelity but also stereophonic by 1958. Stereophonic sound is a method of recording in which two channels are used for recording and playback. Not all stereo is hi-fi, and vice versa. In the 1970s, an attempt was made by the component manufacturers to replace stereo sound with quadraphonic sound, or recording and playback on four channels. Although the quality of sound was superb, sophisticated techniques had to be used for recording and playback. Quadraphonic receivers had to use four channels, which made them more expensive.

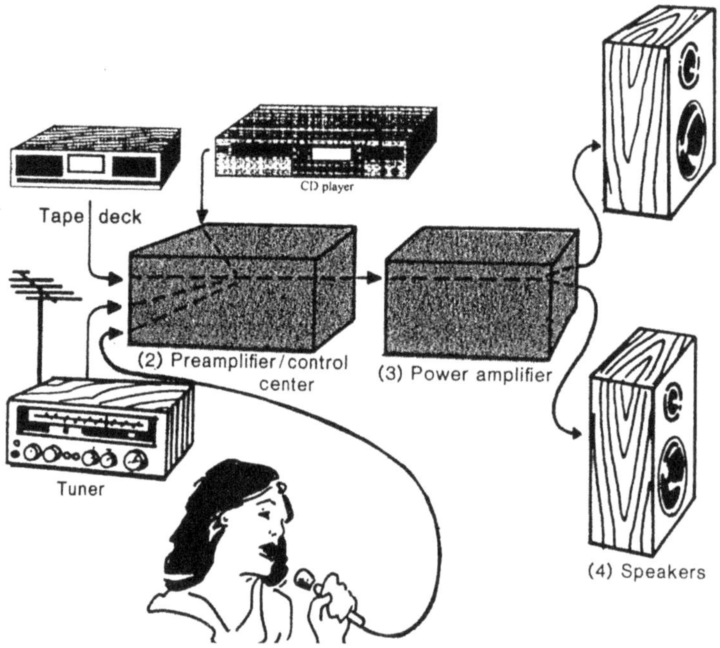

Figure 3.1 Sound sources, preamplifier, power amplifier, speakers
From *Science of Hi-Fidelity* by Johnson, Walker, Cutnell. Reprinted by permission of the authors.

The new approach that is used today is called surround sound, because it gives a spatial quality to music. Surround sound envelops the listener, giving ambience and directionality to sound effects. Usually, five speakers are used instead of the usual two found in stereophonic sound. The center speaker is placed between the left and right speakers at the front of the room, and two speakers are placed at the rear of the room. A special processor is used to decode the channel information and delay the sound to the rear speakers providing the illusion of being surrounded by sound.

In the early 1950s, sound reproduction was essentially monophonic, being produced essentially by one amplifier and played through one speaker. To the listener it seemed that all the sound emanated from one point. Adding a second speaker moved the perceived origin of the sound to halfway between the two speakers. In stereophonic sound, which appeared at the end of the 1950s, the recorded sounds are played over two channels and through two speakers. Now the listener was able to experience the correct placement of instrumental sounds originating from each side of an orchestra. Still, ambience was lacking. All the sound appeared to come only from the front of the room, with no reflections from the back of the room.

Quadraphonic sound in the 1960s used four amplifiers and four discrete channels, providing the listener with an ambient experience. However, it was expensive in that four amplifiers were needed to reproduce the four channels. Special recordings and playback units were necessary. For FM broadcasts two tuners were necessary and tape decks needed four channel audio. The expense was too great.

In the 1970s, **Dolby Surround** sound was developed. Dolby developed a method for recording four channels into a two channel signal using a special chip. Decoded information could be sent to the left and right front speakers, as well as to the right rear and left rear speakers. Ambience was provided by the rear speakers. Now sound effects could not only be right to left, but also rear to front, giving the listener a more realistic experience.

Dolby Pro Logic systems use a more sophisticated decoding chip that adds more separation to channels, creates a new center channel, and gives more directionality to sounds. It has four channels. Three channels play through the left, right and center speakers while a fourth channel of limited band width is shared by two surround speakers. The center and surround channels are actually created from the left and right channel using signal processing techniques. When it is coded, it combines four channels into two for recording and transmission purposes (Fig. 3.2).

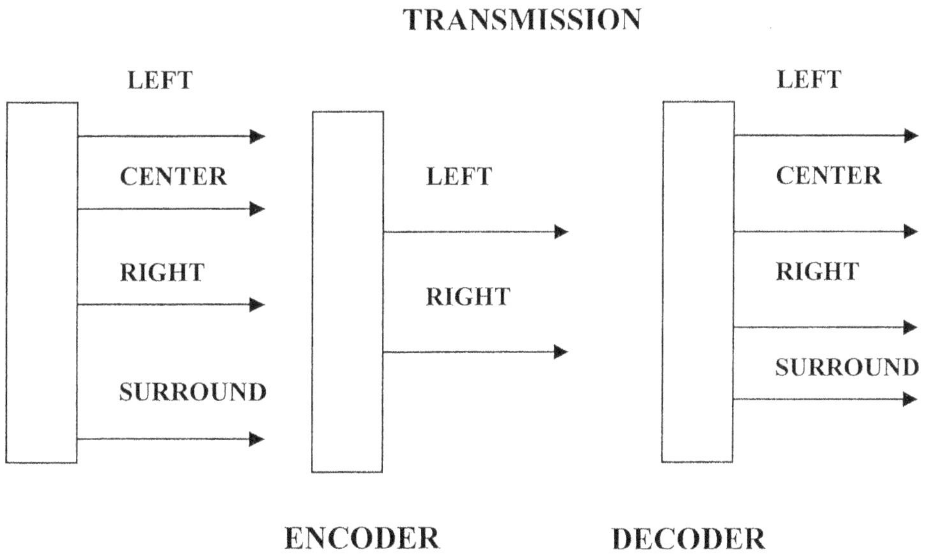

Figure 3.2 Dolby Pro Logic

Recently Dolby has developed a **Dolby Pro Logic II** system which uses a four channel Dolby Surround signal and simulates a five channel surround system, 5:1 audio, using signal processing circuits. In addition to the two full band width channels used to create the sound for the left, right, center, surround left and surround right, it has a low frequency channel. It can also separate the channels encoded in the Dolby Surround coding more distinctly. Using signal processing, a 7:1 channel sound can be created.

Now we have **Dolby Digital** (or 5:1 or 6:1) channel systems. The signal is encoded digitally on five independent and full bandwidth channels and even more directionality and accuracy become possible along with a subwoofer channel for deep bass. In a 5:1 system the Dolby Digital Encoder encodes the left, right, center, surround left, surround right, and low frequency effects from the movie sound track and transmits them to a Dolby Digital Decoder that faithfully reproduces the 5:1 sound sources (Fig. 3.3).

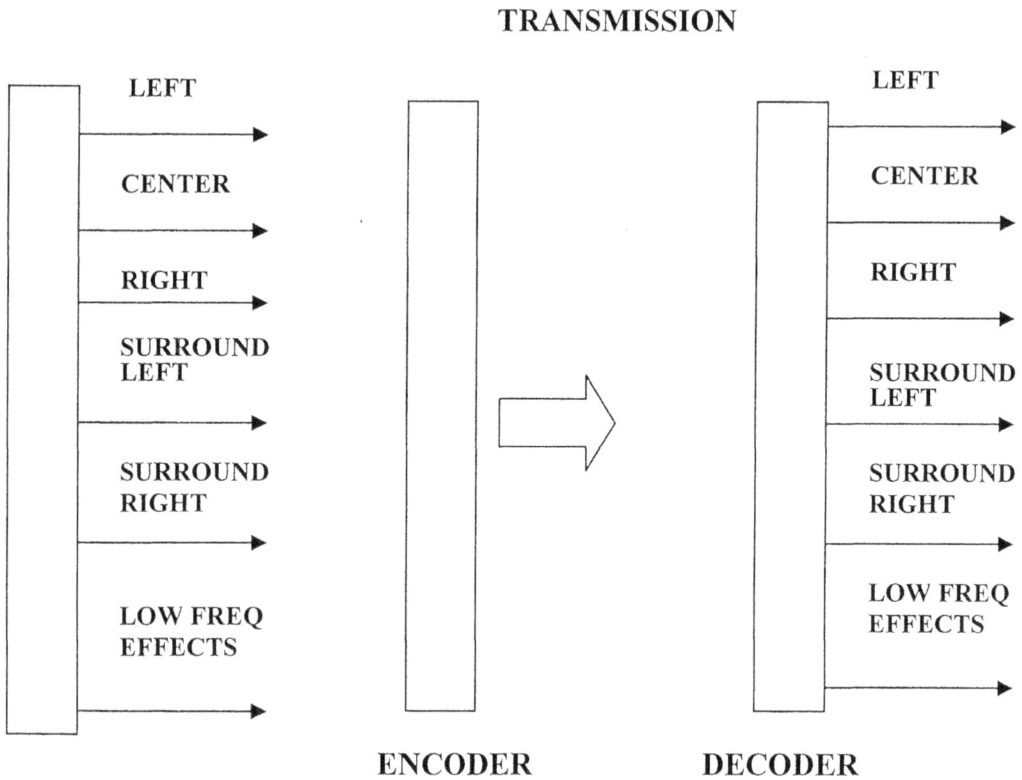

Figure 3.3 Dolby Digital

If we add a rear center channel for one or two speakers for back surround, it is then called **Dolby Digital EX** using six channels. The Dolby Digital EX decoder, when playing source material recorded in 5:1 format, can simulate 6:1 or 7:1 channel sound using signal processing and combining information contained in the 5:1 channel source material.

Sound Sources

In computer music the primary source of sound will usually be from the synthesizer whose signal has been processed by a computer and sent to the audio system for reproduction. The other sound sources such as a CD player, tape deck, or video system would also be sound sources that function in an analogous manner in that their output will be an electrical signal, usually a fluctuating voltage that would be sent to the pre-amp in a continuous fashion. In computer music the primary source of sound will usually be from the synthesizer whose signal has been processed by a computer and sent to the audio system for reproduction.

Traditionally, a turntable having a motor driven platter on which a record rests would be a common source of signal for a sound reproduction system. In this case the signal would be detected by a tone arm equipped with a magnetic cartridge containing a stylus, tracing out the grooves in a record containing the images of sound waves frozen in vinyl and converting them to an electrical signal which is then fed into the audio system. The motor is usually connected to the platter using a flexible belt and a pulley. The flexible belt has the advantage of absorbing any vibrations generated by the motor and thus allowing a less costly motor to be used. Most turntables used this system. A direct drive motor can be used but must be of high quality to maintain a constant speed and prevent any "wow" or "flutter." The cheapest drive mechanism used a motor with a rubber wheel attached to the shaft and had the rubber wheel in contact with the platter under the edge of its rim. It was termed "rim drive". Eventually the rubber wheel wears through friction, and the sound is distorted because of the variation in speed that occurs (Fig. 3.4).

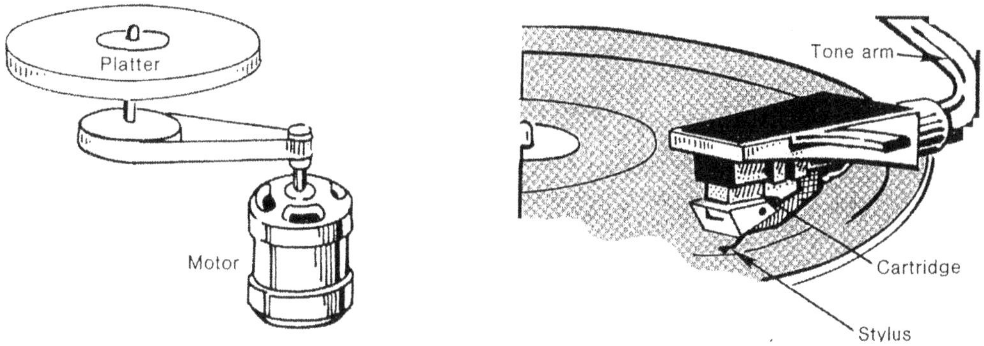

Figure 3.4 Turntable
From *Science of Hi-Fidelity* by Johnson, Walker, Cutnell. Reprinted by permission of the authors.

An FM or AM tuner would be another common sound source. In the case of a tuner, electromagnetic radio waves detected by an antenna are converted by its internal circuitry to an electrical signal that is fed into the preamplifier.

CD players use voltages that have been coded digitally or discretely in numerical fashion as opposed to the continuous mode. These voltages normally must be converted by the CD player to analog by a circuit called a digital-to-analog converter (DAC) before being sent to the preamplifier (Fig. 3.5).

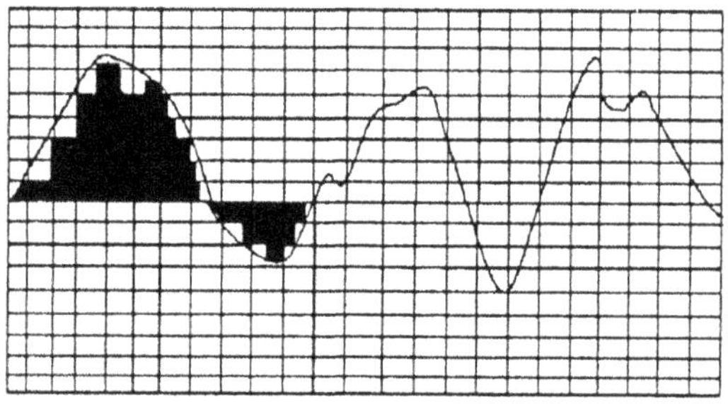

Figure 3.5 Digital to analog

From *Computer Music: Science and Technology of a New Art,* Stanionis and Berberich. Copyright © 1989 by Research Foundation of State University of New York. Reprinted by permission.

Although the signals from the DAC are analog, they follow a series of steps and are smoothed out by passing the signal through a low pass filter and sent to the preamplifier Fig. 3.6).

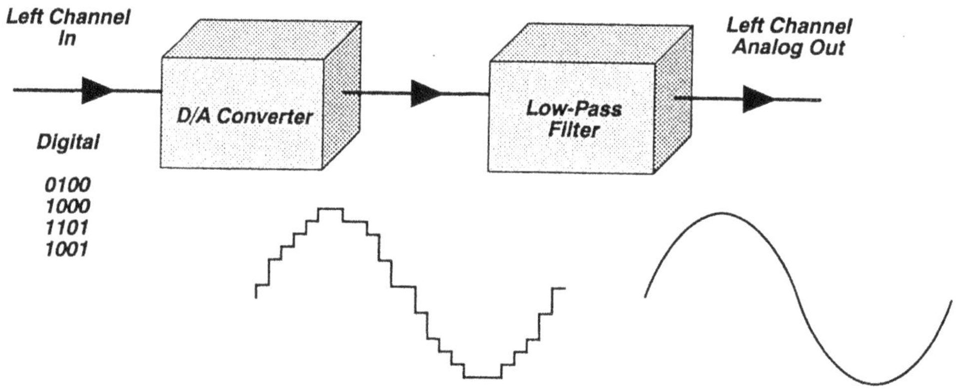

Figure 3.6 DAC and filter

From *Physics of Hi-Fi* by Symko. Reprinted by permission of the author.

A CD disk is made of a thin layer of aluminum covered by a protective lacquer and a printed label. The aluminum surface consists of a series of pits and lands that have been cut into it by a high-powered laser. A low-powered laser shines on the bottom of the disc. If it illuminates a pit, the light is scattered and not received by the sensor. The pit is read as a binary zero. If the laser light reflects from a land and is picked up by the light sensor, it is recorded as a binary 1 (Fig. 3.7). So we see that information and music can be digitized (i.e. converted to binary), recorded on a CD, and decoded. DVDs usually have several reflective layers stacked in between two laminated plastic discs and, as a result of having several layers, store more information than a CD ROM disc.

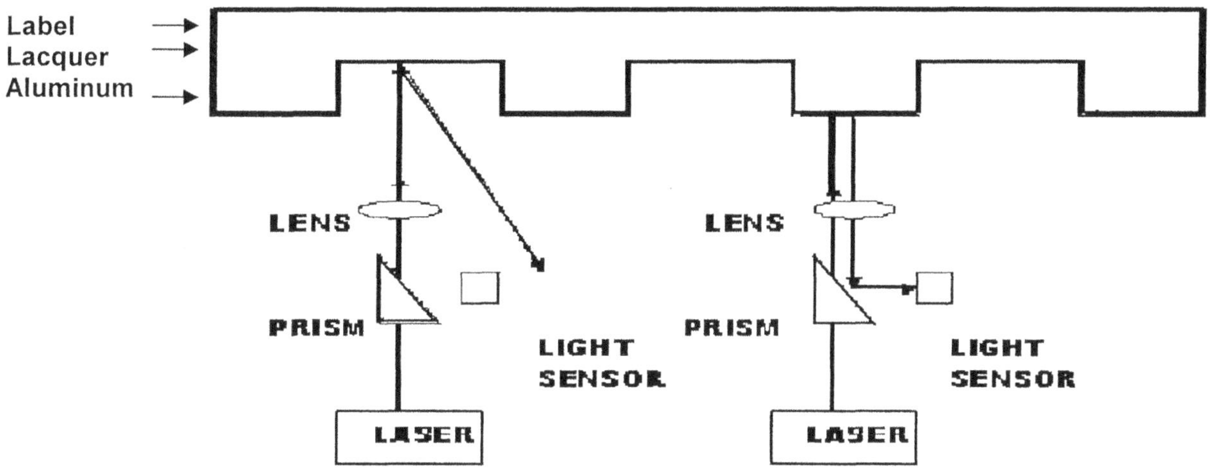

Figure 3.7 CD ROM disc

Other Sound Sources

The Mini-Disc player is another device similar to the CD in that information is stored digitally on a disc that can be played or erased and re-recorded. The digital compact cassette (DCC) is another sound source in which information is stored digitally but on magnetic tape. The DAT or digital audio tape player is similar to the DCC in that information is recorded on magnetic tape.

In addition to reproducing video information, a video cassette recorder or VCR can generate hi-fidelity sound that can be reproduced by an audio system. For a short time, videodisc players using 12-inch CD type discs on which information could be stored digitally were available for video and hi-fidelity sound. Today they have been replaced by DVD, which can provide high quality video and surround sound on a 5 inch disc.

Preamplifier

The preamplifier (see Figure 3.8) is the tonal management center and switching center of the hi-fidelity audio system. When the signal arrives from the sound source, it is still small (approximately 0.5 V) and easily manipulated. The pre-amp contains the volume control, bass, treble, balance, loudness controls. It contains circuits that will filter the sound: high frequency filters, low frequency filters and subsonic filters.

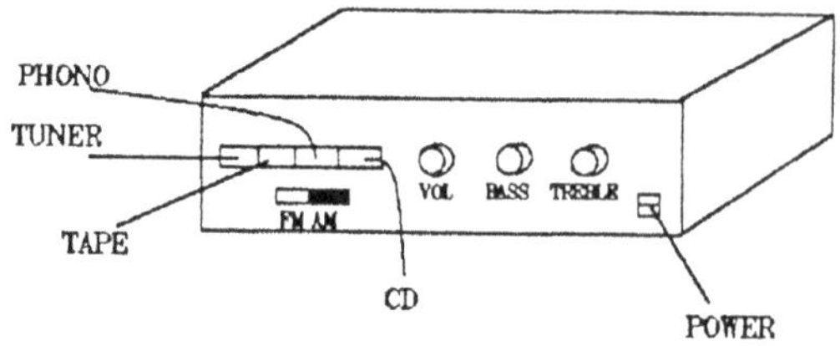

Fig. 3.8 Pre-amplifier

From *Computer Music: Science and Technology of a New Art,* Stanionis and Berberich. Copyright © 1989 by Research Foundation of State University of New York. Reprinted by permission.

It should also contain a stereo/monitor switch and may contain audio/video switching capabilities allowing easy selection from several audio and video inputs by remote control. Some new receivers are called integrated audio/video receivers because they incorporate video switching capability (Fig 3.9). With such a system and a single remote control device it is possible to switch both audio and video from one VCR connected to the system to another, with the audio output sent through a hi-fidelity system, simultaneously copying a tape and viewing it.

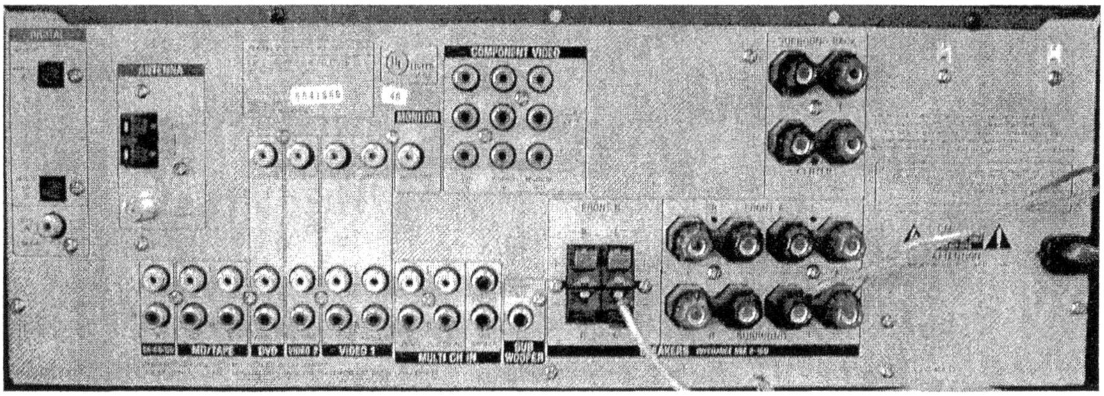

Figure 3.9 Back of receiver

Power Amplifier

The power amplifier's job is to enlarge the signal it receives from the preamplifier in a faithful and true fashion. It must produce a signal sufficiently large to power the speakers that create the pressure variations in the air that we hear as sound. The output of a power amplifier is measured in watts of undistorted power (Fig. 3.10) that it can send to each speaker in a typical stereo system.

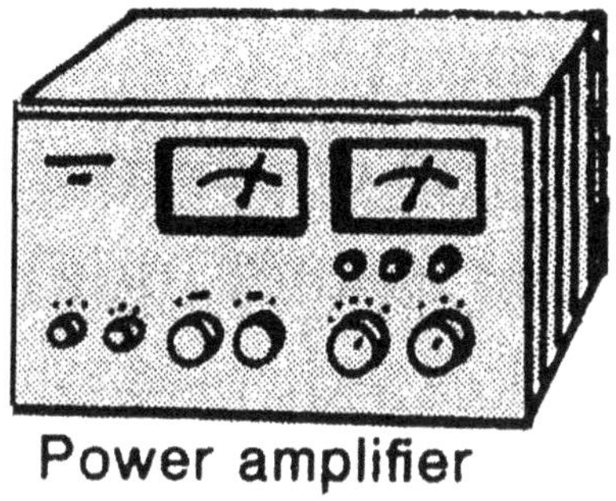

Figure 3.10 Power amplifier

From *Science of Hi-Fidelity* by Johnson, Walker, Cutnell. Reprinted by permission of the authors.

In the early days of hi-fidelity in the 1950s, most consumers assembled their own component hi-fidelity systems with sound sources, preamplifier, power amplifier and speakers all coming from different manufacturers. As electronic design of circuits and components advanced during the 1960s, it was possible to buy a preamplifier and power amplifier both contained in one unit that would have the same quality as the separate units. The main advantage to buying it as one unit, called an integrated amplifier, was that it was cheaper. Manufacturers were able to save on cost by using one cabinet and having them share the same power supply (Fig. 3.11).

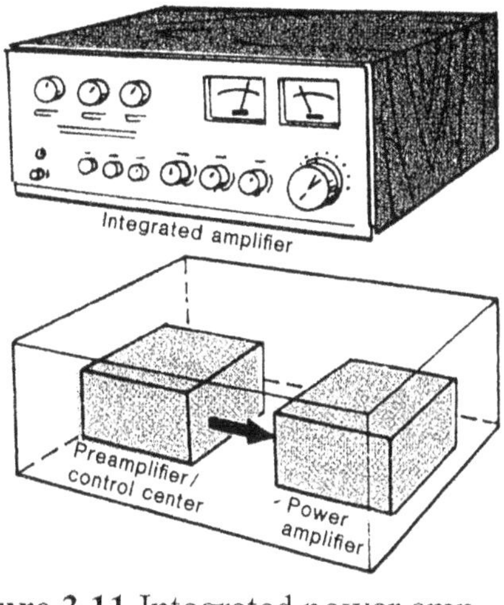

Figure 3.11 Integrated power amp

From *Science of Hi-Fidelity* by Johnson, Walker, Cutnell. Reprinted by permission of the authors.

Sometimes a manufacturer will include a tuner, preamplifier, and amplifier all in one enclosure, called a receiver (Fig 3.12). It is possible to assemble a hi-fidelity system from separate components: tuner, pre-amp, and amplifier, instead of buying it as an all-in-one unit as a receiver. However, by the various units sharing the same power supply and requiring fewer wires, plugs and outlets, and by using the same enclosure, manufacturers are able to reduce their costs and provide affordable receivers with no noticeable sacrifice in quality.

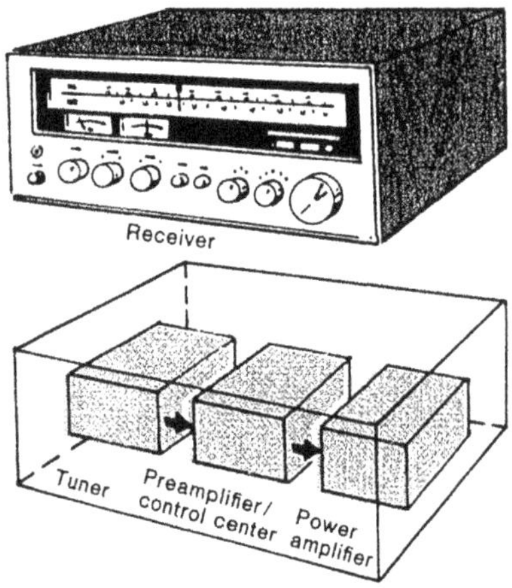

Figure 3.12 Receiver

From *Science of Hi-Fidelity* by Johnson, Walker, Cutnell. Reprinted by permission of the authors.

A typical receiver in the early 1980s was rated somewhere in the range of 35 to 65 watts per channel. Today, receivers are rated upwards of 100 watts per channel are much more common. Why the change? The CD player, which came on the market in the United States in 1983, with its ability to reproduce deep bass, middle range, and high frequencies at equal energy levels, has changed the audiophile's power requirements. The ability of a system to accurately reproduce all frequencies at equal energy levels is measured in terms of its dynamic range in units of decibels.

The dynamic range is the ratio of the softest sound to the loudest sound in a musical instrument or piece of electronic equipment. It is a measure of the difference between the inherent background noise and the loudest peaks on the recording. Decibels are used to measure sound levels from the softest that the ear can hear (0 dB) up to the threshold of pain (130 dB). It is a logarithmic scale and experiment shows that a difference of ten decibels (e.g. 80 to 90 decibels) doubles the apparent loudness of the sound.

CD players have dynamic ranges on the order of 100 dB. To reproduce such a wide range of levels without distorting the sound requires amplifiers of high power capability. Because of the dynamic range of the CD player, a musical piece might generate a high instantaneous power peak at even moderate listening levels, making severe power demands on the amplifier to faithfully reproduce it. If the amplifier is incapable of doing so, "clipping" or buzzing (Fig. 3.13) can occur and the subsequent generation of harmonics that can damage the amplifier and/or speakers may be produced by the electrical circuitry itself.

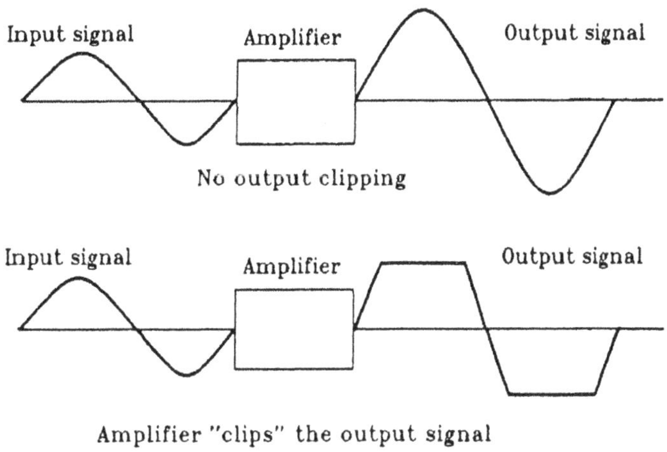

Figure 3.13 Clipping
From *Science of Hi-Fidelity* by Johnson, Walker, Cutnell. Reprinted by permission of the authors.

The dynamic power rating of an amplifier is an important specification in judging whether it can handle a sudden surge. This is especially important in using an amplification system with a MIDI system in which the sounds are created and produced by a synthesizer.

Speakers

The speakers (Fig. 3.14) receive the signal and generate the sound we hear by creating pressure variations in the air around us. Speakers are designed to move large quantities of air and can do this in many different ways as evidenced by the large number of speaker manufacturers and designs. Speakers transform an electrical signal into a physical vibration. Errors are easily introduced at this stage. Speaker design has been modified most recently with the introduction of the CD player and MIDI systems with their large dynamic ranges.

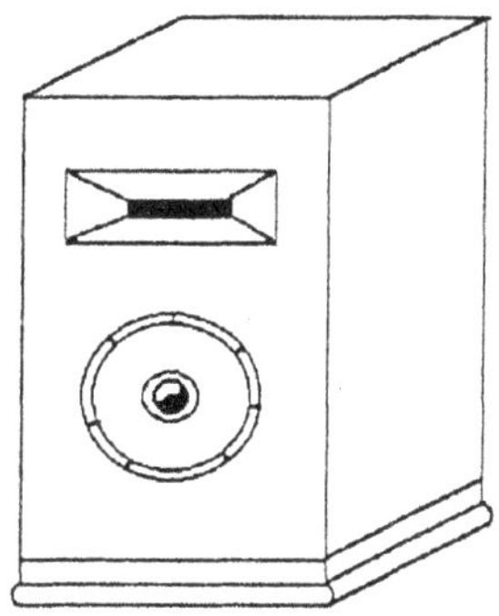

Figure 3.14 Speaker
From *Computer Music: Science and Technology of a New Art,* Stanionis and Berberich. Copyright © 1989 by Research Foundation of State University of New York. Reprinted by permission.

Most speakers operate on the electrodynamic principle which causes a voice coil attached to a moveable diaphragm to move to and away from the pole of a permanent magnet with a motion corresponding to the frequency and amplitude of the alternating current flowing through the coil. Research has revealed that no one speaker can produce all frequencies in the audible spectrum with equal accuracy. Speaker systems are therefore designed with multiple drivers (speakers) that are assigned to reproduce low, middle and high frequencies. Crossover circuits are able to channel high, medium, and low frequencies to the appropriate speaker smoothly, with some overlap in frequencies, thus avoiding abrupt transitions.

Today the typical hi-fidelity system consists of a number of sound sources, a receiver, and a set of speakers as shown in Figure 3.15.

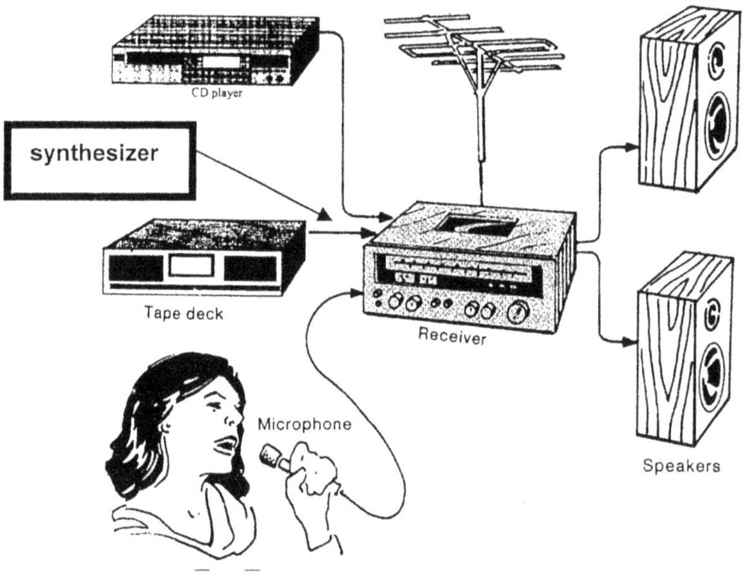

Figure 3.15 Typical component system

From *Science of Hi-Fidelity* by Johnson, Walker, Cutnell. Reprinted by permission of the authors.

Other Common Configurations

There are a number of other common configurations which are popular. The boom box radio is a device that contains all of the components in one enclosure and is portable operating on batteries. Many times it has sound sources such as: a tuner, cassette tape player, and CD player, as well as a preamplifier, power amplifier, and speakers (Fig. 3.16).

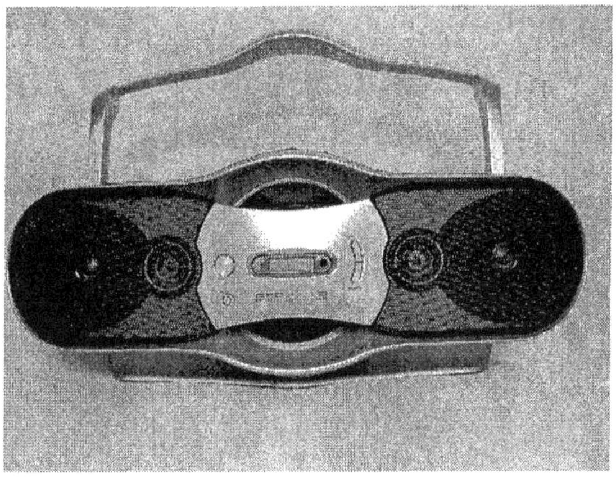

Figure 3.16 Boombox

Many students use compact systems in their dormitory rooms (Fig. 3.17). All the components of the hi-fidelity system are contained in one unit except for the speakers, which can be placed in different areas of the room.

Figure 3.17 Bookshelf system stereo.

Finally, there is the home console system (Fig 3.18) which is sometimes found in a living room disguised as a piece of furniture or credenza. All the components of the hi-fidelity system are contained in a single unit including the speakers. This can pose a problem. Any time speakers are placed on the same bookshelf or in the same cabinet as the rest of the hi-fidelity components there is danger. The vibrations the speakers create can cause a stylus in a phonograph to bounce on the record and skate across it, and in a CD player the vibrations may cause noticeable skips. This problem is easy to correct in a bookshelf system, in that the speakers can be moved from the bookshelf to another location. In the console system the audio engineers usually design the system so that frequencies below 50 Hz are not produced. Most furniture will vibrate sympathetically at frequencies below 50 Hz.

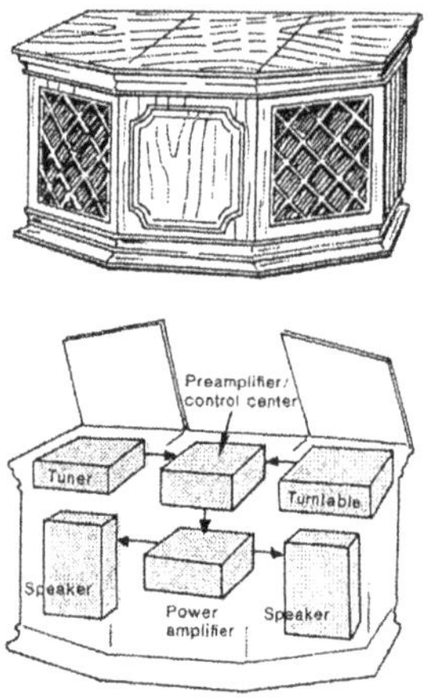

Figure 3.18 Console system

From *Science of Hi-Fidelity* by Johnson, Walker, Cutnell. Reprinted by permission of the authors.

Tape Monitor Switch

Most preamplifiers have a tape monitor switch which is very useful in making a tape recording from a sound source. The signal from the sound source is split into two parts before reaching the tone control section. One half of the signal is sent to the tape out jack (Fig. 3.19) while the other half passes through the tone control section and on to the speakers.

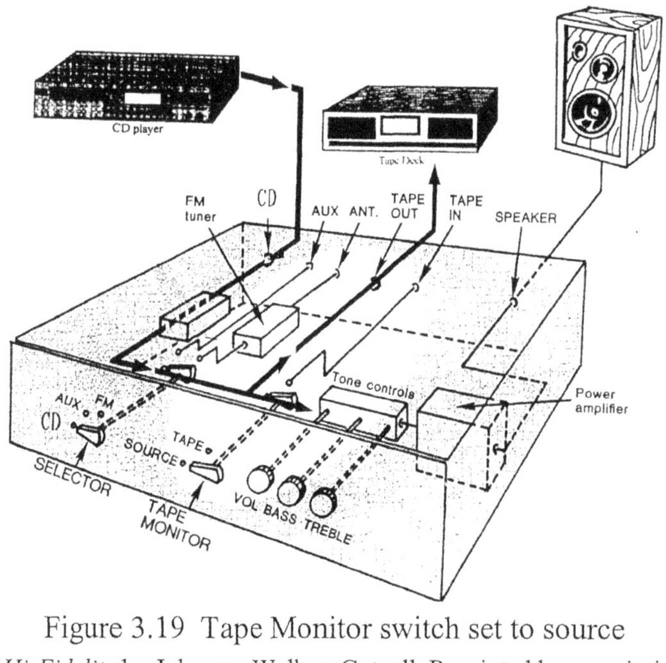

Figure 3.19 Tape Monitor switch set to source

From *Science of Hi-Fidelity* by Johnson, Walker, Cutnell. Reprinted by permission of the authors.

If a tape deck is connected, you can record the music on tape. With output of the tape deck connected to the tape in jack of the preamplifier, the signal recorded on tape is sent to the tape monitor switch. If the tape monitor switch is then turned by the listener from the "source" setting to the "tape" setting the signal passes through the tonal control section and on to the speakers so you can hear the music which was recorded a fraction of a second ago on the tape (Fig. 3.20). So by switching the tape monitor switch from source to tape, and back, you can monitor the quality of the recording. Of course, you can't hear both simultaneously.

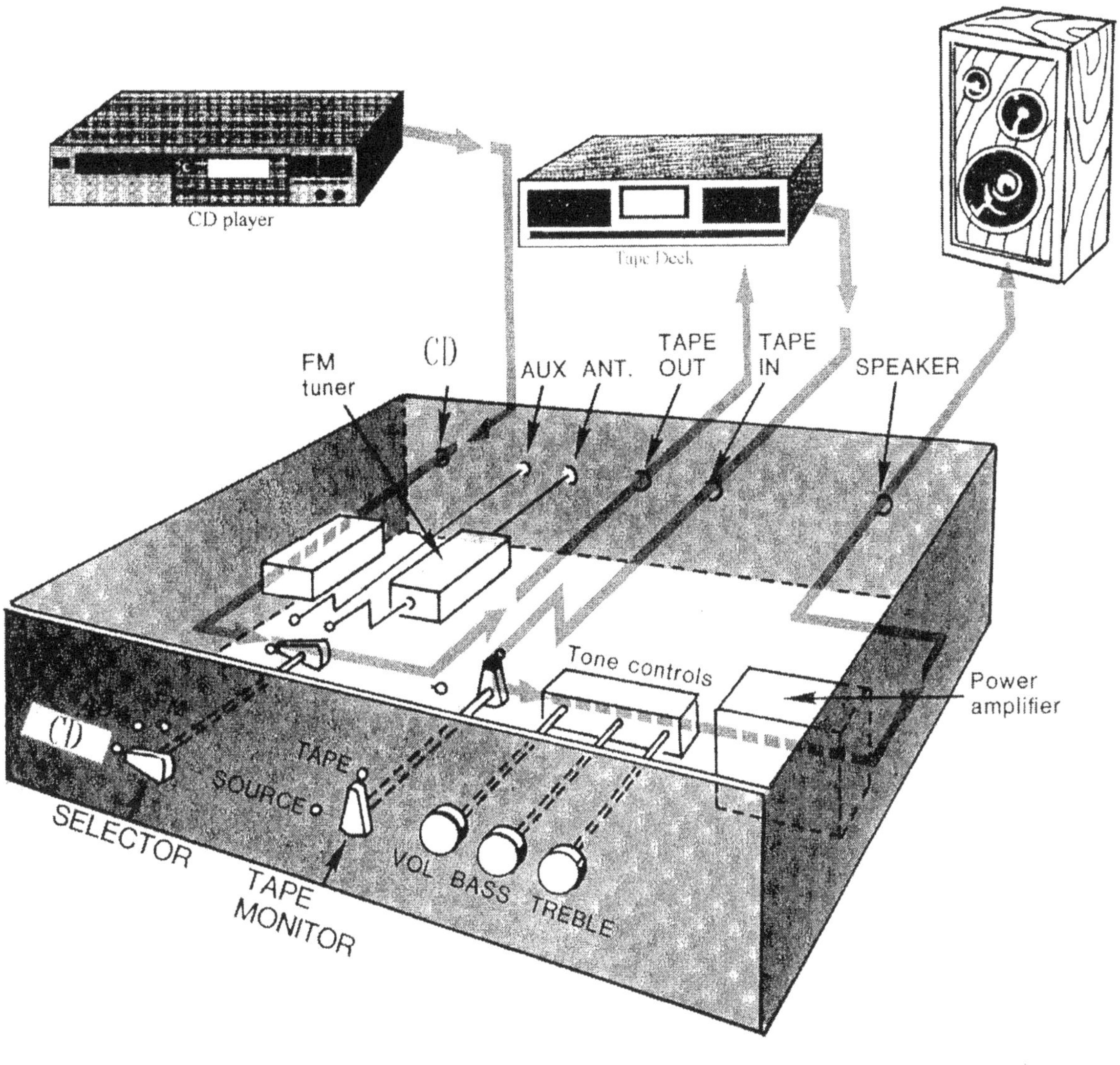

Figure 3.20 Tape monitor switch set to tape
From *Science of Hi-Fidelity* by Johnson, Walker, Cutnell. Reprinted by permission of the authors.

Review Questions

1. What is hi-fidelity?
2. What is stereophonic sound?
3. Define surround sound and describe how it works.
4. What is Dolby Digital Ex?
5. Name five sound sources and describe how they work.
6. What is the role of the preamplifier? What does it do?
7. What is a tape monitor switch and how does it work?
8. What is the main function of the power amplifier?
9. Explain why we need more power in a hi-fi system today than twenty years ago.
10. What is the role of the speakers? How do they work?

Multiple Choice Questions

1. Sound reproduction that uses two channels is called _____ sound.
a. monophonic
b. stereophonic
c. quadraphonic
d. surround

2. Hi-fidelity is defined as ______.
a. stereophonic sound
b. surround sound
c. total faithfulness to the original sound
d. quadraphonic sound

3. In a vinyl disc recording (records) music is stored ______.
a. mechanically in the grooves
b. in digital form
c. in magnetic form
d. electrically in the vinyl

4. A DAC is a _____.
a. type of input jack
b. switch on the preamplifier
c. digital to analog converter
d. digital amplifier converter

5. The tonal management center of a hi-fidelity system is called the _____.
a. power amplifier
b. preamplifier
c. tone control section
d. surround sound processor

6. The largest electrical currents are found in the _____.
a. preamplifier
b. magnetic cartridge
c. power amplifier
d. tonal controls

7. In what order should the four components of a hi-fidelity system be connected?
a. sound source, power amplifier, preamplifier, speakers
b. sound source, preamplifier, tonal controls, speaker
c. sound source, preamplifier, power amplifier, speakers
d. sound source, tape monitor switch, power amplifier, speakers

8. An integrated amplifier is a _____.
a. sound processor and preamplifier
b. preamplifier and power supply
c. sound source and power amplifier
d. preamplifier and power amplifier

9. A _____ has all the components of an audio system in one box.
a. radio
b. integrated amplifier
c. receiver
d. compact system

10. If you wanted to listen to a tape, which combination of audio components would be sufficient?
a. tape deck, power amplifier, speakers
b. tape deck, integrated amplifier, speakers
c. tape deck, speakers
d. tape deck, preamplifier, speakers

11. Surround sound gives a spatial quality or _____ to music.
a. liveliness
b. excitement
c. ambience
d. timbre

12. Dolby Surround sound records a four channel signal on _____ channels.
a. two
b. three
c. four
d. six

13. Dolby Pro Logic uses _____ channels for sound.
a. one
b. two
c. four
d. five

14. In Dolby Pro Logic systems we have _____ speakers.
a. 1
b. 2
c. 4
d. 5

15. Dolby Digital Ex uses ____ channels.
a. 2
b. 4
c. 6
d. 8

16. The frozen image of a sound wave is embedded in transverse form in a ____.
a. CD
b. DVD
c. DAT
d. vinyl record

17. FM broadcasting depends on _____ to carry the information for the broadcast.
a. sound waves
b. electromagnetic waves
c. frequency modulation
d. tuners

18. The main advantage in purchasing a receiver instead of a component system is _____.
a. cost
b. convenience
c. quality
d. space

19. The ratio of the softest sound to the loudest sound that a piece of electronic equipment can reproduce is called _____.
a. frequency response
b. dynamic range
c. output sensitivity
d. total harmonic distortion

20. Another name for a speaker is a _____.
a. harmonic generator
b. frequency analyzer
c. driver
d. baffler

21. _____ circuits channel high, medium, and low frequencies to the appropriate speaker.
a. Resistive
b. Parallel
c. Crossover
d. Series

22.______ due to speakers may cause a tone arm to skate across a record or a CD player to skip.
a. Electrical surges
b. Voltage fluctuations
c. Vibrations
d. Imbalances

23. To record the music from a sound source on tape it is useful to have a ______ on a receiver.
a. tape monitor switch
b. loudness switch
c. filter switches
d. volume control

24. The output voltage from most sound sources reaching the preamplifier is about ______ volts.
a. 0.005
b. 0.05
c. 0.5
d. 5.0

25. The best and most expensive drive for a turntable was ______.
a. belt drive
b. rim drive
c. warp drive
d. direct drive

26. Tone controls like treble, volume, bass are found on the ______.
a. power amplifier
b. preamplifier
c. sound source
d. tuner

27. A device having a preamplifier, tuner, and power amplifier all in one unit is called a/an ______.
a. receiver
b. radio
c. integrated amplifier
d. console unit

28. A device that contains a tuner, pre-amp, power amp and speakers all in one unit is called a/an ______.
a. receiver
b. radio
c. integrated amplifier
d. compact system

29. The unit that produces the large currents is the ______.
a. magnetic cartridge
b. pre-amplifier
c. power amplifier
d. speakers

30. A unit that produces large currents that can drive speakers is called a/an ______.
a. tuner
b. turntable
c. preamplifier
d. integrated amplifier

Bibliography and Web Resources

Johnson, Kenneth, Willard Walker and John Cutnell. *The Science of Hi-Fidelity*. Kendall-Hunt Publishing Company, Dubuque, Iowa, 1995.

Symko, Orest. *Physics of Hi-Fi*. Kendall-Hunt Publishing Company, Dubuque, Iowa, 1995.

Stanionis, Victor and Hugh Berberich. *Computer Music: Science and Technology of a New Art*. Research Foundation of State University of New York, Stony Brook, NY 1989.

Surround sound - **http://hometheater.about.com/od/beforeyoubuy/a/surroundsound_5.htm**

CHAPTER 4

Waves

"I do not think that the wireless waves I have discovered will have any practical application."

Heinrich Hertz, 1885

First person to broadcast and receive radio waves

Why study wave motion? The concept of waves is imbedded in music and its generation. The terminology used in describing the operation of synthesizers is heavily dependent on the notion of wave motion and an understanding of its properties. Sound itself is a form of wave motion. The generation of sound by vibrating objects, the transmission of sound through the air and its reproduction by audio equipment involves understanding waves and how they behave. Plucking a guitar string, blowing a trumpet, or striking a drum head is intimately connected with wave motion. Reading the specifications for any type of electronic equipment reveals that the terminology is related to the study of wave motion. Acquiring an understanding of waves is the most important physical insight needed to fully appreciate the synthesis and generation of computer music.

Goals

After studying this chapter, you should be able to:

- Know why scientists think wave motion is important.
- Define what energy is.
- Define the meaning of work.
- Know what is necessary to produce a mechanical wave motion.
- Know the difference between a transverse and a longitudinal wave.
- Know the definitions of wavelength, amplitude, period, frequency, and wave speed.
- Know what a crest, a trough, condensation, and rarefaction are.
- Know the properties that distinguish a wave from a particle-like phenomenon.
- Know the definitions and meaning of reflection, refraction, interference and diffraction.

Energy

The study of wave motion is concerned with the fact that waves transfer energy from one place to another without mass motion of the energetic particles from one place to the other. Energy, by definition, is the capacity or ability to perform work. In physics, work has a narrow definition that requires a force to act and move an object through a displacement parallel to its line of action (Fig. 4.1).

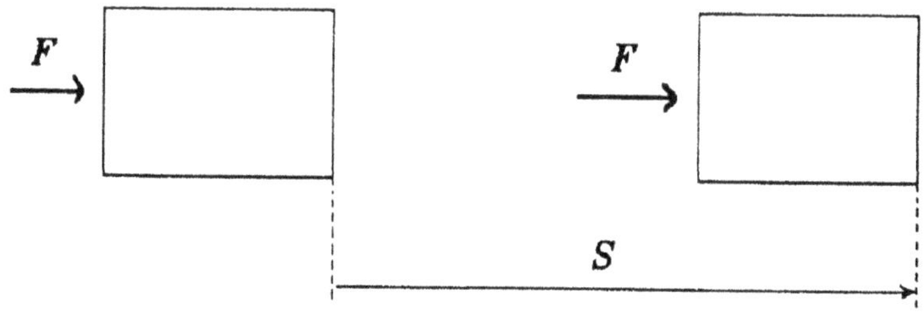

Figure 4.1 Force moving a box through a displacement

From *Computer Music: Science and Technology of a New Art,* Stanionis, Berberich. Copyright © 1989 by Research Foundation of State University of New York. Reprinted by permission.

Pushing on a wall all day would result in no work being done by the force if the wall didn't move. Work = F x S. In the metric system of units force (F) is measured in terms of a unit called the Newton and displacement (S) is measured in meters. So the units for work will be Newton meter (Nm). This combination of units has been given another name, the Joule (J). The renaming of units and combinations of units has been done purposely in science to memorialize the names of great scientists and honor them.

A moving object is said to possess energy because it has the capacity or ability to perform work. A block of wood struck by a speeding bullet will experience a force that will move it through some displacement, and thus have work done on it. A compressed spring also has such a capability.

The word work is used broadly in our everyday language. Having to "go to work" is a relatively recent phenomenon. In the "old days" around 1800 or so, one would awaken, wash, feed the chickens, milk the cows, eat breakfast, and sit on the porch and watch the grass grow. Then all of a sudden a new technology arrived. It required large amounts of capital and had to be housed in a large structure far from home. You had to hop on your horse and commute to work. The so-called muscle amplifier, the steam engine, powered the Industrial Revolution. With the help of the machine, one man could do the work of ten. Industrialists even put it on wheels (steam engine-powered train) to transport workers to the factory. Today, there has been somewhat of a reversal in the new Information Technology Revolution powered by the computer, the so-called brain amplifier. It has turned us all into paper shufflers and some of us don't have to "go to work" because we can telecommute from home. Which do you prefer: commuting or telecommuting?

Mechanical Waves

Suppose an explosion took place a block from your home. A chemical reaction taking place generates a gas at high temperature which then expands, creating a pressure wave moving outward from the focus of the explosion. However, these high velocity air molecules created by the chemical reaction never reach the windows of your home but rather push aside the surrounding air molecules which in turn push others creating a domino-like effect. Finally, the molecules resting against your windows are pushed against the window pane and may break it. The energy has been transferred from the source through a medium, air, to the window and sufficient force may be generated to break it. We see that energy transferred by a wave motion can do work.

Transverse Waves

We are all familiar with waves of one type or another such as sound waves, water waves, or waves on a string. All wave motions have certain properties in common. The study of mechanical wave motion will help in understanding sound. Waves traveling along a string require two things in order to exist: a **source** (the vibrating spring) and a **medium** (the elastic string) (Fig. 4.2 and Fig. 4.3). This need for a source and a medium turns out to be true for all mechanical wave motions such as vibrating strings and sound waves.

Figure 4.2 Mass attached to spring with attached string
From *Computer Music: Science and Technology of a New Art,* Stanionis, Berberich. Copyright © 1989 by Research Foundation of State University of New York. Reprinted by permission.

There are two main types of wave motion that are of interest in computer music: transverse and longitudinal waves. Consider a string attached to an oscillating mass hanging from a spring. Attaching the string to the oscillating mass causes a disturbance, a transverse wave, to travel down its length (Fig. 4.3). As the mass attached to the spring moves up and down, it drags the attached string through a displacement in the vertical direction. This results in a displacement of succeeding particles along the length of the string since each is elastically attached to its neighbor. A wave made up of displacements takes place along the length of the string. The mass attached to the spring constitutes the vibrating source for the mechanical wave and the string provides the elastic medium for propagation of the wave. The motion of the particles of the string is at right angles to the disturbance that travels along the length of the string. Such a wave is said to be a transverse wave.

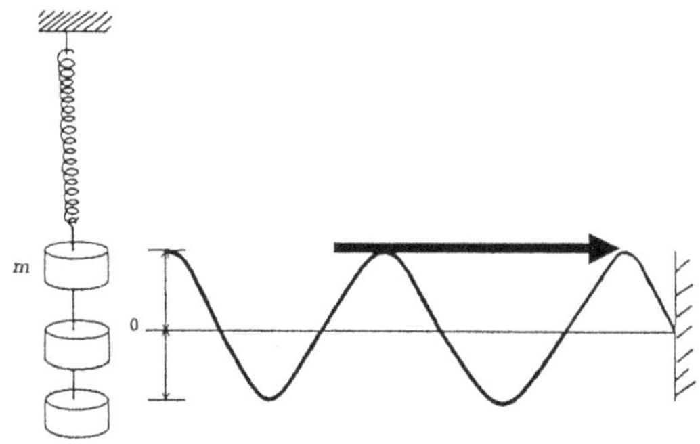

Figure 4.3 Vibrating mass with string attached

From *Computer Music: Science and Technology of a New Art,* Stanionis, Berberich. Copyright © 1989 by Research Foundation of State University of New York. Reprinted by permission.

It is truly amazing that transverse mechanical waves cut into a vinyl record can produce sound and music. Edison recorded his waves on a single play tin foil cylinder and Alexander Graham Bell used a wax cylinder. The master recording was the analog of a sound wave. As the stylus of the magnetic cartridge carried by the tone arm oscillates back and forth it converts the mechanical wave into an electrical signal which is transformed into sound. The music has been captured in a mechanical form on vinyl (Fig. 4.4).

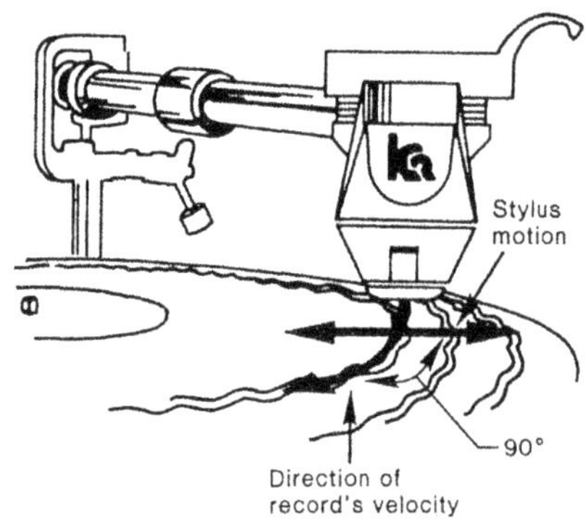

Figure 4.4 Monophonic record and stylus

From *Science of Hi-Fidelity* by Johnson, Walker, Cutnell. Reprinted by permission of the authors.

Wavelength

Suppose the source is vibrating so that the particles of the elastic medium, or in this case the string, reach the same displacements vertically, although at different times. Ignore frictional effects.

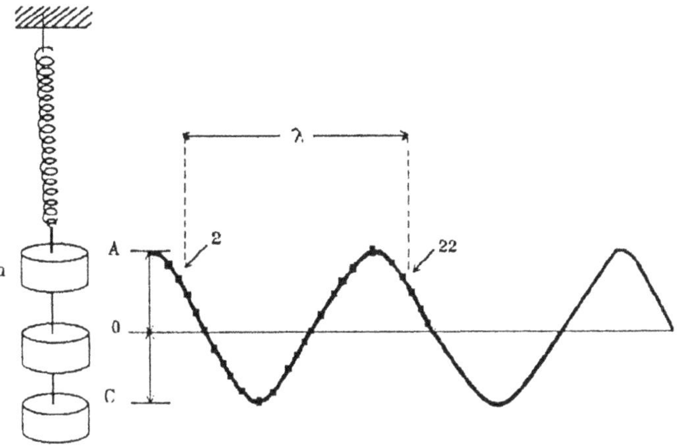

Figure 4.5 Wave traveling on a string

From *Computer Music: Science and Technology of a New Art,* Stanionis, Berberich. Copyright © 1989 by Research Foundation of State University of New York. Reprinted by permission.

Particles 2 and 22 are in phase meaning both are at the same displacement at the same instant of time, although one cycle apart (see Figure 4.5). The distance between these two particles is said to be one wavelength and is denoted by the Greek letter lambda, λ. In fact, the distance between any two successive particles moving in phase is the wavelength of the motion.

Period and Frequency

The time it takes in seconds for the source, or any particle along the length of the string, to execute one complete vibration or oscillation is called its period and is denoted by T. One oscillation would occur when the mass moves from the center to its lowest position, back up through the center to its highest position, and back down to the center. The period tells us how long it takes, in seconds, for the source or any elastic particle, in the string to execute one complete oscillation.

Sometimes we are interested in the number of vibrations that can take place in one second. The number of oscillations that take place in one second is termed the frequency of the vibration and is designated by the letter f. It is simply the reciprocal of the period, i.e. $f = 1/T$, the number of cycles per second. The unit in which frequency is measured is called the Hertz (Hz) and is equal to one cycle per second.

Example:
Suppose you are told that the time or period for one oscillation is 3 sec/cycle and are asked for the frequency.

Since $f = 1/T = 1/(3\ \text{sec/cycle})$,

$f = 1\ \text{cycle}/(3\ \text{sec})$

or

$f = (1/3)\ \text{cycle/sec}$ or $(1/3)$ Hz.

The term frequency is often used in connection with specifications for audio hi-fidelity equipment and is more useful for comparison purposes than the period. For example, you may have seen speakers advertised as capable of reproducing frequencies in a range of 50 to 20,000 Hz. Compare this with the human ear which can be sensitive to frequencies in the range of 20 to 20,000 Hz.

Amplitude

The amplitude of the wave motion is denoted by **A** and corresponds to the maximum displacement of the mass attached to the spring from its equilibrium position. **A** is also the maximum height of the transverse wave traveling along the attached string. A transverse wave is said to consist of a series of crests and troughs that travel along the string (Fig. 4.6).

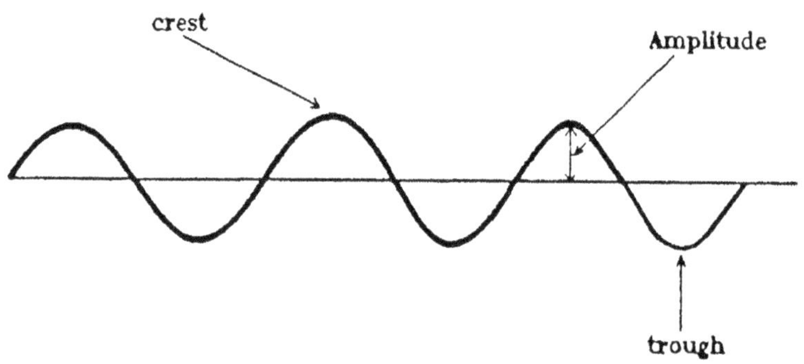

Figure 4.6 Amplitude of a transverse wave

From *Computer Music: Science and Technology of a New Art,* Stanionis, Berberich. Copyright © 1989 by Research Foundation of State University of New York. Reprinted by permission.

The velocity of the wave motion propagating on the string is given by distance divided by time and, since a crest will travel one wavelength in a time equal to one period,

$$v = \lambda/T.$$

The velocity **v** is the phase velocity or wave velocity of a single wavelength. If a wave disturbance consists of a group of waves with slightly different wavelengths, they combine to have a group velocity which may or may not be the same as the phase velocity.

Example:

Suppose a wave has a period of .02 sec and a wavelength of 6.9 meters. What is the phase velocity?

$$v = \text{distance/time}.$$

$$v = d/t.$$

$$v = 6.9 \text{ m}/.02 \text{ sec}.$$

$$v = 345 \text{ m/s}.$$

So, we can write

$$v = \lambda/T = \lambda(1/T).$$

Note that since the frequency is equal to 1/T, it is possible to replace T in the formula and rewrite it in terms of f.

$$v = \lambda(1/T) = \lambda f \text{ or } f\lambda.$$

Since frequency is used more often then period for sound waves, the formula for wave speed will be taken as

$$v = f\lambda.$$

This relationship for v, f and λ is true for all types of wave motion.

Longitudinal Waves

We said that a wave traveling along a string is called a transverse wave because the particles of the medium move at right angles, perpendicular to the direction that the disturbance and energy are traveling (Fig. 4.7).

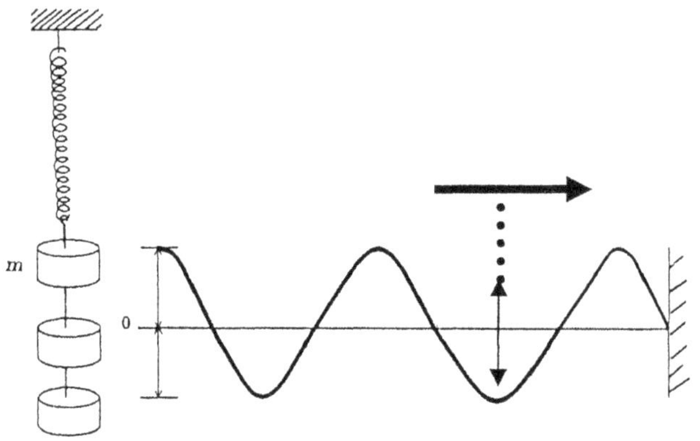

Figure 4.7 Wave moves at right angle to motion of particles on string
From *Computer Music: Science and Technology of a New Art,* Stanionis, Berberich. Copyright © 1989 by Research Foundation of State University of New York. Reprinted by permission.

What happens when a long, slinky-type spring is pushed or pulled at one end by a vibrator? The coils of the spring move back and forth, parallel to the disturbance traveling along the length of the coil (Fig. 4.7). Such a wave is called a longitudinal wave. Motion of the coils (or medium) is parallel to the direction of propagation of the wave motion. Places where the coils (the medium particles) are closer together than normal are called condensations and places where they are farther apart than normal are labeled rarefactions (Fig. 4.8). So a longitudinal wave consists of a series of condensations and rarefactions. Sound waves are longitudinal waves. We shall consider sound as a wave motion in the next chapter.

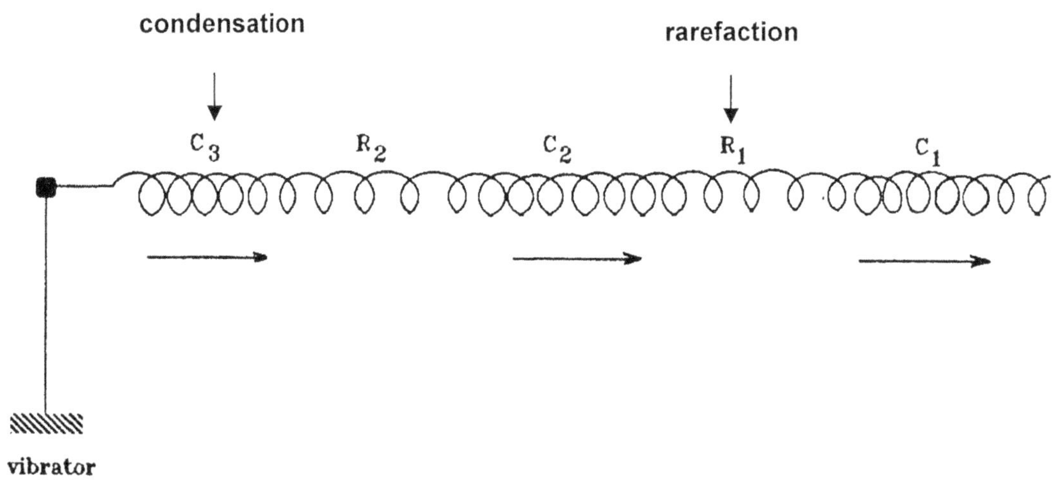

Figure 4.8 Longitudinal waves on a spring
From *Computer Music: Science and Technology of a New Art,* Stanionis, Berberich. Copyright © 1989 by Research Foundation of State University of New York. Reprinted by permission.

MORE ON WAVES

X-rays were discovered accidentally in 1895 by William Conrad Roentgen while experimenting with a cathode ray tube in his lab. Many scientists at this time were experimenting and observing the effects of electron beams in gas discharge tubes. Essentially, they were applying a voltage to two electrodes separated from one another, enclosed in a glass tube, and passed through a gas at extremely low pressure. Some scientists used various gases in the tube and observed colored light and light patterns. In others cases the tube was completely evacuated as the discharge between the electrodes took place (Fig. 4.9).

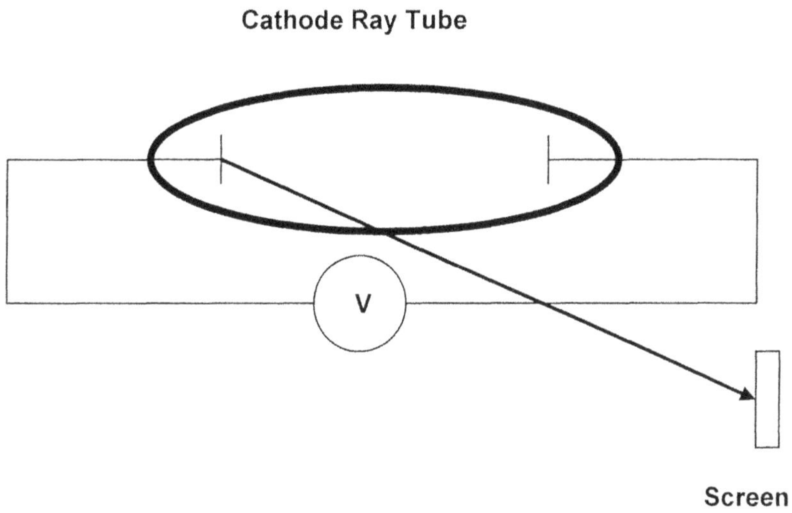

Figure 4.9 Cathode ray tube and fluorescent screen

In one account of Roentgen's discovery, he is said to have noticed that a fluorescent screen in his lab glowed when the electron beam was turned on, in spite of the fact that he had surrounded his tube with heavy black cardboard. He was puzzled. He finally came to the conclusion that the cathode ray tube was giving off some kind of rays that were affecting his fluorescent screen. What is the name given to the unknown variable in algebra? Roentgen called the rays X-rays to indicate what he was observing was an unknown type of radiation.

A week later Roentgen took an X-ray photograph of his wife's hand (Fig. 4.10) with her wedding ring. The stunning photograph showed the bones in her hand and her wedding ring, impressing the general public and exciting the scientific world. Lawmakers tried to legislate against the use of X-rays because they were afraid scientists would be able to use them to see through people's clothes. They did not understand that only bones and dense objects would be seen through the clothing. Conrad Roentgen was the first recipient of the Nobel Prize in physics in 1901.

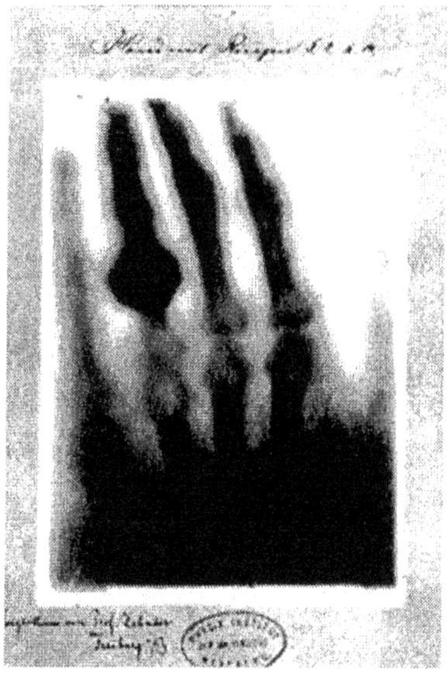

Figure 4.10 X-ray of Roentgen's wife's hand
http://imagers.gsfc.nasa.gov/ems/xrays.html

Scientists subdivided physical phenomena at this time into two main categories. A newly discovered phenomenon was either considered to be a particle-like phenomenon or a wave-like phenomenon. What were X-rays? How did scientists determine whether they were a beam of particles or some sort of wave?

Technology and Consumers

As consumers, one of the thing we have to keep in mind are the side effects that might be associated with a new technology. X-rays were thought of as one of the greatest advancements in medical science, allowing physicians to see bone structures, broken bones and even swallowed objects, through human tissue without having to cut through layers of flesh. However, sometimes there is a time lag between the introduction of a new technology and the documentation of harmful side effects associated with it. In the case of X-rays it took about fifty years for researchers to discover that X-rays can produce mutations in living cells and exposure to X-rays are the source of many varieties of human cancer. We always have to weigh the benefits of a new technology versus the risks it presents.

X-ray Shoe Machines

During the late 1940s and early 1950s practically every shoe and department store had a shoe-fitting X-ray unit which used a fluoroscope (Fig. 4.11). A fluoroscope consisted of an X-ray tube mounted near the floor in a partially or wholly shielded box and a fluorescent screen. The consumer placed feet with the new shoes into the opening at the bottom of the machine. An image of feet outlined by the shoes was formed by the X-rays penetrating the shoes and feet and projected on a fluorescent screen. The image was viewed simultaneously by the wearer, salesman, and usually your mother. Mothers have always been concerned with obtaining the right fit for shoes so that their children's' feet would not be malformed as they grew.

Figure 4.11 Typical x-ray shoe machine
Permission granted by Oak Ridge Associated Universities

Many times the machines were not shielded properly or out of adjustment. By the early 1950s, the radiation hazard associated with these machines became known. Each state by the 1970s had banned the use of these shoe x-ray machines but it is reported that such a machine was still operating in West Virginia as late as 1981.

Color TV sets

During the 1960s color TV sets became a popular consumer item. The general public switched from their black and white sets to full color TV. The color picture tubes operated at a higher operating voltage than in the black and white sets, which meant that high voltage power supplies accelerated electrons that struck and activated the phosphor coating at the front of the picture tube, simultaneously producing X-rays. The electron beam also produced soft (low energy) X-rays when it struck the front of the tube.

People, especially young children, were warned to sit at least six feet away from the front of the color set. Some consumers thought that the warning would help prevent eye disorders and were not aware of the harmful effects of X-rays. Children are especially vulnerable to the possibilities of suffering from X-ray exposure since human cells are most susceptible to damage while dividing. On the average cell division in children is most rapid as they are growing. So, the meaning of this warning wasn't exactly clear to the general public.

The highest emission levels from color TV were reported in 1968 and 1969. The production of X-rays could be limited by using thick glass in the walls of the picture tube and using solid state components in the power supplies. In 1968, the U.S. Congress passed into law the legal limits for radiation of 0.5 milliroentgens per hour at a distance of 5 cm from any external surface on the set. The law is enforced by the Federal Food and Drug Administration (FDA). The annual exposure since the 1970s for the US population is about 1 millrem per year.

Manufacturers of TV sets complied, so that by 1972 all manufactured sets had thicker glass on the front and sides of the picture tube. As a result, most CRTs have glass envelopes made from heavily leaded glass. The leaded glass presents an environmental hazard upon disposal because of the possible leaching of the lead in landfills in the presence of acid rain. The phosphor coating is also toxic and may be released upon implosion. Today, disposal of TV sets and computer monitors is governed by local laws because of the hazards they present to the environment.

Other X-ray sources

Video terminals and computer display screens are essentially the same as color TV screens and the X-ray emissions from such devices are the same as present day TV screens. Airport X-ray security devices that subject luggage to low doses of X-ray radiation protect the flying public from potential harm from terrorists and pose a low risk versus the benefit derived. The amount of radiation exposure is smaller than from natural exposure in the environment. Although the luggage being screened has been radiated, there is no residual contamination.

Electromagnetic Radiation

Another concern consumers have had to face recently deals with electromagnetic radiation from power lines and radio broadcasting stations, microwave ovens and cell phones. The radiation from the transmitting tower of a radio station is caused by accelerating electrons up and down a large vertical antenna tower (Fig. 4.12).

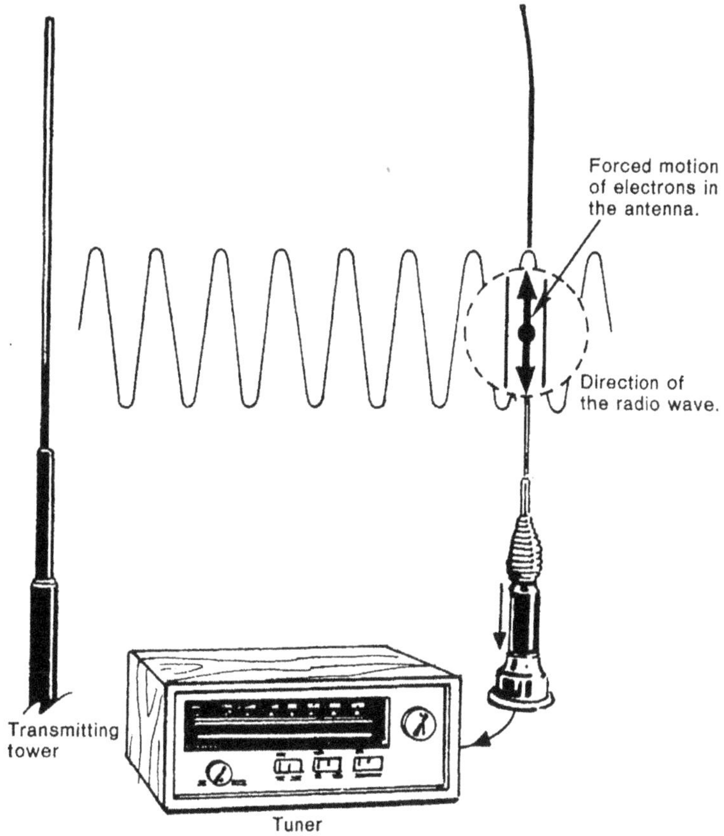

Figure 4.12 Radio waves
From *Science of Hi-Fidelity* by Johnson, Walker, Cutnell. Reprinted by permission of the authors.

As the electrons oscillate up and down the transmitting tower they generate an electronagnetic wave that travels through space. When the wave strikes the metal receiving antenna, it causes electrons in the antenna to oscillate. This produces a small voltage in the antenna which is sent by the tuner to the preamplifier and is ultmately converted into the sound we hear. Since the 1920s, we have been exposed to a greater amount of long wavelength electromagnetic waves produced in radio transmission, TV transmission, microwave transmission, etc.

These same oscillating electrons are found in overhead high voltage transmission lines that some people believe cause living cells to malfunction and develop defects causing all sorts of diseases. Experience and history has taught us to be wary of new technologies. The fact that relatively little scientific research has been done on the effects of electromagnetic waves on humans is a cause for concern.

Some research seems to indicate that electromagnetic waves are useful in promoting healing in bone fractures by stimulation with magnetic fields. Bone is a living tissue and has the ability to heal itself when broken or fractured. Occasionally, the fracture fails to heal in the normal period of time and the danger of gangerene exists. Applying electromagnetic fields to the fracture seems to promote bone growth and healing.

One study involved two groups of rats that had their legs bound for a period of time. One group was exposed to electromagnetic radiation and the other was not. When released from their bonds, those that were exposed ran around normally and the others were crippled.

Russian astronauts in the early days of earth orbiting were found to have large calcium losses in bone tissue after being in orbit for several months and had to be carried from their landing module. They recuperated slowly on earth as they absorbed more calcium. It seems that the gravitational field of the earth triggers electrical impulses in humans as they move under the influence of gravity in carrying out their daily activities. The pounding effect seems to stimulate the nervous system and promote calcium absorption. As a result, U.S. astronauts when orbiting in space have to perform a rigorous set of exercises in space while weightless to promote the electrical stimulation for calcium absorption.

What about those microwave ovens? When they first were produced, everyone thought they were technological marvels. By the way, how does a microwave cook? Why is it that a cup of coffee when placed in a microwave will get very hot and a paper plate always remains cool? It turns out that it is the water molecules in the material that oscillate when struck by a certain frequency of microwave radiation and its their oscillations which generate friction in the food causing a heating or cooking effect. The paper plate doesn't have any water molecules and therefore doesn't heat up.

However seals around microwave oven doors age like all plastic materials. What do you first notice when you enter a brand new car? Yes, the new car smell is evident. What causes it? Basically it is the plasticizers evaporating form the plastic materials in the interior of the car. Eventually, they will dry and the new car smell will be gone. Without the plasticizers and with exposure to UV radiation the plastic becomes brittle and cracks. Time to buy a new car? Well the same thing happens to the seals on a microwave oven and this allows microwave radiation to leak out. Microwave radiation can cause cataracts and sterility. So, when cooking with a microwave, it is a good idea to put the food in and walk away. Watching the food go round-and-round is not a good idea.

In the case of cell phones it is known that all cell phones by their nature radiate electromagnetic waves from their antennas which are usually in close proximity to your brain when using a cell phone. What are the effects? One neurosurgeon is suing because he developed brain cancer. He claims he developed the cancer from frequently using an analog cell phone in the early 1990s. A study done in Scandanavia in 2003 showed that digital cell phone radiation does effect human tissue and the report suggests using an earphone to keep the cell phone antenna at a distance, just in case. More studies are presently being done to ascertain the effects of cell phone radiation on humans. **What does all this say about being the first on the block to adopt a new technology?**

Back to the Properties of Wave Motion

How do we distinguish between a particle-like phenomenon and a wave-like phenomenon? Roentgen faced this problem when trying to determine the nature of X-rays. Wave-like phenomena are characterized by possessing certain properties such as: reflection, refraction, interference, and diffraction.

Reflection

In order for mechanical waves to transfer energy from one place to another, a medium is necessary. Reflection of a wave motion is a measure of the amount of energy that is returned to the original medium when a wave motion encounters an interface between two different media (Fig. 4.13). If a long taut string attached to a wall is snapped a pulse (a crest), will travel down the length of the string until it reflects from the wall. It is interesting to observe that the crest will be reflected as a trough and travel back to the source.

The wave, as it travels through the less dense medium, and reflects from the more dense medium back into the less dense medium, undergoes a phase shift from a crest to a trough. Observation reveals that when the wave travels as a trough down the string and comes to the string-air interface, it also is reflected back, still as a trough. Reflections play an important role in room and concert hall acoustics, which we shall examine in a later chapter.

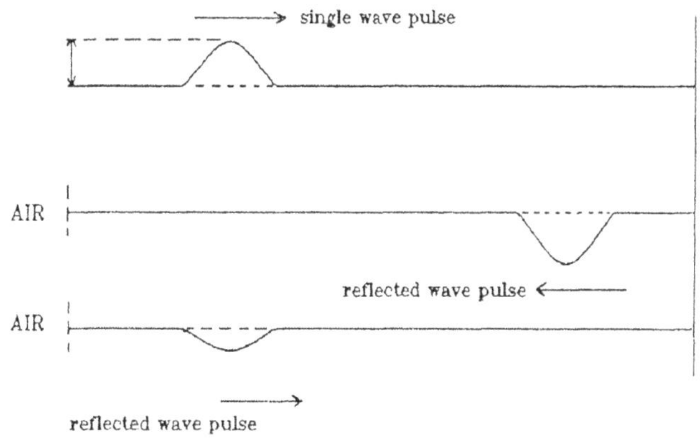

Figure 4.13 Transverse wave pulse on a string reflecting from a wall
From *Computer Music: Science and Technology of a New Art,* Stanionis, Berberich. Copyright © 1989 by Research Foundation of State University of New York. Reprinted by permission.

In reflecting from a more dense medium onto less dense and then back to more dense, there is no phase shift (Fig. 4.11). Experiment reveals that the effect of reflection on longitudinal waves is the opposite; when a condensation travels in the less dense medium and reflects from a more dense medium back into the less dense medium, no phase shift occurs.

Refraction

If waves travel from one medium into another, there will be some reflection whether the medium is more dense than the original or less dense. Some of the energy carried by the wave motion will be transmitted and some reflected back into the original medium. Refraction of a wave motion is defined in terms of the change in velocity it experiences as it travels from one medium into another (Fig. 4.14).

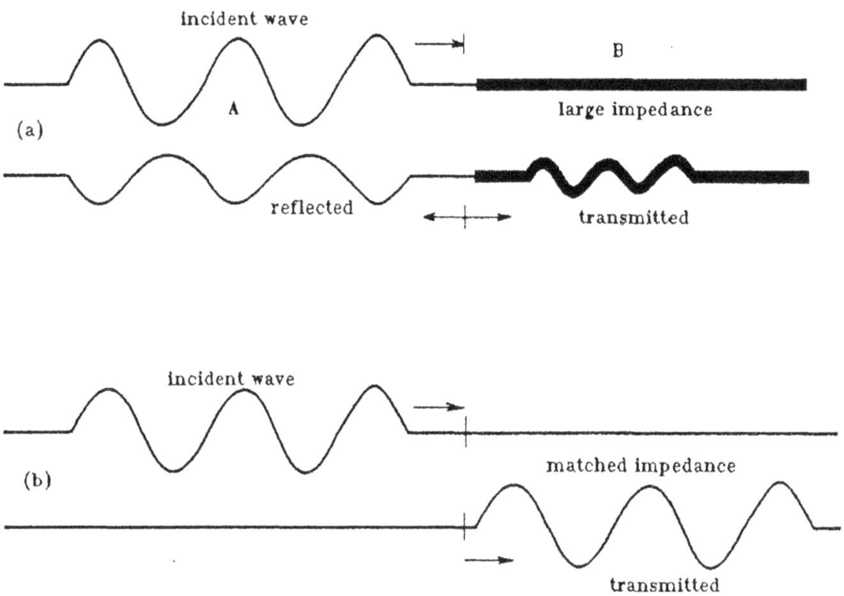

Figure 4.14 Wave partially reflected and partially transmitted at interface with change in velocity
From *Computer Music: Science and Technology of a New Art,* Stanionis, Berberich. Copyright © 1989 by Research Foundation of State University of New York. Reprinted by permission.

If a pulse is traveling down the length of a string, and all of the energy it carries is transmitted at the junction with another string, we say that their impedances are matched. Impedance is concerned with the response of a system when forces are applied to it. If the impedances are mismatched, none or only a portion of the energy may be transmitted (Fig. 4.12).

Unlike the waves on a string sound waves travel in three dimensions. When sound waves reflect from a smooth surface, the angle of incidence and the angle of reflection are found to be equal. When listening to a lecture in an auditorium, part of the sound wave from the speaker travels directly to you while part reflects off the walls before it reaches you. Each time a wave is reflected some of its energy will be absorbed by the reflecting medium. Consequently, the reflected sound has a lower intensity then the original sound. This is true because each time a sound wave is reflected, part of the sound power is refracted into the material, where it undergoes a change in velocity and is partially absorbed.

Interference

Interference of wave motions, which is used extensively in synthesizers of all types depends on the Principle of Linear Superposition. This principle basically states that when two wave motions are traveling through a medium simultaneously, each wave proceeds through the medium and produces its effect independently of the presence of the other wave. In any region where two waves are present simultaneously, the resulting disturbance is made up of a sum of the two individual disturbances. The y value (amplitude) of a particular particle in the resultant wave disturbance is computed by adding the y values in the two independent motions for the same values of x. This could lead to total destructive interference, total constructive interference, or partial constructive and partial destructive interference ("**R**" in Fig. 4.15).

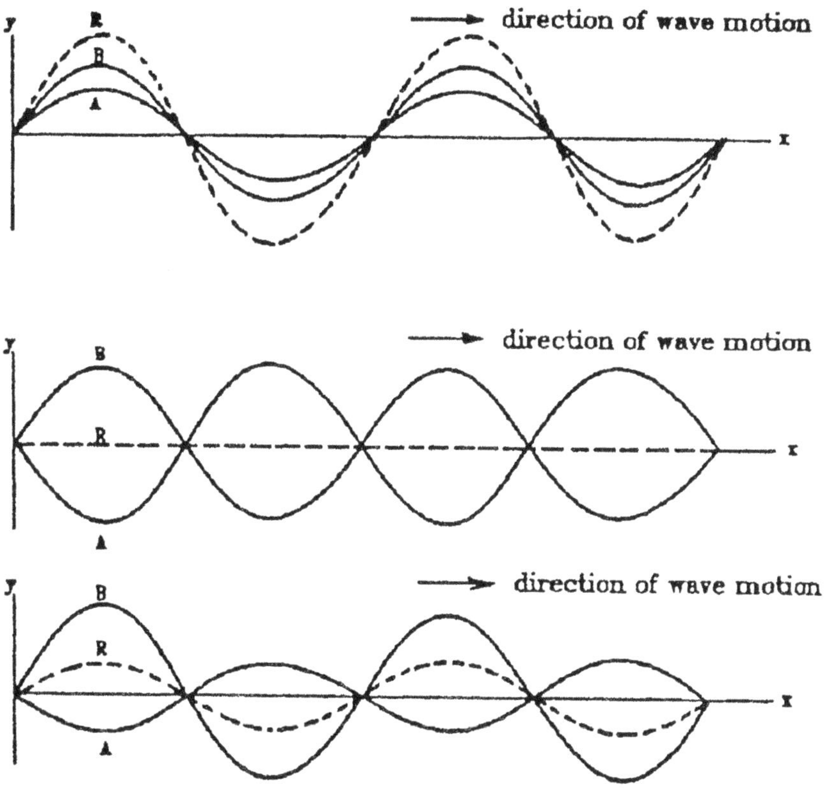

Figure 4.15 Interference: constructive, destructive, partial constructive and destructive
From *Computer Music: Science and Technology of a New Art,* Stanionis, Berberich. Copyright © 1989 by Research Foundation of State University of New York. Reprinted by permission.

Standing Waves

Standing waves in a medium occur because of interference and reflection effects. The phenomenon of standing waves may occur when there are two waves that have the same frequency and amplitude and are traveling in the same medium in opposite directions. Such a wave combination sets up a system of stationary waves where there will be points along the string where there is almost no movement, nodes, and points of maximum vibration called loops or anti-nodes (Fig. 4.16). As we shall learn in the next chapter, standing waves may result from how loudspeakers are arranged in a room.

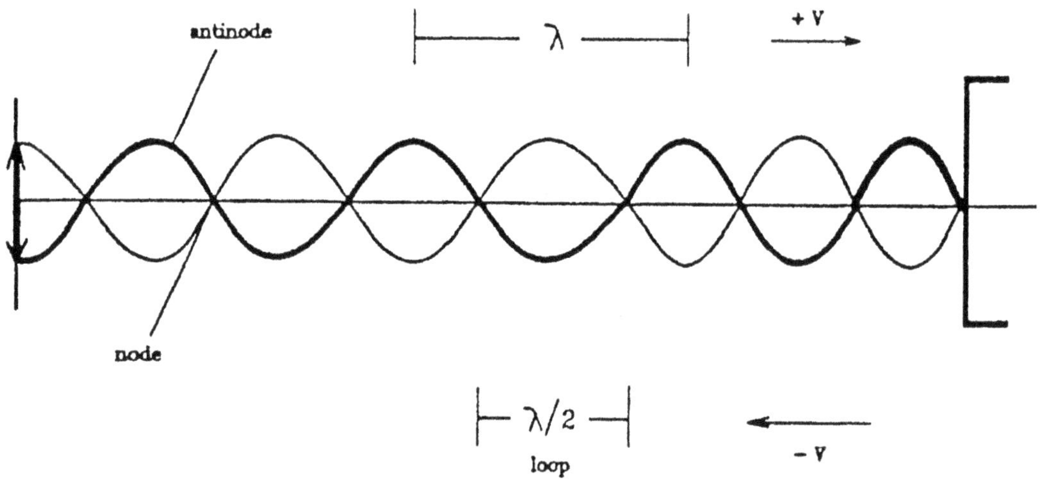

Figure 4.16 Standing waves

From *Computer Music: Science and Technology of a New Art,* Stanionis, Berberich. Copyright © 1989 by Research Foundation of State University of New York. Reprinted by permission.

Standing waves are important in gaining an understanding of how musical instruments produce sound. For string instruments, the medium is a stretched string. Wind instruments utilize air as the medium and percussion instruments use various media. A drum uses a stretched membrane and a bell its hollow metal form.

By plucking a string we may set up standing waves. If we pluck a string under tension at its center, it's possible to excite the string to vibrate in one segment (i.e. two nodes and one anti-node). It is interesting to write the frequency for this mode of vibration in terms of the length of the string, L.

$$v = f\lambda$$

$$f = v/\lambda$$

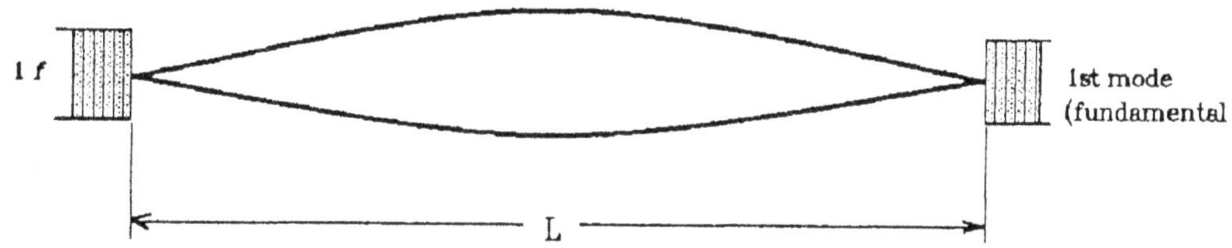

Figure 4.17 First harmonic (fundamental mode of vibration)
From *Computer Music: Science and Technology of a New Art,* Stanionis, Berberich. Copyright © 1989 by Research Foundation of State University of New York. Reprinted by permission.

$L = \lambda/2$ as seen from Figure 4.17. Therefore $f = v/2L$. The string is vibrating in one segment, or, its fundamental mode. Sometimes this is called the first harmonic.

If we pluck the string carefully to begin with, we can have it vibrate in two segments. This is the second harmonic, or, first overtone (Fig. 4.18).

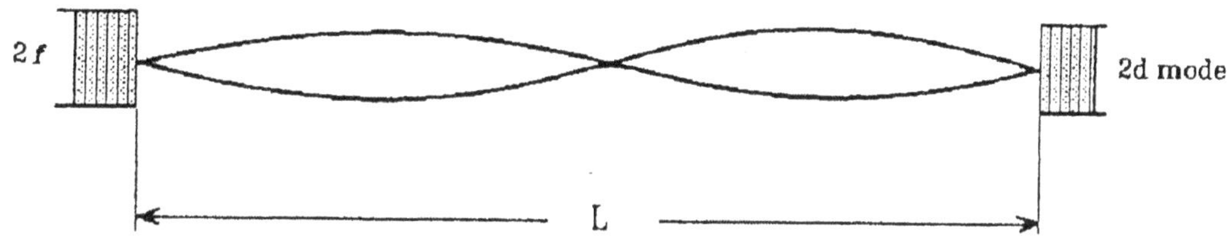

Figure 4.18 Second Harmonic (first overtone)
From *Computer Music: Science and Technology of a New Art,* Stanionis, Berberich. Copyright © 1989 by Research Foundation of State University of New York. Reprinted by permission.

Now $L = \lambda$ and $f = v/L$ or $f = 2(v/2L)$. Note that this is twice the frequency of the fundamental mode.

When the string vibrates in three segments (Fig. 4.19)or in its third harmonic (or second overtone), $L = 3\lambda/2$ and $f = 3(v/2L)$.

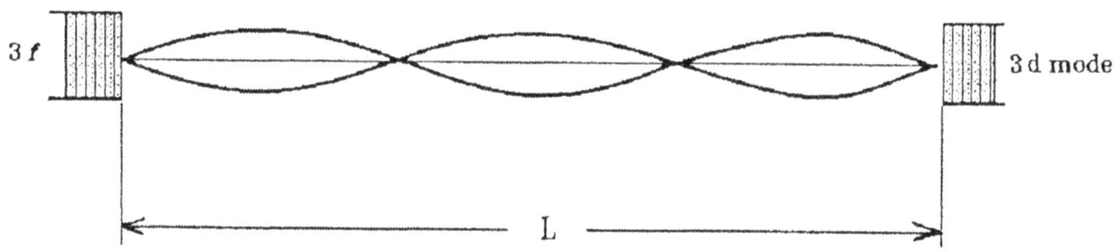

Figure 4.19 Third harmonic (second overtone)
From *Computer Music: Science and Technology of a New Art,* Stanionis, Berberich. Copyright © 1989 by Research Foundation of State University of New York. Reprinted by permission.

For a string vibrating in eight segments (Fig. 4.20), or, in the eighth harmonic (seventh overtone) f = 8v/2L.

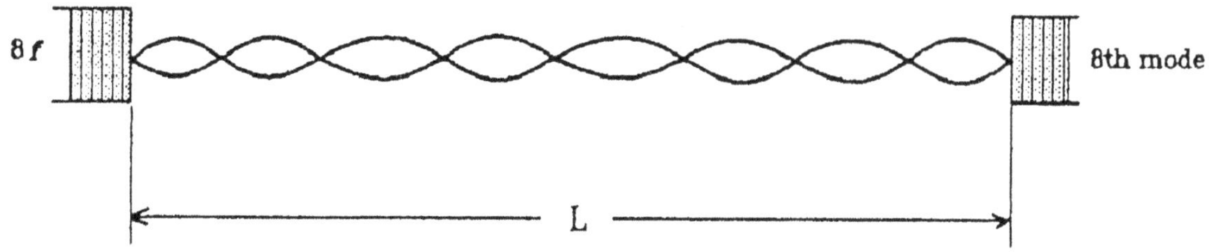

Figure 4.20 Eighth harmonic (7th overtone)
From *Computer Music: Science and Technology of a New Art,* Stanionis, Berberich. Copyright © 1989 by Research Foundation of State University of New York. Reprinted by permission.

So for the fundamental or first harmonic

$$f = 1(v/2L).$$

For the second harmonic,

$$f = 2(v/2L).$$

For the third harmonic,

$$f = 3(v/2L).$$

For the eighth harmonic,

$$f = 8(v/2L)$$

The pattern is clear.
For the nth harmonic,

$$f = n(v/2L).$$

It is possible for the plucked string to vibrate in several of its frequencies at (Fig. 4.21). Ordinarily, when a string is plucked, the fundamental frequency and several harmonics are excited. The loudness of each frequency is different but when all are heard simultaneously, we hear the musical note characteristic of the guitar.

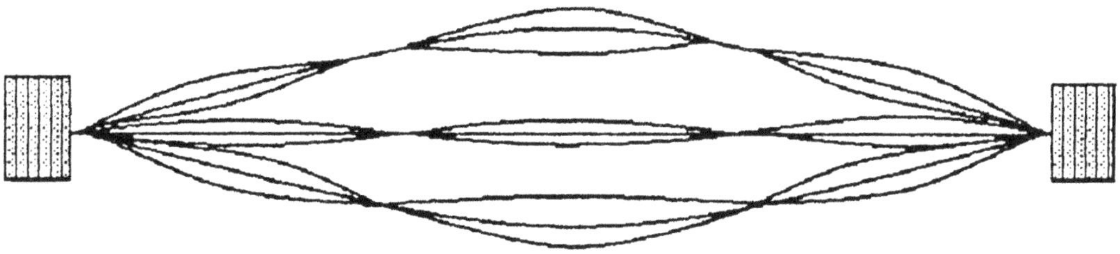

Figure 4.21 String vibrating with several frequencies at once
From *Computer Music: Science and Technology of a New Art,* Stanionis, Berberich. Copyright © 1989 by Research Foundation of State University of New York. Reprinted by permission.

When the observer wants to play another note, the allowed standing wave frequencies can be altered by changing the wave velocity, which is done by adjusting the tension, or density, of the string.

The player can easily change the length of the string with his fingers. Remember, $f = n(v/2L)$. As the length changes so do the allowed frequencies for standing waves. It can be shown that for transverse waves traveling on the string that the velocity can also be written as $v = (T/\rho)^{1/2}$, where T is the tension in the string and ρ is the density of the string. If we substitute this formula for v into $f = n(v/2L)$, we see the relationship for allowed standing wave frequencies depends on the tension and density.

$$f = n(T/\rho)^{1/2} /2L.$$

Tightening a string on the guitar increases its tension and, as learned from the formula, raises its frequency. Plucking a heavier or thicker string produces a lower frequency.

Diffraction

Diffraction of waves describes the phenomenon in which a wave disturbance, under certain circumstances, can bend around obstacles and spread out as it passes through openings. For diffraction of a wave motion to occur, the size of the obstacles and openings must be small in comparison with the magnitude of the wavelength of the wave. If the size of the object or opening is large compared with the wavelength of the disturbance, the wave will appear to travel in straight lines (Fig. 4.20). For example, we know sound waves can have wavelengths measured in feet. Windows, doors, chairs, buildings and other similar structures would also be measured in terms of feet and are capable of producing diffraction effects for sound waves (Fig. 4.22).

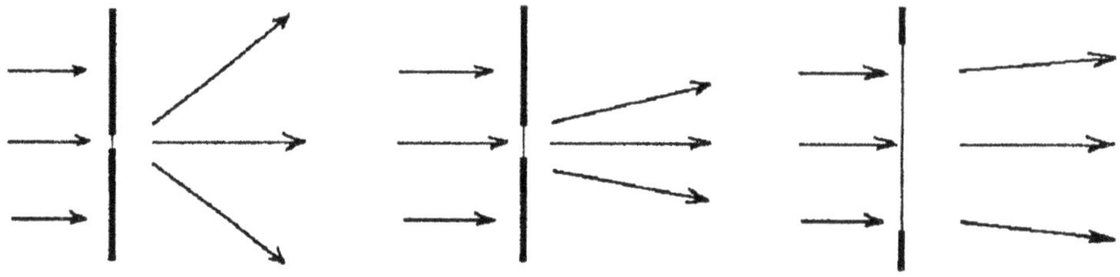

Figure 4.22 Waves being diffracted through different sized openings

From *Computer Music: Science and Technology of a New Art,* Stanionis, Berberich. Copyright © 1989 by Research Foundation of State University of New York. Reprinted by permission.

Roentgen conducted many experiments using X-rays to determine if he could refract and reflect them. He thought they had many of the characteristics of light, which was believed to be a wave-like motion. In 1912, a definitive observation was made by William Henry Bragg and William Lawrence Bragg (father and son) that X-rays showed interference effects when interacting with crystals and being diffracted in the small openings between atoms. This demonstrated that X-rays were electromagnetic waves.

Polarization

If the amplitude vector of a transverse wave traveling down the length of a string is completely symmetrical around the direction of propagation, the wave is said to be unpolarized (Fig. 4.23).

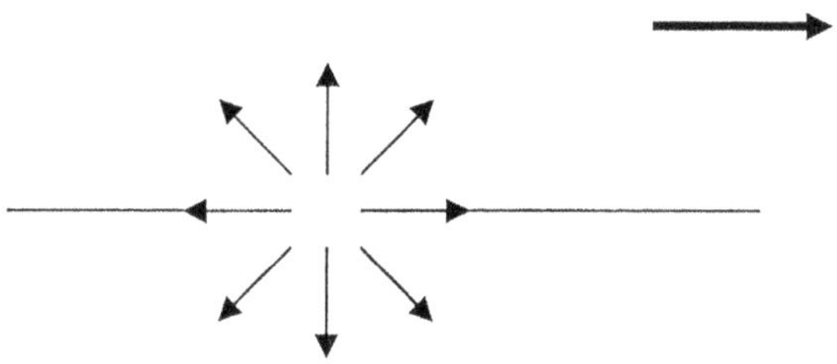

Figure 4.23 Amplitude vector of a transverse wave traveling to the right

If we thread the string through a gate with vertical stiles, the amplitude vector can only vibrate in the up and down direction in the plane of the paper. We say the wave is now plane polarized (Fig. 4.24).

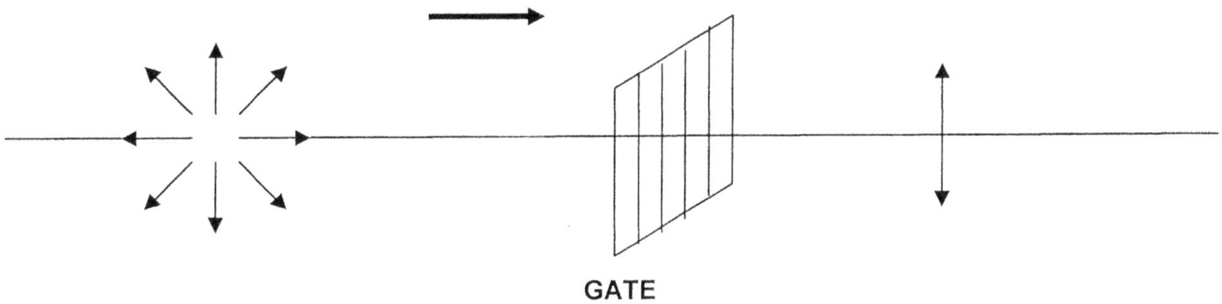

Figure 4.24 Unpolarized wave is plane polarized

If we send a longitudinal disturbance in the same direction, the back and forth vibrations easily fit between the stiles of the gate without change and the wave is unpolarized. So polarization allows us to test a wave motion to see if it is transverse. Polarization is a phenomenon that allows us to distinguish between transverse and longitudinal waves.

We have all observed how light can be polarized by pieces of Polaroid material or by Polaroid sunglasses. Scattering also polarizes light. Scattering occurs when light is reflected sideways out of a beam. Fog droplets will scatter light from a headlight so that it doesn't shine as far ahead as on a clear night. In fog you are able to see the beam emanating from the headlight because of scattering. Sunlight is scattered as it passes through the atmosphere of the earth, partially polarizing the light. This scattering, however, produces a more noticeable effect.

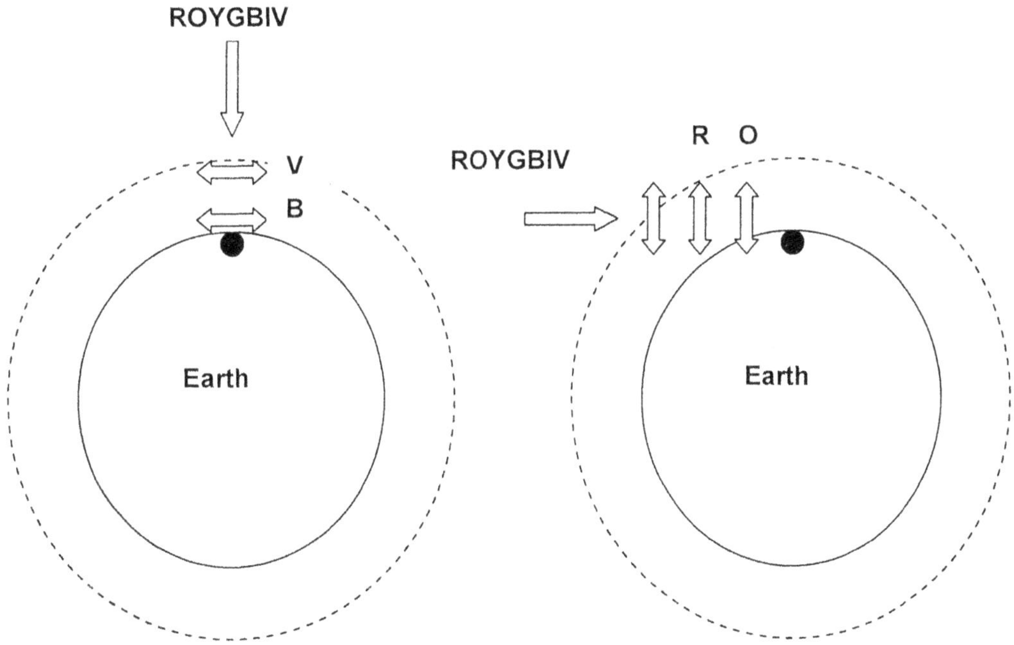

Figure 4.25 Scattering in the atmosphere

Has anyone ever asked you why the sky is blue? White light (as it comes from the sun) is made up of the colors of the spectrum: red, orange, yellow, green, blue, indigo, and violet – (ROYGBIV). At noon, when the sun is more or less directly overhead, the light is scattered sideways as it enters the atmosphere, with the short wavelengths being scattered first. When the light reaches us at the surface, the blue wavelengths are being scattered and the sky appears blue. At sunset, with the sun low on the horizon and not overhead, the light has to travel through a larger amount of atmosphere and more wavelengths are scattered out of the beam before it reaches us. As a result, we see the yellows and reds scattered at sundown (Fig. 4.25).

Doppler Effect

A train at rest blows it whistle once each second for five successive seconds. The frequency of the sound is the same in all directions. However, as the train moves in the forward direction, an observer ahead will hear an increase in frequency and one behind the train will hear a lower frequency. This is a case of the Doppler effect, which occurs because of the relative motion between the source of a sound and the observer of the sound (Fig.4.26). Doppler demonstrated this phenomenon using a band of musicians playing on a train approaching a group of observers.

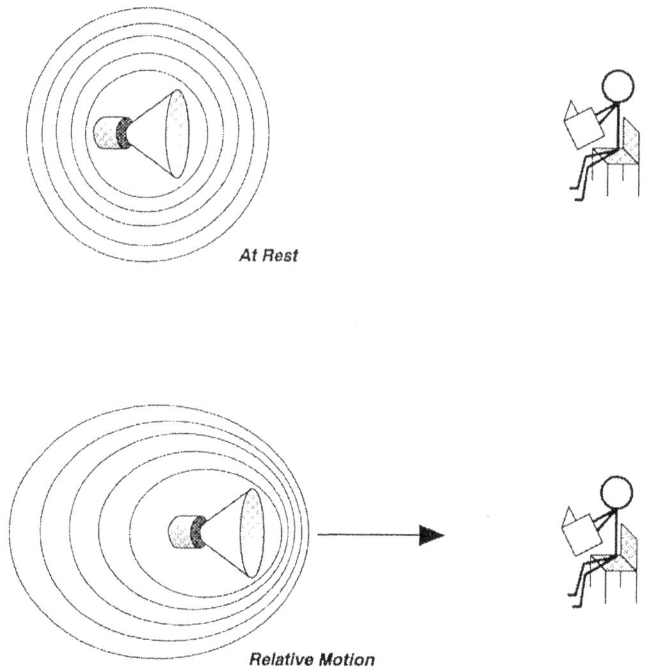

Figure 4.26 Doppler effect
From *Physics of Hi-Fi* by Symko. Reprinted by permission of the author

At a race car track, you notice the high pitch of the car engines as they approach, and the low roar as they pass you in the stands. This is a common example of the Doppler effect. The Doppler effect turns out to be important when considering the motion of the diaphragm of a loudspeaker. When the speaker is generating two tones of different frequencies simultaneously, the Doppler effect will cause a blurring of the tones. So a speaker vibrating at 100 Hz and 1000 Hz will be such as to cause a Doppler frequency shift in the 1000 Hz tone.

Police departments also use the Doppler effect to determine the speed of moving vehicles. They use a RADAR source sending out a steady beam of microwave radiation which then reflects back to the source from a moving vehicle. The waves are shifted in frequency by the Doppler effect. The beat frequency between the original wave and the reflected wave allows them to determine the speed of the vehicle.

Review Questions

1. What is energy?
2. Why study wave motion?
3. What is work?
4. What two things are necessary in order to have a mechanical wave motion?
5. Define what is meant by a transverse wave.
6. Define what is meant by a longitudinal wave.
7. What is reflection?
8. What is refraction?
9. What is meant by constructive and destructive interference?
10. Define diffraction of a wave.
11. What is meant by polarization and how is it used?
12. What is the Doppler effect?

Multiple Choice Questions

1. Waves transfer _____ from one place to another.
a. particles
b. frequencies
c. energy
d. material

2. An object is said to possess energy because it can _____.
a. do work
b. generate power
c. deliver jewels
d. generate force

3. For mechanical waves to exist, we need _____.
a. a mechanic and energy
b. a medium and a source
c. energy and power
d. amplitude and frequency

4. In a transverse wave the particles of the medium _____ to the direction of propagation of the wave.
a. vibrate parallel
b. vibrate perpendicular
c. vibrate randomly
d. act opposite

5. In a longitudinal wave the particles of the medium _____ to the direction of propagation of the wave.
a. vibrate parallel
b. vibrate perpendicular
c. vibrate randomly
d. act opposite

6. The maximum displacement of the particles from the equilibrium position is termed the _____ of the particle.
a. frequency
b. period
c. amplitude
d. wavelength

7. The distance between two successive particles in a wave executing the identical motion is called the _____.
a. frequency
b. period
c. amplitude
d. wavelength

8. The time it takes for the source to execute one complete oscillation is called the _____.
a. frequency
b. period
c. amplitude
d. wavelength

9. The number of vibrations the source makes in one second is called its _____.
a. frequency
b. period
c. amplitude
d. wavelength

10. The velocity of a wave as it travels in a medium is given by _____.
a. frequency x wavelength
b. wavelength/frequency
c. period/wavelength
d. period x wavelength

11. The Greek letter λ represents the _____ of the wave.
a. wavelength
b. period
c. frequency
d. velocity

12. Roentgen discovered _____.
a. wave motion
b. gamma rays
c. x-rays
d. cathode rays

13. When a wave motion is incident at the interface of two media and the energy is returned to the original medium, we call the phenomenon _____?
a. reflection
b. refraction
c. interference
d. diffraction

14. The ability of a wave motion to bend around obstacles and through openings is called _____.
a. reflection
b. refraction
c. interference
d. diffraction

15. When a wave motion changes speed as it passes from one medium into another, we call this _____.
a. reflection
b. refraction
c. interference
d. diffraction

16. The example of two wave motions traveling simultaneously through a medium and canceling each other out is called _____.
a. reflection
b. refraction
c. interference
d. diffraction

17. Standing waves occur when two waves which have the same _____ are traveling in opposite directions through a medium.
a. amplitude and frequency
b. frequency and period
c. wavelength
d. frequency

18. When a fixed string of length L vibrates in one segment, we have excited the _____.
a. first overtone
b. second overtone
c. first harmonic
d. Doppler frequency

19. When a fixed string of length L vibrates in four segments, we have excited the_____.
a. third harmonic
b. third overtone
c. fifth overtone
d. fifth harmonic

20. The sky is blue mainly due to _____.
a. reflection
b. refraction
c. scattering
d. polarization

21. When a wave's amplitude vector lacks symmetry around the axis of propagation, we say the wave is _____.
a. polarized
b. diffracted
c. refracted
d. scattered

22. When a fire engine is approaching you at an intersection, the frequency of its siren will be _____.
a. deeper
b. higher
c. lower
d. the same as when at rest

23. As a race car passes you, the pitch of its engine will be _____.
a. shriller
b. higher
c. lower
d. the same as when at rest

24. The change in frequency that occurs when there is relative motion between a source and observer is termed the ____.
a. Hertz effect
b. frequency effect
c. speaker dispersion
d. Doppler effect

25. When police car with a siren is approaching an observer, the pitch of the siren will be _____ when at rest.
a. lower than
b. higher than
c. the same as
d. no change as

26. If sound travels at 344 m/s at room temperature, the wavelength associated with a tuning fork vibrating at a frequency of 200 Hz is _____.
a. 344 m
b. 688 m
c. 1.72 m
d. 0.580 m

27. If the period of a wave is 5 s, then its frequency is _____ Hz.
a. 0.5
b. 5
c. 1/5
d. 2

28. If the frequency of a wave is 300 Hz and its wavelength is 5 m, its speed is _____ m/s.
a. 1500
b. 60
c. 1/60
d. 1.67

29. If 20 waves pass by a point in 4 s, its frequency is _____ Hz.
a. 5
b. 80
c. 1/5
d. 24

30. A string of length 2 m is vibrating in its 5th harmonic. Its wavelength is _____ m.
a. 10
b. 5
c. 2
d. 0.8

Bibliography and Web Resources

Johnson, Kenneth, and Willard Walker and John Cutnell. *The Science of Hi-Fidelity.* Kendall-Hunt Publishing Company, Dubuque, Iowa, 1995.

Stanionis, Victor and Hugh Berberich. *Computer Music: Science and Technology of a New Art*. Research Foundation of State University of New York, Stony Brook, NY 1989.

Symko, Orest. *Physics of Hi-Fi*. Kendall-Hunt Publishing Company, Dubuque, Iowa, 1995.

X- rays -
http://imagers.gsfc.nasa.gov/ems/xrays.html

Shoe machine
http://www.orau.org/ptp/collection/shoefittingfluor/shoe.htm

Shoe machines -
http://www.mtn.org/quack/devices/shoexray.htm

Radiation laws -
http://www.google.com/search?hl=en&lr=&q=Adrian+x-rays+shoe+machines+gov

X-rays as electromagnetic waves -
http://www.bae.ncsu.edu/research/blanchard/www/465/textbook/otherprojects/2003/group6/cyeago/info/

Electromagnetic waves and bones
http://www.ncbi.nlm.nih.gov/entrez/query.fcgi?cmd=Retrieve&db=PubMed&list_uids=2075809&dopt=Abstract

Bone healing
www.telecomlab.gr/2002/oct/rhodes/pap3rs/N%20031%20_p231%20-%20p240_.pdf

E&M waves and effects
http://copradar.com/preview/chapt8/ch8d1.html

Chapter 5

Musical Coding

"ut [now "doh"], re, mi, fa, sol, la"

Guido of Arezzo, 1025
Benedictine Monk

The origins of musical coding and musical notation are found in the work of the ancient Egyptians, Greeks, and Chinese. During the Middle Ages such a large number of chants had been developed that it was no longer possible to transmit them all orally and the monks began using a system of dots, dashes and squiggle lines called "neumes" to code the chants. This notation gradually evolved into the modern staff and notes we use today to indicate pitch and duration.

Goals

After studying this chapter, you should be able to:

- Know the basic history of musical notation and how it evolved.
- Distinguish between the treble and bass clefs and know their significance.
- Determine the names of notes by their position on the staff.
- Identify the white and black keys on a keyboard with the names of notes on a staff.
- Determine the duration of notes by their shape and know the symbols for silence.
- Understand something about the organization of notes in a musical piece.

The Ancients

The Egyptians are thought to have used primitive forms of musical notation around 3000 B.C. A more sophisticated form was developed in China as early as the second-century B.C. At that time, ideograms were used to symbolize notes of the scale and finger charts for instruments. The origin of modern musical notation seems to have been developed by the ancient Greeks who took their ideas from ancient Mesopotamia and Egypt.

Pythagoras (sixth-century B.C.) noted that when a blacksmith struck his anvil, different notes or musical tones were produced according to the weight of the hammer. He used a monochord consisting of a string stretched between two supports to develop his ideas.

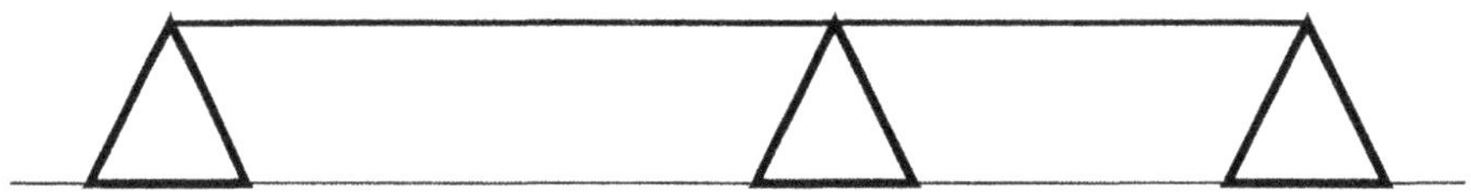

Figure 5.1 Monochord

Using a moveable bridge, the string can be subdivided into different lengths (Fig. 5.1) and when plucked resulted in different musical tones. He discovered a definite relationship between the length of the string (number) and the musical tone. Pythagoras hypothesized that each of the seven planets (according to its distance from the Earth, which was stationery) produced a distinct musical tone much like the subdivisions of the string when plucked.

Pythagoras believed that all relations could be reduced to number relations. This generalization stemmed from Pythagoras' observations in music, mathematics and astronomy. He noticed that vibrating strings produce harmonious tones when the ratios of the lengths of the strings are whole numbers, and that these ratios could be extended to other instruments. He founded a school of mathematics and philosophy in which he and his followers believed all reality could be expressed in terms of whole numbers and their ratios, rational numbers. Most of us are familiar with the Pythagorean Theorem which was discovered by Pythagoras or one of the members of his school in which he demonstrated that the lengths of the sides of a right triangle were related by the equation:

$$a^2 + b^2 = c^2$$

The discovery of irrationals went against Pythagoras's philosophy that all things are numbers, since by a number he meant the ratio of two whole numbers. Consider a right triangle whose sides are one unit in length. When Pythagoras's theorem is applied, the third side would have length $\sqrt{2}$ which cannot be expressed as the ratio of two whole numbers.

The Middle Ages

For a long time, music was passed down by memory from person to person and as a result was often changed, or even forgotten. There was no standardized form of keeping track of music. For example, in keeping with Jewish tradition, the Psalms were sung. The Gregorian Chant of Pope Gregory the Great (590-604) was built on existing chant. It was simply an improvement on Roman Plainchant. Chant is defined as prayer that is sung, or as liturgical music. Pope Gregory I created the Schola Cantorum (Song School) to train choirs in the chant and developed a musical notation, which was basically an improvement on the existing notation. He was not the creator of the lined staff of modern musical notation.

During the liturgical year of the Catholic Church different chants were sung according to their usage during Proper Time (Feasts of the Lord) or Ordinary Time (Feasts of the Saints). The early music of the church was monophonic, having one melodic line without an accompaniment. Around the 1200s, chant consisted of the blending of two or more sounds or polyphony.

By the 760s, a large number of chants had been developed to such an extent that it was no longer possible to transmit them all orally. The first attempts consisted of little reminders written down by the monks. Towards the end of the nineth-century the monks began using a system of dots, dashes, and squiggle lines called neumes. The neumes were graphic signs that triggered the memories of the monks as to the contour of the melody. The word "neume" comes from the Greek word "Pneuma" which means breath. They enabled the monks to distinguish between high and low tones. When a tone was to be sung at a high pitch, the neume was written higher on the page than a tone that was to be sung at a lower pitch

Still the notation was vague. The Greeks had used letters to represent notes of the scale and this system of associating music with the letters of the alphabet was passed on in Europe by Boethius, a Roman scholar and statesman (A.D. 500). He used the first fifteen letters of the Roman alphabet to indicate notes that were in use at the end of the Roman period. Boethius's *De Institutione Musica*, was written in the early sixth-century and was an authoritative guide for medieval authors in the ninth-century. Soon a horizontal red line was introduced, which represented the note F. Writing the notes at certain distances above or below the red line showed how high or low it should be sung. Soon after, a yellow or green horizontal line was introduced to represent middle C.

Neumes arranged above and below a line made the melody more discernible. By the end of the twelfth-century the staff was perfected by Guido d'Arezzo, a monk at the Benedictine abbey at Pomposa. Around A.D. 1025, he devised a four line staff (Fig. 5.2). He started with the F line colored red and the C line yellow, corresponding to the first usage of clefs. Two more lines were added in between so exact pitch could be determined. Guido of Arezzo placed letters on certain lines to designate their pitch.

Figure 5.2 Four line staff

He also used the space between lines to denote pitches. At this time the neumes took on square shapes instead of dot, dashes and squiggles. The scribes added gothic letters to the staff to indicate the names of the lines. It became the foundation for the modern G and F clefs. The word "clef" comes from the word "clavis" meaning key. The G clef, or treble clef, is used for music written for the soprano voice and the F clef, or bass clef, for low voices or instruments.

By the thirteenth-century musical notation was fairly accurate in designating pitch but not duration. Signs for single notes were developed: long, breve, double long, semibreve. By the fourteenth-century time signatures and (white and multi-colored) note heads were used to designate duration.

In the fifteenth-century paper became available to the scribes to record musical notation instead of using the traditional vellum which was made out of animal skins. However, since the paper was made out of fibers, the ink they used to enter the square notes had a tendency to spread out, ruining the shape of the note. At that point they started to outline the notes and they became more rounded. The notes became even more rounded in the sixteenth and seventeenth-centuries and the basic unit, the quarter note, became widely accepted. Various symbols were developed for bar lines, articulation, ties, dots, tempo and Italian phrases for dynamics and expression.

Modern Musical Coding

In order to designate pitches we use the five line staff, consisting of five equally spaced horizontal lines and four spaces (Fig. 5.3). It helps the musician or singer to recognize the pitch of a note but the staff must be qualified by a clef symbol to accurately determine the pitch of a note.

Figure 5.3 Five line staff

We saw that square or rectangular symbols (Fig. 5.4) were used to symbolize notes in the past. Their position on the staff was a code for their pitch or sound. If it was placed on an upper line, it would have a higher pitch than one placed below it. So position was the key element.

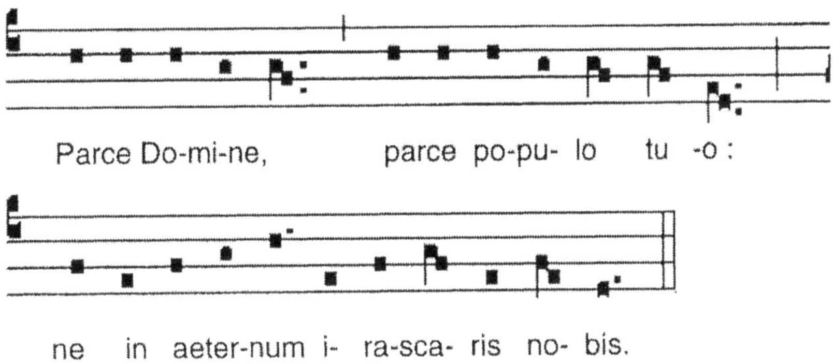

Figure 5.4 Four line staff with square notes
http://comp.uark.edu/-rlee/missal.html

Today because of the large range of pitches that we have to deal with in modern music, another structure called the Great Staff has been created. It consists of two five line staffs connected by a vertical line. Each five line staff has a special symbol at the front called a clef symbol. There are four different clef symbols that are commonly used. We shall consider two of the most common: G clef (or treble clef) that is associated with medium to high pitches. The F clef (or bass clef) is used for low and medium pitches (Fig. 5.5). For most musical instruments notes are written on one staff using one clef symbol. For example, the G clef would be used for the violin, flute, clarinet, and trumpet while the F clef for trombone, tuba, and bassoon.

Figure 5.5 Two staffs

Piano music is written on the Great Staff (Fig. 5.6). The right hand is used to play music on the treble clef, usually above middle C on the keyboard, while the left hand is used to play the lower pitches written on the lower staff, designated by the F clef.

Figure 5.6 Great Staff

If more lines are needed to designate higher or lower pitches than available on the Great Staff, additional lines may be added above or below the staff. These are called ledger lines (Fig. 5.7). Modern musical notation uses rounded symbols to designate pitches.

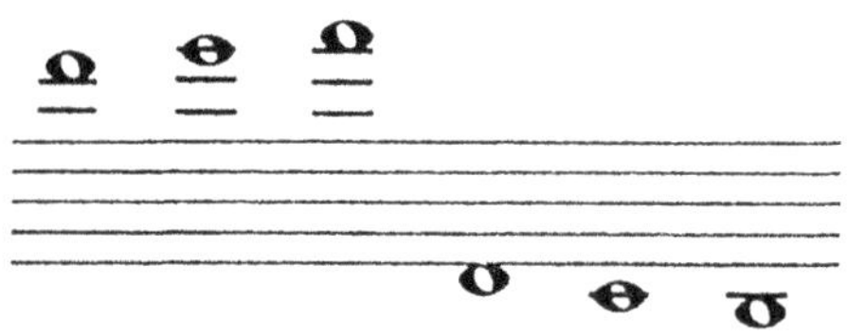

Figure 5.7 Ledger lines

Historically, we saw that the letters of the alphabet have been used since Greco-Roman times to name the pitches. Today, we use the first seven letters of our alphabet: A, B, C, D, E, F, G. The most important or key letter in the code is called middle C (Fig. 5.8).

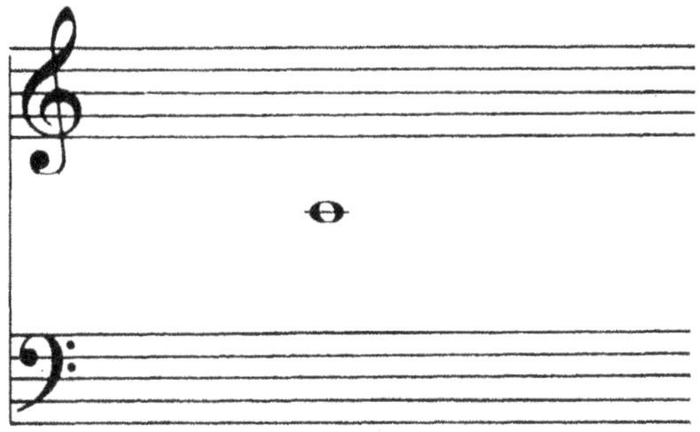

Figure 5.8 Middle C

Knowing the location of middle C allows us to determine the name of any note on the staff by starting with middle C and following with the letters of the alphabet until we reach the unknown note.

The lines on the Treble clef are assigned the names **E G B D F** or "Every Good Boy Does Fine." and the spaces are: **F A C E** (Fig. 5.9).

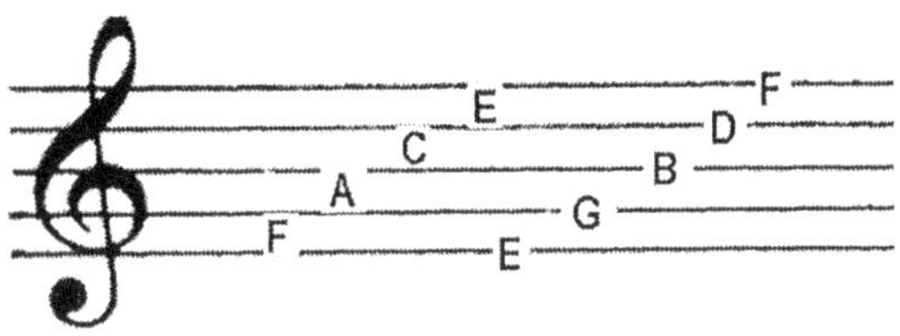

Figure 5.9 Lines and spaces on treble clef

The names of the pitches on the bass clef can be determined by starting with middle C and proceeding backward through the alphabet starting with C (Fig. 5.10).

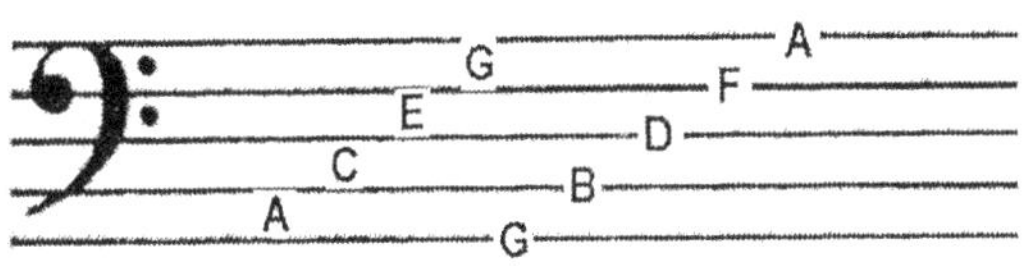

Figure 5.10 Lines and Spaces on bass clef

The Keyboard

One of the most common MIDI input devices and controllers is the piano keyboard. Therefore, it is important that we learn how musical coding is connected to the white and black keys on the keyboard. One of the patterns that we shall have to learn is to associate the names of the pitches on the Great Staff with the names of the black and white keys on a piano keyboard. Theoretically, we can have a piano keyboard with 128 black and white keys (Fig. 5.11). Each key is identified by a number from 0 to 127. Ordinarily, an acoustical piano keyboard has 88 keys while an electronic keyboard usually has 61 keys. However, using a computer sequencer program we can access any of the 128 possible pitches by shifting the range of the electronic keyboard.

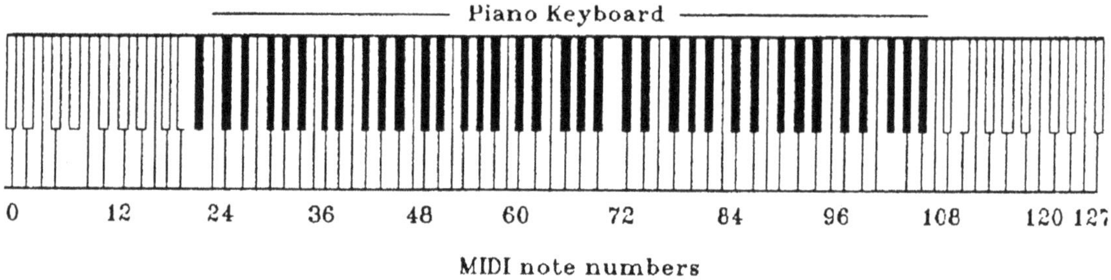

Figure 5.11 Keyboard and key numbers

From *Computer Music: Science and Technology of a New Art,* Stanionis, Berberich. Copyright © 1989 by Research Foundation of State University of New York. Reprinted by permission.

Notice the pattern of white and black keys on the keyboard. Note that the black keys appear in groups of twos and threes. The white key to the left of a group of two black keys is always a C. The C key near the middle of the keyboard is called middle C. The various pitches on the Great Staff are associated with the black and white keys on the keyboard (Fig. 5.12).

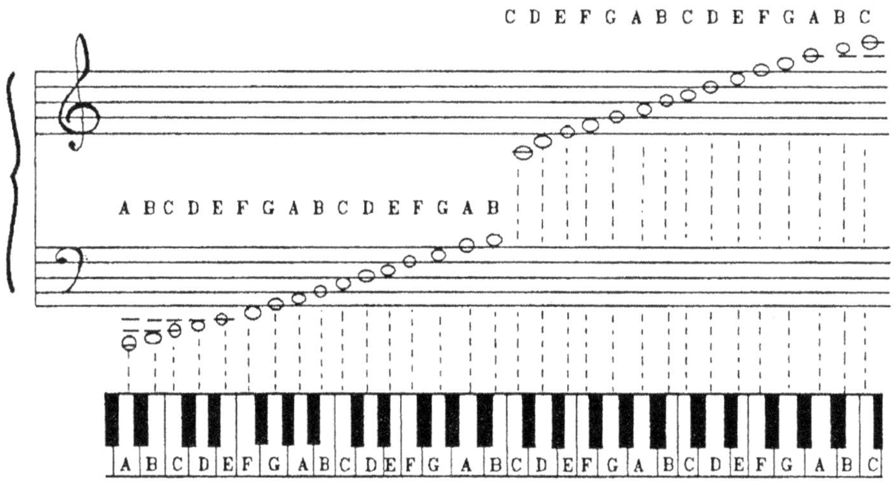

Figure 5.12 Great Staff and keyboard

From *Computer Music: Science and Technology of a New Art,* Stanionis, Berberich. Copyright © 1989 by Research Foundation of State University of New York. Reprinted by permission.

An octave is a musical interval of eight tones. We find that as we move from C to C on the piano keyboard, we span a range of one octave (Fig. 5.13).

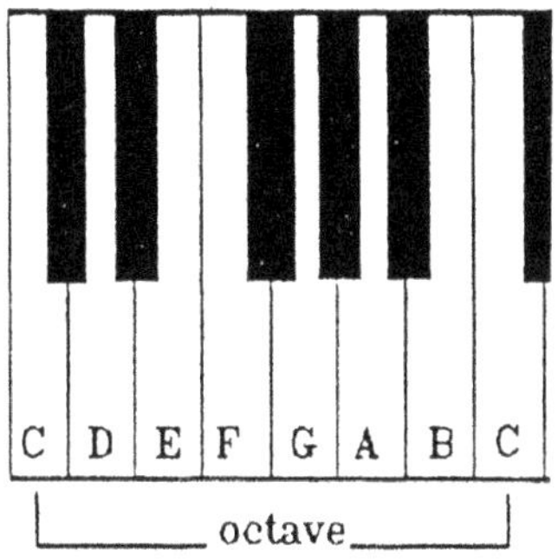

Figure 5.13 The octave

From *Computer Music: Science and Technology of a New Art,* Stanionis, Berberich. Copyright © 1989 by Research Foundation of State University of New York. Reprinted by permission.

In Western music the smallest distance between pitches is called a half-step. So as we proceed from C to C, counting white and black keys, we have 12 half-steps or half-tones. Therefore in an octave we have 12 half-steps. Each of the half-steps is designated by a different name. If we start with C, the first black key to the right is C#, or C-sharp. The next white key is D and the next black key is D#, or D sharp. E is the next white key but instead of E# the next key is F. Following this are we have F, F#, G, G#, A, A#, B, B# or C. This is as we move from left to right pressing successive keys.

When we move from right to left, the black keys are given different names. For example, if we move from E to the black key at the left it is called E*b*, or E flat. Then continuing we have D and D*b*, or D flat, and finally C. So, C# and D*b* are the same halftones but have different names depending on our direction (Fig. 5.14).

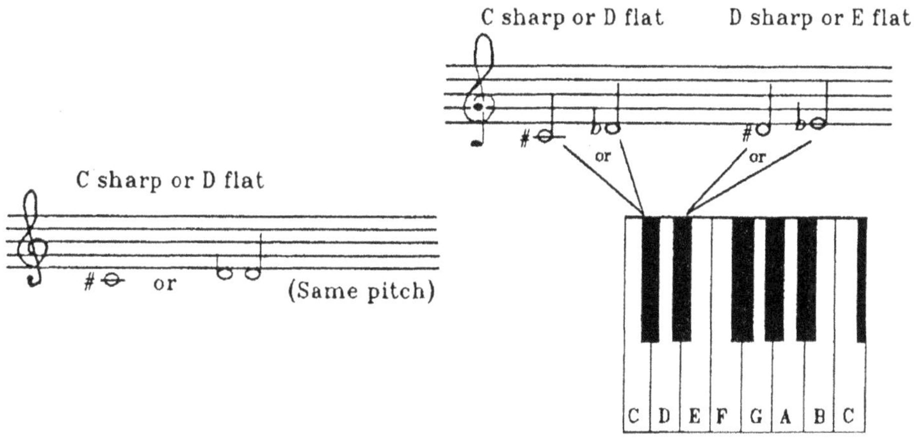

Figure 5.14 Enharmonic spelling

From *Computer Music: Science and Technology of a New Art,* Stanionis, Berberich. Copyright © 1989 by Research Foundation of State University of New York. Reprinted by permission.

Durations

The shape of a note determines how long it lasts and we can count how long it should sound. A whole note lasts for four beats (Fig. 5.16).

Figure 5.15 A whole note

So we count: ***1 and, 2 and, 3 and, 4 and***

If we add a stem to the whole note, then it becomes a half note and lasts one-half as long as a whole note or two beats (Fig. 5.16).

Figure 5.16 A half note

Now we count: ***1 and, 2 and***

If we fill in the head of the note making it completely black, it is known as a quarter note and has half the duration of a half note or one beat (Fig. 5.17).

Figure 5.17 A quarter note

So we count, ***one and***

If a flag is added to the stem, the note has one half the duration of a quarter note and is called an eighth note (Fig. 5.18).

Figure 5.18 An eighth note

It duration is given by the count: ***one***

The notation for sixteenth notes and thirty-second notes is symbolized in a similar fashion.

Dots

When a dot is placed directly to the right of the note head, it increases the duration of the note by one-half its original value in 2/4, 3/4, or 4/4 time (Fig. 5.20).

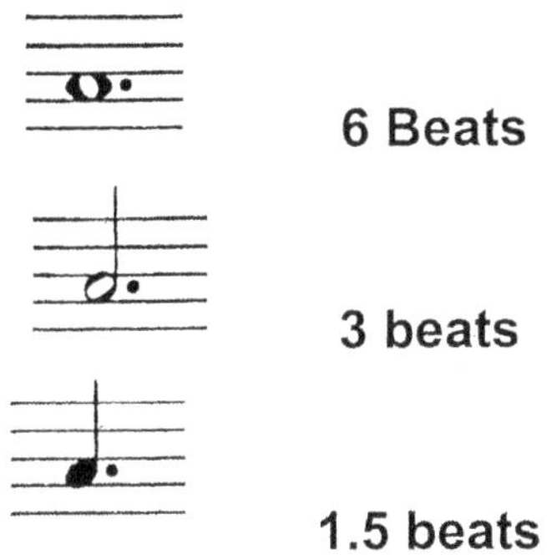

Figure 5.19 Dotted notes

Tie Lines and Slurs

When the addition of dots to the note is insufficient to specify the duration, tie lines are used between notes to obtain the required duration. Both notes are of the same pitch but the duration is the sum of both (Fig. 5.20).

Figure 5.20 Tie line

So the two notes with their heads connected by the tie line have a duration of two beats. This means when striking this note on the keyboard, it is held down for two beats. If there were different notes with different names in between, the curved line would not be a tie line but rather a "slur." This means that the notes between the first and last connected by the curved line or slur should be played smoothly (Fig. 5.21).

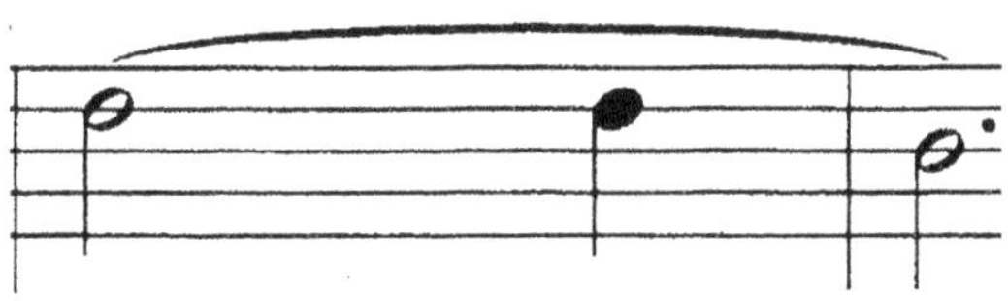

Figure 5.21 Slur

Rests

There are times in a musical piece when there is silence. The symbols for silence are called rests (Fig. 5.22).

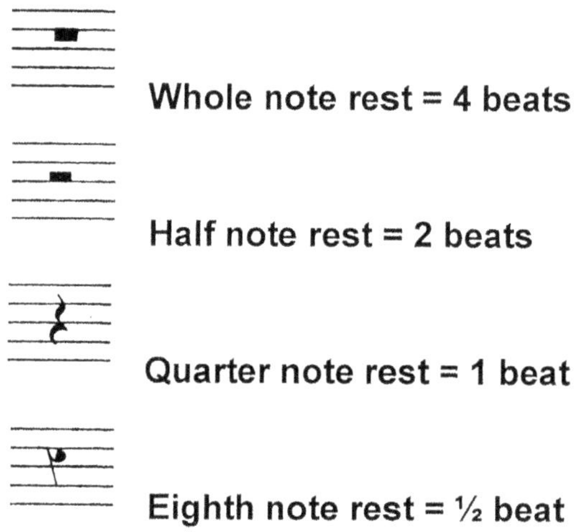

Figure 5.22 Rests

Bar Lines

The staff is divided up into sections called bars. The space between two bar lines is called a bar or measure. The number of beats in each bar is the same. A double bar line marks the end of the piece of music (Fig. 5.23).

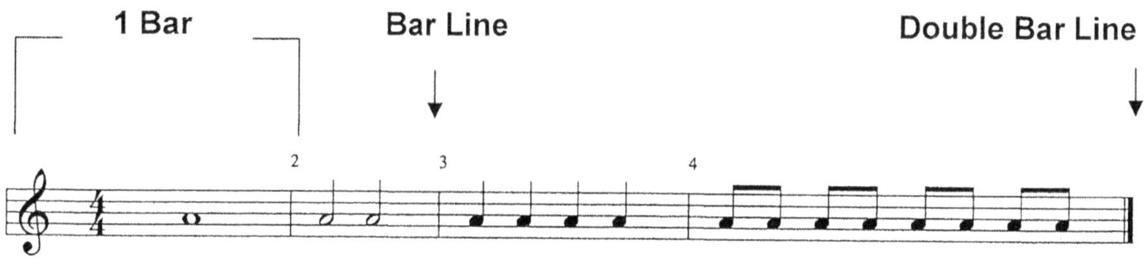

Figure 5.23 Bars

Repeat Signs

When repeat signs appear in a song (Fig. 5.24), the song is played from the beginning through the first ending and is then played from the beginning again, but with a second ending now substituted for the first.

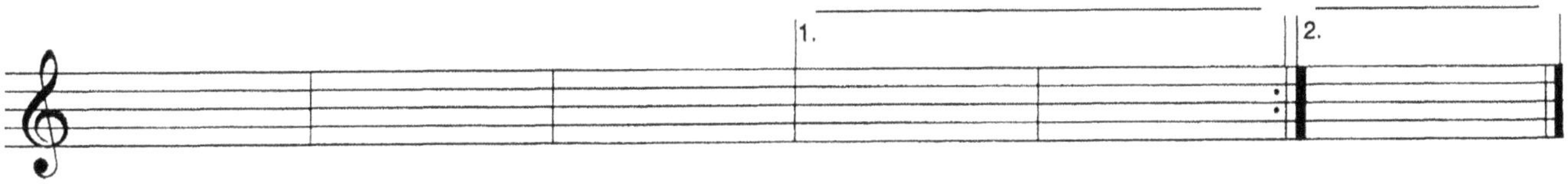

Figure 5.24 Repeat symbols

If the repeat symbols appear in complementary pairs (Fig. 5.25), the portion of the song bounded by the set of repeat symbols is repeated with the second ending again substituted for the first.

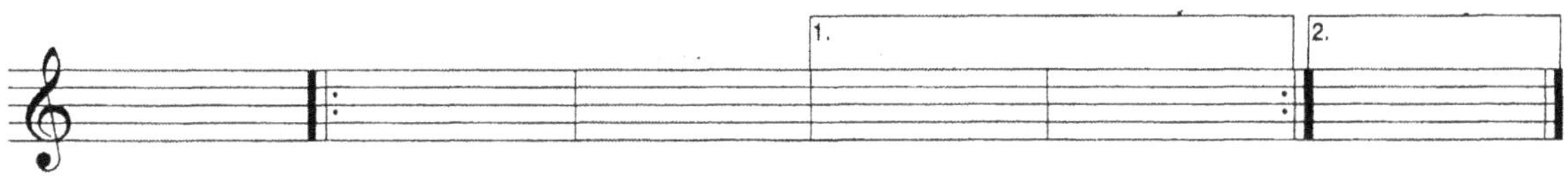

Figure 5.25 Repeat symbols in the middle of a song

Time Signatures

At the beginning of a piece of music following the clef symbol are two numbers that look like a fraction or a capital "C" (or 4/4). These are called time signatures. The top number tells us how many beats are in each bar and the bottom note tells us what note length is equivalent to a beat, the type of note that is beating (Fig. 5.26).

Figure 5.26 Time signatures

The first time signature tells us there are 4 beats in a bar and each quarter note is a beat. The second time signature tells us there are 3 quarter beats in each bar. A time signature of 3/8 would tell us there are 3 beats in a measure and each eighth note is a beat. The time signature tells us the length of a bar. So a ¾ time signature tells us that each bar is exactly as long as four quarter notes.

Sometime the notes at the beginning of a song just before the first bar line do not contain the full number of beats indicated by the top number in the time signature. These notes are called "pick-up" notes. The last measure in the song is then incomplete and contains the number of beats required to complete the first measure and make it complete. They complement one another.

We have considered some of the basic symbols in musical coding which should allow us to enter music into the computer accurately. The next few chapters will be concerned with learning more about the basic knowledge needed to understand music, and the underlying scientific principles that govern the operation of different components of a computer music system. You will apply what you have learned in this chapter to create and orchestrate music during the laboratory activity sessions.

Review Questions

1. Who was Pythagoras?
2. What was Pope Gregory's contribution to musical coding?
3. What is monophonic music?
4. What is polyphonic music?
5. What were neumes?
6. What was Boethius's contribution to musical notation?
7. Who was Guido d'Arezzo?
8. What is a five line staff used for?
9. How do we designate the name of a pitch on the five line staff?
10. What are clef symbols and how do they function?
11. What are ledger lines?
12. What is the relationship between the names of the pitches and the white and black keys on the piano keyboard?
13. What is an octave?
14. What is a half-step and how many are in an octave?
15. What are flats and sharps and how are they used?
16. How are the durations of notes symbolized in musical coding?
17. What are bars?
18. What are time signatures and what is their function?
19. Distinguish between ties and slurs.

Multiple Choice Questions

1._____ is prayer that is sung.
a. Chant
b. Aria
c. Neume
d. Popular music

2. The early music of the Church was _____.
a. multiphonic
b. polyphonic
c. monophonic
d. stereophonic

3. "Pneuma" is a Greek word meaning _____.
a. song
b. notation
c. breath
d. chant

4. Neumes was a system of ____ used by the monks.
a. musical notation
b. breathing exercises
c. prayers
d. songs

5. Boethius was a ____ who used letters of the Roman alphabet for musical notation.
a. monk
b. priest
c. bishop
d. Roman scholar

6. The four line musical staff for musical notation is attributed to _____.
a. Pope Gregory the Great
b. Boethius
c. Guido d'Arezzo
d. Nero

7. The G clef is used for music written for the _____ voice.
a. soprano
b. alto
c. bass
d. mezzo

8. Scribes, in writing notes on the staff, initially shaped the notes as small _____.
a. circles
b. cubes
c. squares
d. ovals

9. When paper started to be used for musical notes, the ink used by the scribes tended to _____.
a. be too dark
b. spread out
c. be invisible
d. too light

10. The number of lines used in the modern musical staff is _____.
a. three
b. four
c. five
d. six

11. The staff is qualified by a ______ symbol.
a. sharp
b. flat
c. clef
d. ledger

12. If a musical note is placed high on a five line staff, its pitch is _____.
a. high
b. low
c. medium
d. unknown

13. If you are told that the four spaces in a five line staff are FACE, then the clef symbol at the beginning of the staff is _____ clef.
a. C
b. E
c. F
d. G

14. The extra lines that are sometimes added to the top or bottom of a staff are called _____ lines.
a. extra
b. ledger
c. alto
d. neumes

15. The letters of the alphabet that we use to name pitches are _____.
a. FACE
b. EGBD
c. ABCDEFG
d. ABC DEF HIJ

16. On the Great Staff the name of the note half way between the two five line staffs is called _____.
a. concert A
b. middle C
c. bottom B
d. elegant E

17. The most common MIDI input device and controller is called the _____.
a. MIDI connector
b. sound card
c. piano keyboard
d. computer mouse

18. Theoretically, a piano keyboard can have _____ keys.
a. 61
b. 88
c. 99
d. 128

19. On a piano keyboard the white key to the left of two black keys next to each other is a _____ key.
a. A
b. B
c. C
d. D

20. An octave on the piano keyboard consists of ________ white keys adjacent to one another.
a. 12
b. 10
c. 8
d. 6

21. In western music, the smallest distance between pitches is a ________.
a. octave
b. half-tone
c. period
d. whole step

22. The number of half-steps or half-tones in an octave is _____.
a. 8
b. 10
c. 12
d. 14

23. C# is the same key as _____.
a. Db
b. Cb
c. D#
d. E

24. The duration of notes in standard musical notation is determined by their _____.
a. color
b. shape
c. size
d. length

25. The name of the note shown when written on the treble clef is _____.

a. F
b. A
c. C
d. E

26. The duration of the note shown above is _____.
a. one beat
b. two beats
c. three beats
d. four beats

27. The following note lasts for an interval of _____ beats.

a. 2 beats
b. 4 beats
c. 6 beats
d. 8 beats

28. The total duration of the two notes shown is _____.

a. 2 beats
b. 3 beats
c. 4 beats
d. 5 beats

29. The symbol shown represents a ______.
a. quarter note rest
b. half note rest
c. whole note rest
d. eighth note rest.

30. A staff is usually divided into sections called _____.
a. sequences
b. neumes
c. bars
d. groups

Bibliography and Web Resources

History of Music
http://www.stevenestrella.com/composers/index.html?ancientmusic.html

Boethius -
http://www.musicbizadvice.com/who_invented_musical_notation.htm

Boethius -
http://www.angelfire.com/indie/anna_jones1/boethius.html

Neumes -
http://library.thinkquest.org/15413/history/history-med.htm?tqskip1=1

Neumes -
http://www.skypoint.com/~waltzmn/Neumes.html

Neumes -
http://www.schuyesmans.be/gregoriaans/EN/ENmu_04.htm

Neumes -
www.bartleby.com/61/1/N0070100.html

Pythagoras -
http://www.aboutscotland.com/harmony/prop.html

Pythagoras -
http://www-groups.dcs.st-and.ac.uk/~history/Mathematicians/Pythagoras.html

Chapter 6

Sound

"Mr. Watson, come here; I want you."
(First sentence transmitted by the telephone, March 10, 1876)

Alexander Graham Bell
Inventor of the Telephone

Philosophers wonder whether a tree will make a sound when it falls in a forest if there is no one around to hear it. Psychologists study the role of the senses and the brain in understanding sound while physiologists are concerned with the ear, its parts, and the tissues and nerves involved in its operation. We shall define sound as any physical disturbance capable of producing a hearing sensation in the human ear.

Goals:

After studying this chapter, you should be able to:

- Define what is meant by sound to a scientist.
- How to determine the speed of sound.
- Define the subjective characteristics of sound: pitch, loudness, quality.
- Know the relationship between the pitch of a sound and its frequency.
- Know the relationship between the loudness of a sound and its intensity.
- Know the relationship between the quality of a sound and its harmonics.
- Define octaves and half steps on a keyboard.
- Discuss the decibel scale.
- Know how loudness affects hearing sensitivity.
- Know how hearing varies with age in men and women.

Sound

Sound is a longitudinal wave that normally travels through air as a series of condensations and rarefactions. It has all the characteristics of wave motion we have defined: wavelength, frequency, period, speed, and amplitude. It is produced by a source that is vibrating, such as vocal chords, a guitar string, a tuning fork, speakers, etc. The medium that the vibrations from the source normally travel through is air. In an elementary science class a classic demonstration is to put a ringing doorbell into a bell shaped glass container and pump out the air. When the air is removed, the hammer is seen striking the gong but the ringing bell goes silent, demonstrating the importance of air as the medium to carry the sound waves (Fig. 6.1).

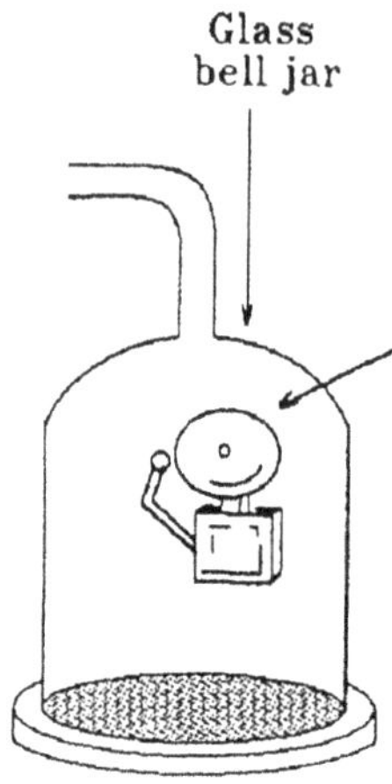

Figure 6.1 Bell jar and ringing bell

From *Computer Music: Science and Technology of a New Art,* Stanionis, Berberich. Copyright © 1989 by Research Foundation of State University of New York. Reprinted by permission.

The pressure variations in the sound wave generated by ordinary conversation are very small, varying by one millionth of a pound per square inch (psi) in the case of normal conversation. The human ear is a remarkable device in that it can detect such small variations. This implies that with normal air pressure, about 14.7 psi, due to the weight of the atmosphere around us, the ear can detect pressure variations in the air of about one millionth of a pound per square inch.

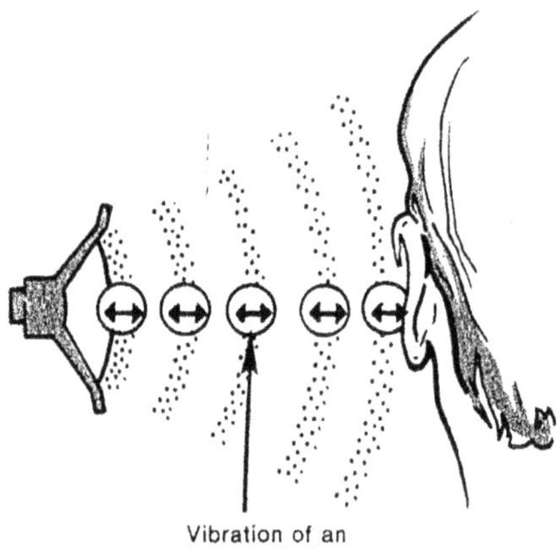

Figure 6.2 Ear and sound waves

From *Science of Hi-Fidelity* by Johnson, Walker, Cutnell. Reprinted by permission of the authors.

It is these small pressure variations that the ear detects as sound as long as the pressure variations occur in a range of about 20 times per second to 20,000 times per second. Variations outside these limits are not sensed as sound (Fig. 6.2).

What do we mean when we say normal air pressure around us is 14.7 psi? Wouldn't such a weight on each square inch of our body (Fig. 6.3) put us flat on our backs starring at the sky?

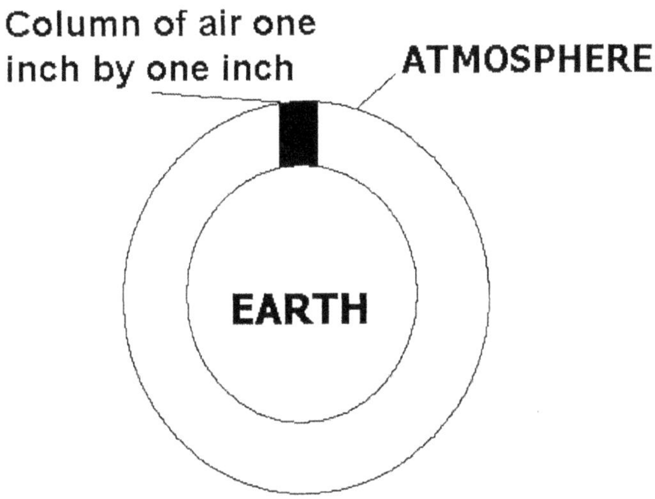

Figure 6.3 Weight of column of air 1 inch by 1 inch by height of atmosphere

Sometimes when deep sea fishermen reel in a fish from deep in the ocean where the pressure is high, they find that it explodes upon reaching the surface. The internal pressure seems to be much larger than the external pressure. As a result of their physical anatomy, fish at great depths balance out the external pressure of the water with an equal internal pressure. On earth, humans function in a similar fashion, as the internal pressure balances out the external pressure due to the weight of the column of air extending to the top of the atmosphere. Thus, we can move under the action of such a great weight. This column of air extending to the top of the atmosphere is made up of gas molecules each of which experiences the gravitational attraction of the earth. Thus each one has an associated weight. The sum of the weights of all the air molecules acting over one square inch of area is termed air pressure and measured in pounds per square inch.

Speed of Sound

The speed of sound in air increases with a rise in temperature but is about 769 mi/hr (or 344 m/s or 1128 ft/s) at room temperature. Sound waves move very rapidly from the source to the listener. For example, consider how long it would take a sound to travel from a speaker to an observer 5 m away?

speed = distance/time.

time = distance/speed.

$t = d/v$.

$t = 5\text{ m}/(344\text{ m/sec})$.

$t = .015\text{ sec}$

In liquids and solids, sound travels even faster.

Characteristics of Sound

There are three main subjective characteristics that are assigned to all sounds: pitch, loudness, and quality. They are associated with certain objective properties of sound that can be measured in the laboratory. Like the hotness or coldness of an object, individuals may disagree on the subjective characteristics of sound.

Pitch

The pitch of a musical tone is determined by its frequency. Men have lower pitched voices then women. A police siren has a high pitch and the sound of bass drum is low pitched. Scientists use the number of vibrations per second (or frequency) to quantify pitch while musicians express pitch using symbols called notes. Musical notes on a staff (Fig. 6.4) provide the musician with an indication of the pitch of the sound and also its duration.

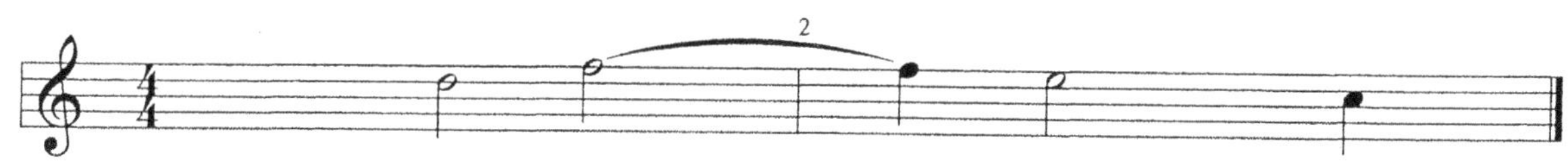

Figure 6.4 Musical staff

When a tuning fork is struck (Fig. 6.5), it emits a sound that is fairly loud at first but weakens with time. The frequency or number of vibrations/sec remains the same throughout, no matter how weakly or strongly the fork has been struck. Suppose we have a tuning fork that can vibrate at 262 Hz. The frequency of middle C on the piano is 262 Hz. This frequency will be associated with the pitch of middle C no matter how it is produced. A singer vocalizing a middle C and a factory whistle sounding middle C will have the same pitch as the piano playing a middle C note. It should be noted that most sounds and musical tones contain the fundamental frequency of vibration associated with their source, as would be emitted by a tuning fork at that frequency, but also multiples of the fundamental frequency, called its harmonics. If we were to remove the harmonics from the sound of the piano playing a middle C note and do the same for the factory whistle, with only the fundamental frequency present they would sound identical. Striking tuning forks of different frequencies will reveal that higher pitch is associated with higher frequency.

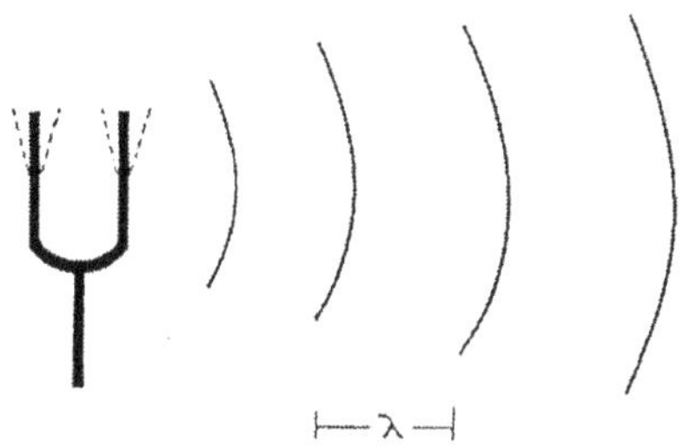

Figure 6.5 Vibrating tuning fork
From *Computer Music: Science and Technology of a New Art,* Stanionis, Berberich. Copyright © 1989 by Research Foundation of State University of New York. Reprinted by permission.

Suppose we examine the frequencies of all of the notes in the middle octave of a piano and list them as in the chart below (Figure 6.6):

A	440	D#	622	A	880
A#	466	E	659		
B	493	F	698		
C	523	F#	740		
C#	554	G	784		
D	587	G#	831		

Figure 6.6 Octave beginning at middle A
From *Computer Music: Science and Technology of a New Art,* Stanionis, Berberich. Copyright © 1989 by Research Foundation of State University of New York. Reprinted by permission.

The frequency of the last note in an octave is twice the frequency of the first. The octave corresponds to a 2:1 ratio of frequencies. The octave (Fig. 6.7) is fundamental to all music. Doubling the frequency raises the pitch by an octave. Further study of the table reveals that each note is higher than the previous by about six percent or .06. For example, 440 x 1.06 = 466.4, which is close to 466. The ratio of frequencies of two adjacent notes is 1.05946. The rise in the whole octave would then be 1.05946 x 1.05946 x ...x 1.05946 = (1.05946) x 12 which is two times or double the initial frequency.

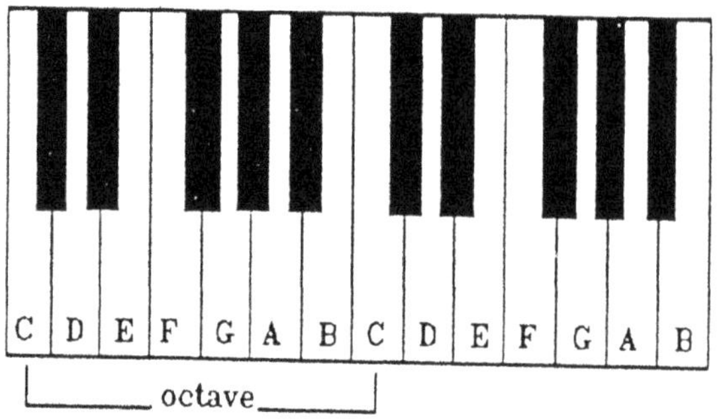

Figure 6.7 Octave on keyboard
From *Computer Music: Science and Technology of a New Art,* Stanionis, Berberich. Copyright © 1989 by Research Foundation of State University of New York. Reprinted by permission.

The smallest interval that traditional Western music uses is the half step. If the frequency of one pitch is 1.05946 times the frequency of a second pitch, the pitches are a half-step apart. If one pitch is twice the frequency of a second pitch, the two pitches will then be one octave apart and there are twelve half-steps in an octave. A black key on the keyboard and the adjacent white key are separated by a half-step.

As we learned in Chapter 5, letter names are assigned to the pitches in the octave. Starting with C in order: C, C#, D, D#, E, F, F#, G, G#, A, A#, B for one octave. For the next octave they repeat. C#, D#, F#, G#, A# are the names of black keys. The # sign stands for the word sharp. C# is read as C sharp which means that this key has a pitch a half-step higher than C. The flat symbol, *b*, is used to designate a pitch which is a half-step lower. B*b* means the key generates a pitch a half step lower than B. The same key can also be interpreted as sounding the note A#, which means the key generates a pitch a half-step higher than A (Fig. 6.8).

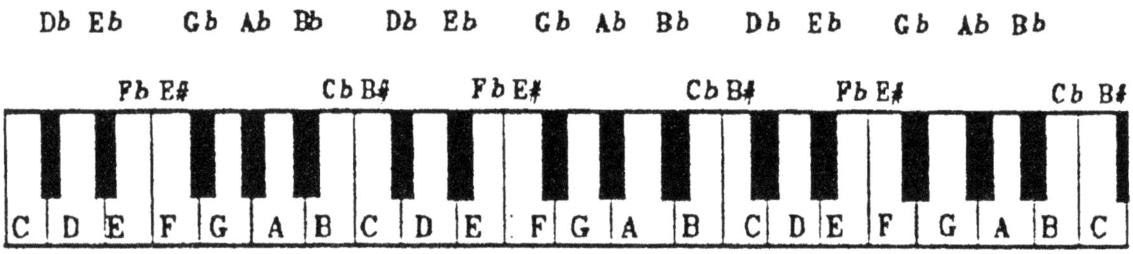

Figure 6.8 Half steps on keyboard
From *Computer Music: Science and Technology of a New Art,* Stanionis, Berberich. Copyright © 1989 by Research Foundation of State University of New York. Reprinted by permission.

Loudness

Remember, the reason we are interested in waves has to do with their ability to transfer information in the form of energy from one place to another. The energy content of the vibrating source, and therefore of the wave motion itself, depends on the amplitude squared (A^2). Intensity of a wave motion is defined as the amount of energy transported per unit time, per unit area ($I = (Work/time)/A^2$). We know energy is the capacity or ability to perform work. When we take into account how fast the work is done we use the term power, which is work or energy expended over a time interval.

$$\text{Power} = \text{work/time or } P = \text{Work}/t.$$

In the metric system the unit for energy is the Joule and therefore the unit for power would be Joules/sec (J/s), which has been renamed the watt (W). Light bulbs are rated in terms of their power consumption in watts. Receivers in a hi-fidelity system are also rated in terms of how many watts of power they can deliver to a speaker, e.g. 50 watts per channel. So intensity, or energy per unit area, over a time interval can be interpreted in terms of power per unit area and measured in watts per meter squared, or W/m^2.

Why is it that just about every country in the world, except for about four (including the United States), uses the metric system of units? How can we expect to sell our products with labels that other people cannot understand? What is the driving force in a capitalist oriented nation? Money? Yes, there are tremendous costs in switching to the metric system. To do it overnight is a revolutionary move and is too costly. However, the path we are following is more evolutionary. Check out the dual weights listed on bags of potato chips and other food products. Check the speedometer in your car for km/h. On January 1, 2005, Ireland switched listing its speed limits on signs and speedometers to km/h from mi/h. The biggest complaint seemed to be that the km/h markings on their speedometers were too small. After January 1, all new cars would be sold only with the km/h scale. Do you know how to convert Fahrenheit temperatures to Celsius? If you travel to a foreign country, it turns out to be useful. The few businesses that have totally converted are liquor manufacturers who sell their products in Liters and fractions of Liters.

Sound travels through the air as a series of condensations and rarefactions. For a 1000 Hz tone the pressure of the surrounding air will alternately vary from 14.700001 to 14.699999 psi one thousand times each second. When these pressure variations reach the ear, they cause the sensation of hearing. Because the pressure fluctuations are so rapid, the ear hears the sound as a steady tone.

The power carried to the ear by these waves is related to the size of the pressure fluctuations or amplitude. The intensity of a sound wave is proportional to the square of the sound pressure amplitude. Doubling the amplitude does not double the power. It multiplies it by a factor of four.

Sound intensity is associated with the amount of power spread out over an area such as 1 m^2. Suppose the power is spread out over a 4 m^2 area is 8 x 10^{-6} watts. See Figure 6.9.

$$I = \text{sound intensity} = \text{sound power/area.}$$

$$I = \text{intensity} = 8 \times 10^{-6} \text{ watts}/4 \text{ m}^2 = 2 \times 10^{-6} \text{ W/m}^2.$$

Sound intensity allows one to compare two sound levels when the areas are different.

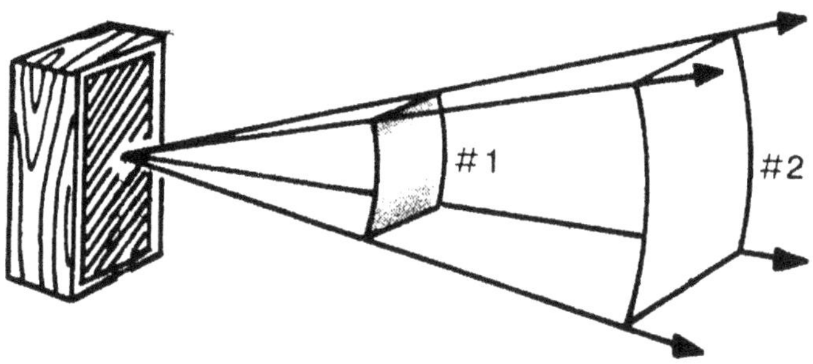

Figure 6.9 Intensity of sound waves decreases farther from speaker
From *Science of Hi-Fidelity* by Johnson, Walker, Cutnell. Reprinted by permission of the authors.

Interestingly, doubling the sound intensity of a wave does not double the loudness of a sound that the ear hears. The ear is sensitive to sounds that range from an intensity of 10^{-13} W/m^2, the faintest sound that can be heard, up to 1 W/m^2, the threshold of pain. A relationship that approximates the ears response to sound is the Weber-Fechner law. It says that the response of a sense organ to a stimulus is proportional to the logarithm of the magnitude of the stimulus. The law is only approximate, but the use of the logarithmic scale matches the wide range of intensities that the ear can sense. The bel is a logarithmic unit of sound intensity named after Alexander Graham Bell but is too large for ordinary calculations. Instead we use the decibel (1/10 of a bel) to assign sound levels on the logarithmic scale. Thus, the intensity of sound on the decibel scale is defined by the equation:

$$dB = 10 \log_{10} (I/I_o),$$

I_o is the reference level, and is taken as 10^{-13} W/m^2.

So a sound of 10^{-13} W/m^2 would have a decibel rating of:

$$dB = 10 \log_{10} (10^{-13})/(10^{-13}).$$

$$dB = 10 \log_{10} 1.$$

$$dB = 0 \text{ decibels.}$$

On the decibel scale (Fig. 6.10) the faintest sound that the ear is sensitive to would have a sound intensity of 0 decibels. On the other hand a sound intensity of 1 W/m^2 would give:

$$dB = 10 \log_{10} (1\ W/m^2)/(10^{-13}\ W/m^2).$$

$$dB = 10 \log_{10} 10^{13}.$$

or

$$dB = 130 \log_{10} 10.$$

$$dB = 130 \text{ decibels.}$$

Source	Noise Level (dB)
Threshold of pain	130
Rock concert	120
Unmuffled car	100
Riveter	95
Elevated train	90
Hair dryer	80
Street traffic	70
Ordinary conversation at 3 ft	65
Whisper	20
Rustle of leaves	10
Threshold of hearing	0

Figure 6.10 Sound intensity levels in decibels

From *Computer Music: Science and Technology of a New Art,* Stanionis, Berberich. Copyright © 1989 by Research Foundation of State University of New York. Reprinted by permission.

Sound Level Meter

A sound level meter (Fig. 6.11) allows us to directly measure the sound intensity in decibels. It measures the sound power per unit area and, using electronics, converts it to decibels which can be read from the meter. A little experimentation will reveal that the smallest audible change in loudness that most people can detect is about one decibel. Further experiment reveals that a 10 decibel change in sound level doubles the loudness of a sound. Hearing a sound at 110 decibels means it sounds twice as loud as the one at 100 decibels and four times as loud as one at 90 decibels.

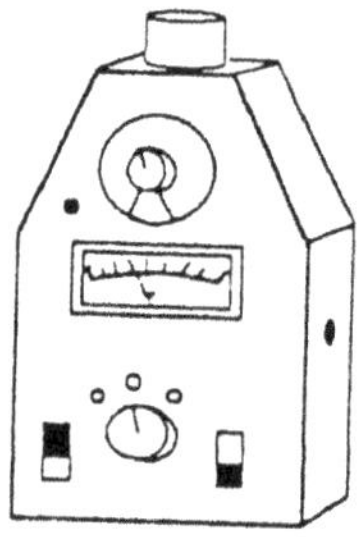

Figure 6.11 Sound level meter

From *Computer Music: Science and Technology of a New Art,* Stanionis, Berberich. Copyright © 1989 by Research Foundation of State University of New York. Reprinted by permission.

Hearing

Research has also shown that the ear's sensitivity to sound varies with frequency. The ear's sensitivity drops off at the low and high ends of the frequency scale, i.e. 20 Hz to 20,000 Hz respectively. It is most sensitive to a tone of about 1000 Hz (Fig. 6.12). Each curve represents the amount of sound intensity required to keep the sound loudness constant as the frequency changes. The loudness level in decibels is shown for each curve. Note that the curve is nearly horizontal at the 100 dB level, which implies that the ear will hear the low, middle and high frequencies with almost equal loudness. Lowering the decibel level, as shown by the 20 dB curve, reveals that the ear is less sensitive to the low and high frequencies. It doesn't hear them at the same level of loudness as the middle frequencies. Most sound level meters have a means of switching to a scale that takes the ear's sensitivity into account when measuring the loudness of a sound.

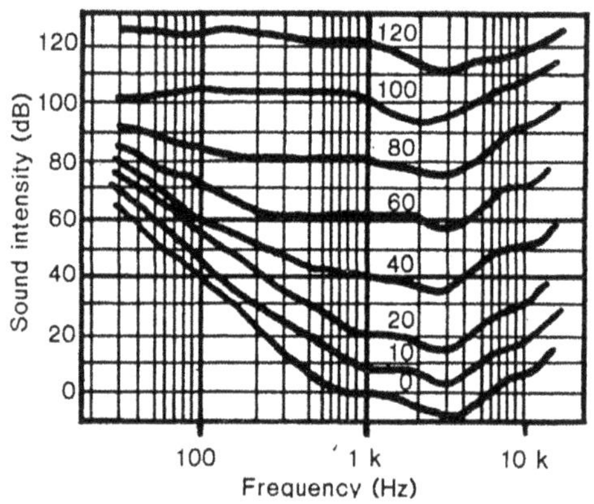

Figure 6.12 Fletcher-Munson curves

From *Science of Hi-Fidelity* by Johnson, Walker, Cutnell. Reprinted by permission of the authors.

When auditioning speakers in a consumer electronics outlet, the ear's sensitivity change with loudness must be kept in mind. If one speaker is playing louder than another, it will sound better because the ear is able to hear more frequencies equally at the higher volume level. This also explains why some people like to listen to music at high volume levels. Why? They hear all frequencies equally well at the higher volume. Fortunately for others, manufacturers of receivers usually include a loudness switch that when turned on activates an electric circuit that allows more power to flow to the low and high frequencies at low volume levels.

Age also plays a role in hearing sensitivity, with young people having a greater sensitivity on the average than old people (Fig. 6.13). Notice the difference between men and women. It appears that women lose their ability to hear low frequencies faster than men as they age and men lose the ability to hear high frequencies faster than women. Maybe this is nature's method for helping couples have long marriages. However, exposure to sound levels over 75 decibels for prolonged periods of time will produce hearing damage. Many rock musicians, when called for military service in the past, were rejected for active duty because of hearing impairments, incurred from years of concerts.

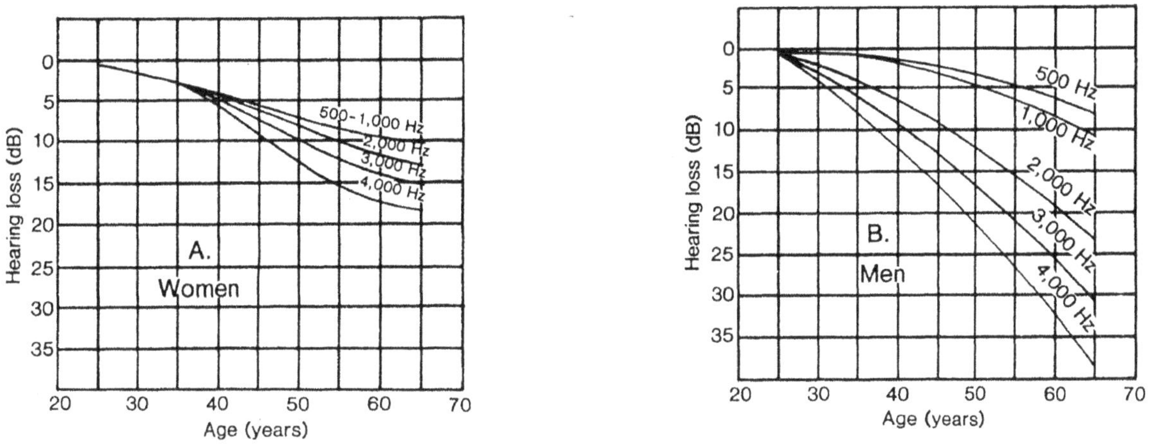

Figure 6.13 Loss of hearing with age

From *Science of Hi-Fidelity* by Johnson, Walker, Cutnell. Reprinted by permission of the authors.

Quality

What allows one to distinguish between a piano sounding middle C, a soprano singing middle C, and a factory whistle sounding a middle C note? As previously discussed, the fundamental frequency of all three is the same as a tuning fork vibrating at the middle C frequency. It is the presence of the harmonics (multiples of the fundamental) and their relative intensities that modify the pure tone of a tuning fork.

Harmonics are multiples of the fundamental that allow the sound to take a complex form, giving rise to a characteristic sound that allows us to distinguish between different sounds at the same frequency. Scientists refer to this as the quality of the sound while musicians use the term timbre (sounds like *tam-burr*) to describe the complicated mixture (color) of frequencies (Fig.

6.14). The more harmonics, the "fatter" is the sound. A "thin" sound would be a sound incorporating few harmonics and a simpler waveform.

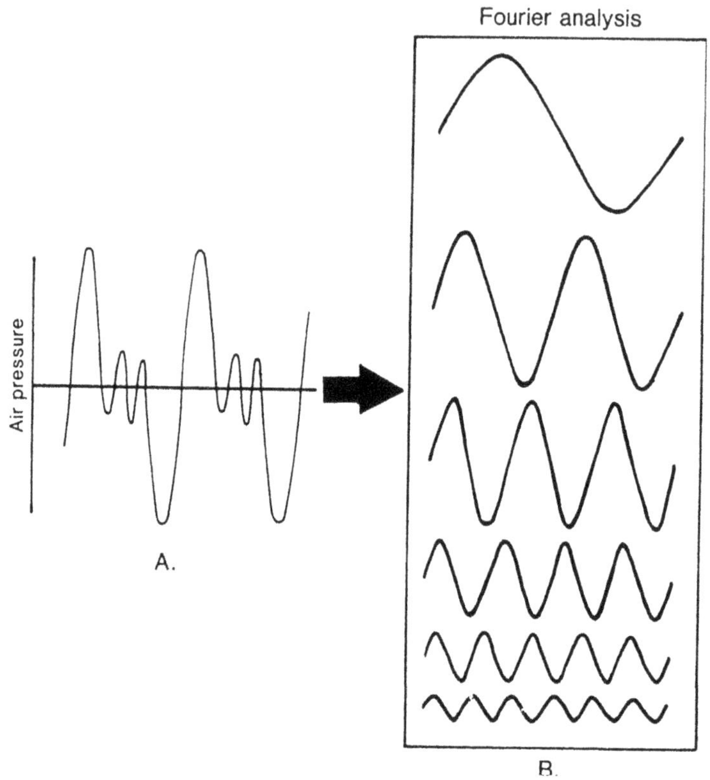

Figure 6.14 Harmonics

From *Science of Hi-Fidelity* by Johnson, Walker, Cutnell. Reprinted by permission of the authors.

A continuous tone generated by an electronic tone generator, as in an analog synthesizer, is characterized by its frequency, intensity, and waveform. A musical note has additional transient characteristics called attack, sustain and decay (Fig. 6.15). These intensity variations vary in length and allow a synthesizer to change the sound of the tone generated from a wind instrument sound to a string instrument sound.

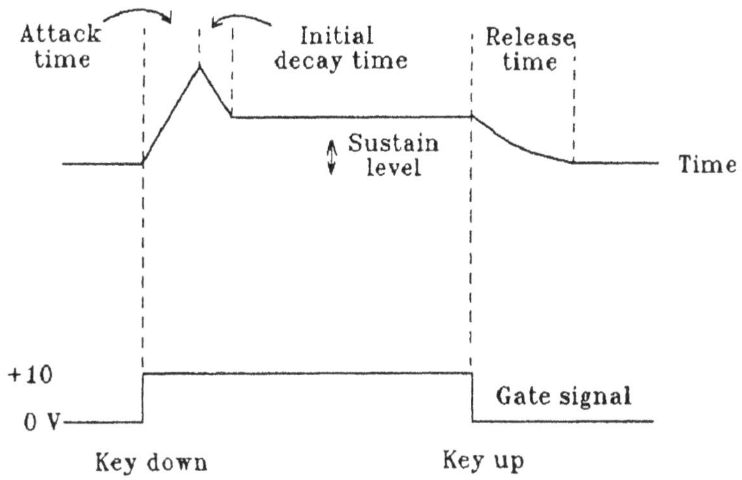

Figure 6.15 Attack, sustain, decay

From *Computer Music: Science and Technology of a New Art,* Stanionis, Berberich. Copyright © 1989 by Research Foundation of State University of New York. Reprinted by permission.

Reflection, Refraction, Interference of Sound Waves

We know that when a wave motion is incident on a hard surface, most of the energy that is incident is returned to the original medium. When light hits a smooth surface like a mirror, it is reflected. If we draw a line perpendicular (normal) to the mirror, the angle the incident ray makes with the normal as it strikes the mirror is called the angle of incidence. Similarly as the light ray leaves the mirror, the angle the reflected ray makes with the normal to the mirror is called the angle of reflection. When sound is reflected from a smooth surface, we find that the angle of incidence equals the angle of reflection just as in the case of a light ray striking a mirror (Fig. 6.16).

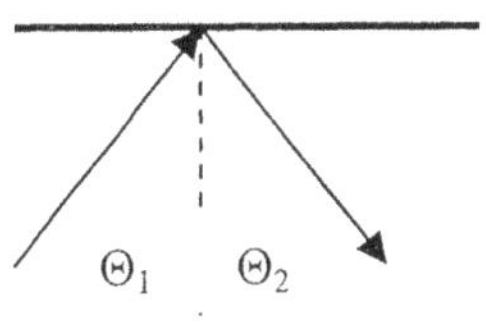

Figure 6.16 Refection of sound from a smooth surface

Reflection of sound from the walls of a room gives it what is known as presence. When speakers are moved outdoors, the decrease in the loudness of the sound is easily noticeable due to the lack of reflections. The loudness of the sound also drops off rapidly as one moves away from the speakers. Measurement reveals that the sound intensity and therefore the loudness drops off as $1/\text{distance}^2$. In a concert hall, the listener hears the direct sound, which orients the listener to the source, while the reflected sounds are related to the acoustic properties of the hall and its contents. The reflected sound gives the hall presence or ambience (Fig. 6.17). The hall adds to the quality of the sound.

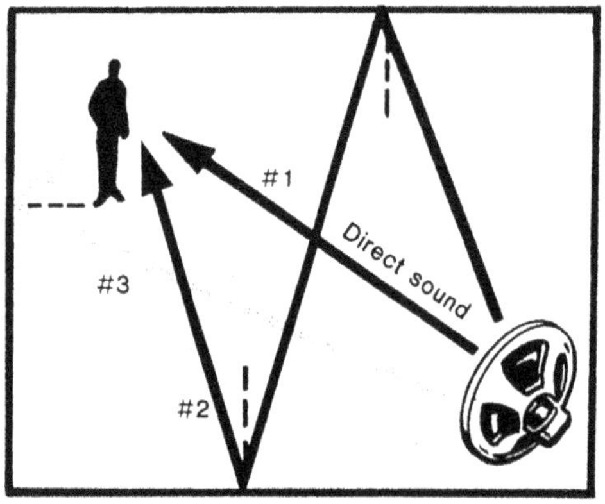

Figure 6.17 Reflection of sound
From *Science of Hi-Fidelity* by Johnson, Walker, Cutnell. Reprinted by permission of the authors.

It is interesting to note that a person's ability to distinguish the relative position of instruments in an orchestra is partially due to the sound from the instruments reaching his ears at different times. This is so because of the spatial separation of the ears on a person's head as one ear is closer to a particular instrument than another (Fig 6.18).

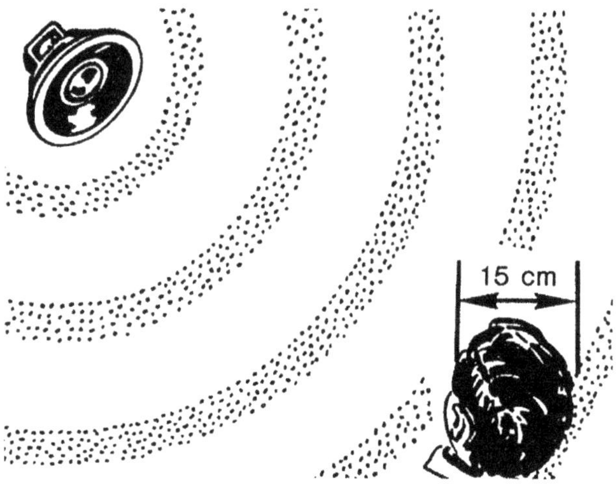

Figure 6.18 Sound directionality
From *Science of Hi-Fidelity* by Johnson, Walker, Cutnell. Reprinted by permission of the authors.

Acoustical tiles by their very design trap sound waves through a series of multiple reflections and partial absorptions, quickly diminishing the intensity of any sound waves that escape. Rooms that have ceilings or furnishings that act on sound waves in a similar fashion (i.e. carpets, curtains, upholstered furniture, etc.) tend to be very effective in absorbing sound and are referred to as being "dead" (meaning they have short reverberation times). "Live" rooms on the other hand tend to have smooth surfaces that absorb very little of the sound waves' energy (Fig. 6.19).

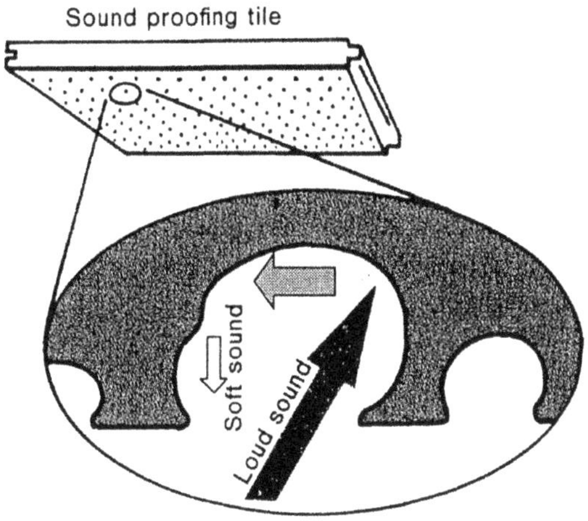

Figure 6.19 Reflection of sound

From *Science of Hi-Fidelity* by Johnson, Walker, Cutnell. Reprinted by permission of the authors.

Reverberation Time

Since reflected sound travels over a longer distance, it may reach the listener after some delay causing an echo. The reflected sound is sometimes referred to as a reverberation. The reverberation time is the time required for the reverberated sound to decay to one-millionth of its original power level after the sound source has been turned off (Fig. 6.20). For opera a reverberation time of about 1.4 seconds would be suitable, while for Gregorian chant in a cathedral 2.4 seconds to 4 seconds is not unusual. Singers of the National Anthem at New York's Shea Stadium have claimed the reverberation time to be 5 seconds. Perhaps this accounts for some unusual renditions of the "Star Spangled Banner" by singers in a stadium caused by confusion they suffer because of the echoes.

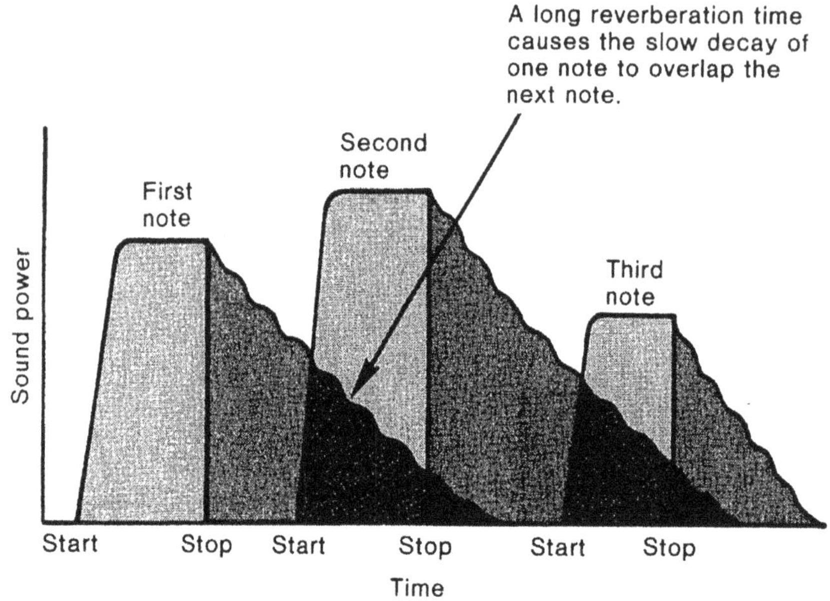

Figure 6.20 Reverberations

From *Science of Hi-Fidelity* by Johnson, Walker, Cutnell. Reprinted by permission of the authors.

Rooms with tile floors and brick walls reflect sounds for long periods of time and create long reverberation times. When sounds overlap because of long reverberation times, there tends to be a loss of clarity and sharpness. If the sounds reverberate too long, they may become unintelligible. Legend tells us that the Tower of Babel was probably constructed of tile, which would give rise to long reverberation times so that even when learned men spoke, their speech was unintelligible.

Prior to the opening of the Philharmonic Hall in New York City its architect, Dr. Leo L. Beranek, wrote in his book, *Music, Acoustics and Architectural Design*, "Lady Luck has finally been supplanted by careful analyses and the painstaking application of new but firmly grounded acoustical principles." The new Phiharmonic Hall was to be the best example of the principles of acoustic science and technology applied to the design of a concert hall. Before this new concert hall was built, there was a move by developers to tear down Carnegie Hall which was world famous for its acoustics. It was to be replaced by the Carnegie Office Tower.

Lincoln Center's Philharmonic Hall in New York City (Fig. 6.21) opened in 1962 where critics panned it as one of the acoustic disasters of the twentieth century, "a $16,000,000 lemon."

Figure 6.21 Avery Fisher Hall

Critics detected echoes possibly due to resonating air conditioning ducts and the modern building materials used to cover them. Low frequencies seemed to be lost. Large reflecting panels in the ceiling were apparently not large enough and low frequencies bent around them and were lost. Parts of the orchestra could not be heard nor performers on stage. Absorption of sound by the audience in the front, because of the low stage and the fact they brought their coats into the hall instead of checking them in the lobby, left the people in the cheap seats (in the back) with very little sound. Several attempts over subsequent years were made to correct the flaws with no success. Music lovers thanked their lucky stars that the movement to "Save Carnegie Hall" was successful.

On September 18, 1973, Philharmonic Hall was renamed Avery Fisher Hall to honor him for the large donation he made to rebuild the Hall. Cyril Harris, a Professor at Columbia University, was hired for the reconstruction, and he gutted the Hall, removing the reflectors, stripping the carpeting and absorbing materials from the walls and floors, adding sound-proofing to the AC and using traditional materials like plaster instead of sheetrock. Avery Fisher Hall reopened on October

19, 1976 to rave reviews. Great concert halls follow a sort of Law of Natural Selection. Only the great ones are saved from the wrecker's ball over time and survive. It is the survival of the fittest once more. Today concert hall designers study these survivors in addition to using their computers and physics when designing a new hall.

Interference

If we were to strike two identical tuning forks about ten feet from each other and there were no sound reflections, it is possible that at certain positions in the room there would be no sound due to total destructive interference. At other points the sound would be much louder. Because of reflections in a real room, the effect might appear as a slight difference in loudness at various points (Fig. 6.22). As previously discussed in Chapter 5, we found that when two wave motions travel through a medium simultaneously, each wave proceeds through the medium and produces its effect independently of the other. It was also noted that in any region where the two waves were present simultaneously, the resulting disturbance is made up of the sum of the two individual disturbances. This phenomenon is known as interference and is exhibited by sound because of its wave nature.

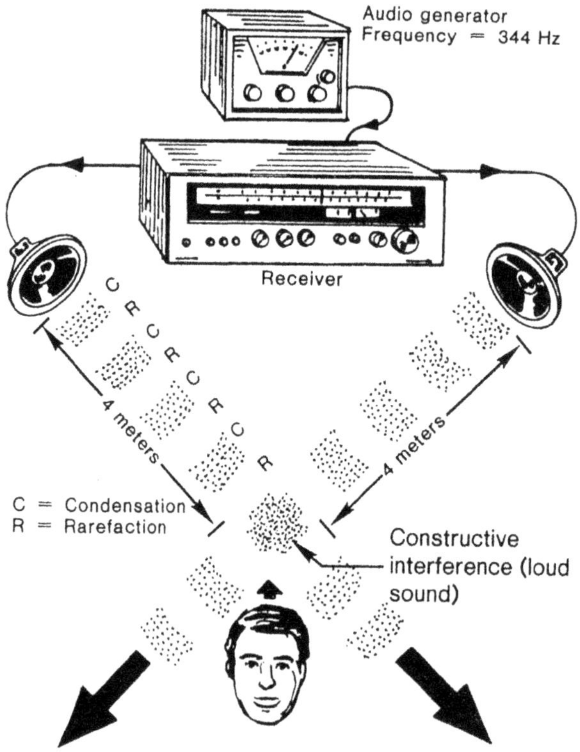

Figure 6.22 Interference of sound waves

From *Science of Hi-Fidelity* by Johnson, Walker, Cutnell. Reprinted by permission of the authors.

Another type of interference results when two waves of slightly different frequencies travel in the same direction through the same medium. A pulsating interference (called beats) is setup, modulating the wave disturbance as it travels through the medium. In the case of sound waves the ear can detect such pulses if the frequency of the two waves is within about ten cycles per second of one another. It is quite noticeable (Fig. 6.23). Imagine a large flexible sheet of tin being waved in the air generating a "wah wa, wah wa" sound.

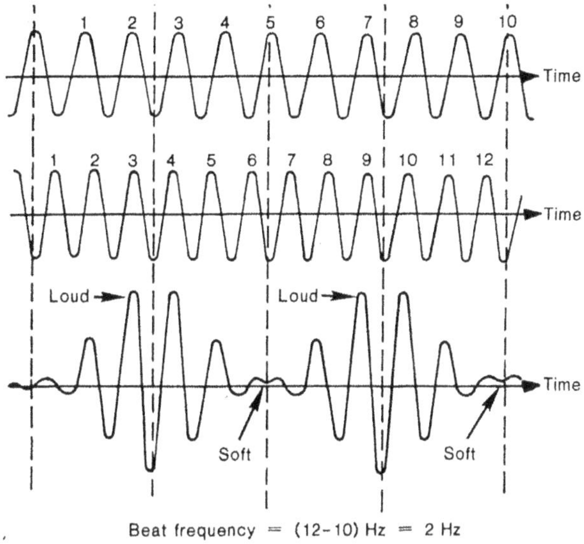

***Figure 6.23** Beats*

From *Science of Hi-Fidelity* by Johnson, Walker, Cutnell. Reprinted by permission of the authors.

Standing Waves

Wind instruments setup standing waves in columns of air. By striking a tuning fork and holding it over a variable length air column closed at one end we can determine the length of the column that will resonate with the fork corresponding to the fundamental frequency. Resonance is a reinforcement of the sound heard. It is an interference phenomenon in which the pressure wave generated directly by the fork is reinforced by the pressure wave reflected from the bottom of the tube, and producing a standing wave (Fig. 6.24).

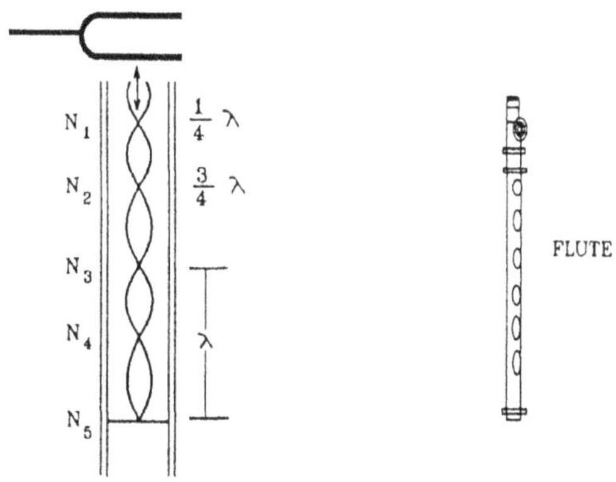

Figure 6.24 Interference in a musical instrument

From *Computer Music: Science and Technology of a New Art,* Stanionis, Berberich. Copyright © 1989 by Research Foundation of State University of New York. Reprinted by permission.

Changing the length of the air column normally changes the natural frequencies of a wind instrument. Long air columns correspond to low frequencies and short ones to high frequencies. Exposing higher or lower openings by lifting the keys over the holes would change the effective length of a flute. Opening a hole shortens the effective length. Percussion instruments, such as a drum, set up standing waves on a membrane. The xylophone's standing waves are analogous to those produced on a string of different lengths.

If we pluck the string on a guitar, we can change the frequency of the sound produced. By changing the length of the string we allow it to vibrate, just like the monochord that Pythagoras used (Fig. 6.25).

Figure 6.25 Different length of string is excited

From *Science of Hi-Fidelity* by Johnson, Walker, Cutnell. Reprinted by permission of the authors.

Each musical instrument has a characteristic sound generated by the standing waves it forms. The quality of the sound we found is determined by the relative amplitudes of its harmonics. The frequency of the fundamental that each instrument generates varies over a range larger than that of human hearing (20 - 20000 Hz). When we call upon a synthesizer to insert the sound of a particular instrument, we may have to adjust the frequency range over which the instrument plays in order to make it sound more realistic (Fig. 6.26).

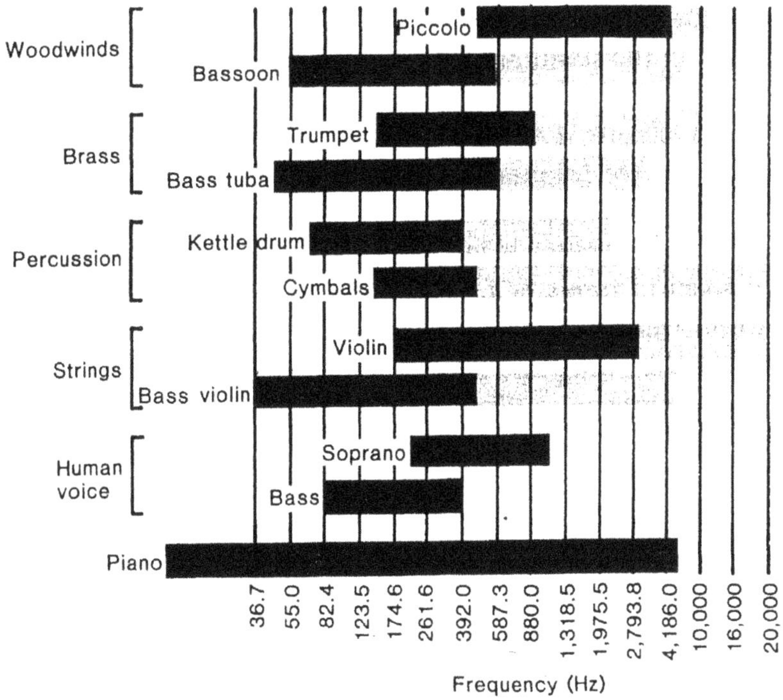

Figure 6.26 Frequency ranges of musical instruments
From *Science of Hi-Fidelity* by Johnson, Walker, Cutnell. Reprinted by permission of the authors.

To avoid interference effects when using a microphone, would it be better to place it flat on a table or in a microphone stand (Fig. 6.27)? Which configuration is more likely to suffer from standing waves? The microphone on the stand receives a direct sound and one that is reflected from the table. The two different path lengths that the two waves travel allow for interference. The microphone on the table receives only a direct sound wave.

Figure 6.27 Two microphones
From *Science of Hi-Fidelity* by Johnson, Walker, Cutnell. Reprinted by permission of the authors.

It is also possible to produce standing waves with the placement of loudspeakers in a room. Depending on the sound frequency and the length of the room it is possible to produce a standing wave for this frequency with its nodes and antinodes. A person standing in the room would notice changes in the loudness of the sound as they moved parallel to the room's length

(Fig. 6.28). By placing the speakers at a slight angle with the wall and not placing them flush against it, this problem can be avoided.

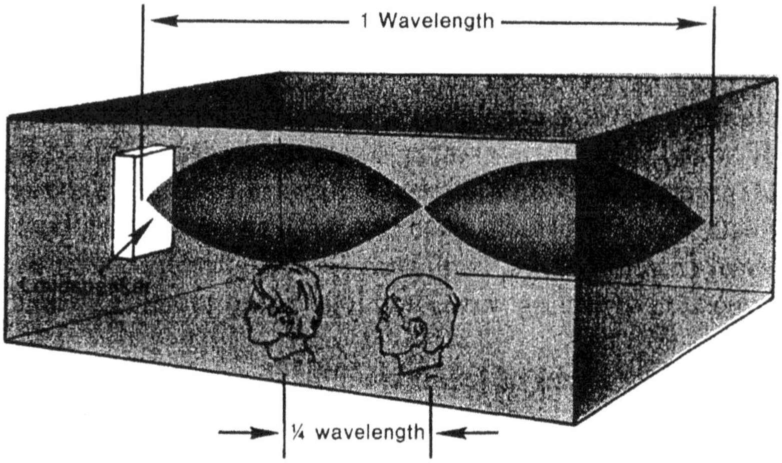

Figure 6.28 Standing waves in a room
From *Science of Hi-Fidelity* by Johnson, Walker, Cutnell. Reprinted by permission of the authors.

More on Resonance

All systems in nature have a natural frequency, which is the frequency at which they will vibrate when excited. This fundamental frequency depends on the materials the systems are made of, their size and shape. These natural frequencies are called resonant frequencies. When you push a child on a swing and you want the swing to move through a large arc, you don't push at any old time, but push just as it reaches it maximum displacement. Pushing at the wrong time may lead to an appointment with the dentist. You have probably seen a famous commercial for magnetic tape in which a soloist sings and is recorded on tape which is then played back so as have the sound generated by his/her voice shatter a wine glass. Energy is absorbed by the glass when the singer sounds its natural or resonant frequency (Fig. 6.29).

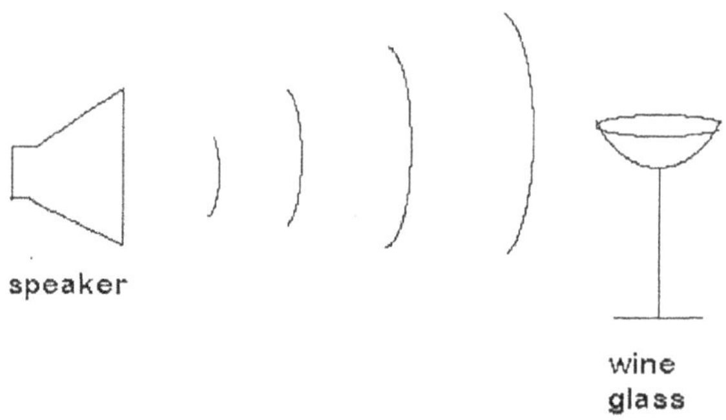

Figure 6.29 Sound waves striking a wine glass

Probably the most dramatic example of resonance and the forming of standing waves took place in the collapse of the Tacoma Narrows Bridge in Tacoma Washington in 1940. The engineers had designed the bridge to be somewhat flexible so that it could withstand high winds and dissipate energy in its supports much as a tree bends with the wind and then returns to its original position after the wind stops blowing. On the day it collapsed, the wind was only blowing with gusts of about 42 mph. The bridge had a 30 Hz transverse vibration and then also began a twisting motion at 14 Hz. As a result it absorbed energy from the wind. its amplitude of oscillation increased. It seems that the gusts produced vortices as they blew horizontally across the deck of the bridge. As the vortices spun off they contributed to the oscillations of the bridge. The collapse of the bridge was due ultimately to resonance brought about by a complicated series of events.

Diffraction and Speakers

A neighbor's or a child's noisy stereo is easily heard outside their rooms because of diffraction effects. Interestingly enough, it's the boom of the bass which is most discernible. Diffraction is frequency dependent. Low frequency sounds spread out over a wide area and have a wide dispersion. High frequency sounds have a narrow dispersion. Setting up speakers in a room for listening to music would require that the listener be positioned as close to the speaker's axes as possible if the high frequencies as well as the low frequencies are to be heard. When sound waves pass through an opening, they are diffracted and bent. Consider the wave front in Figure 6.30.

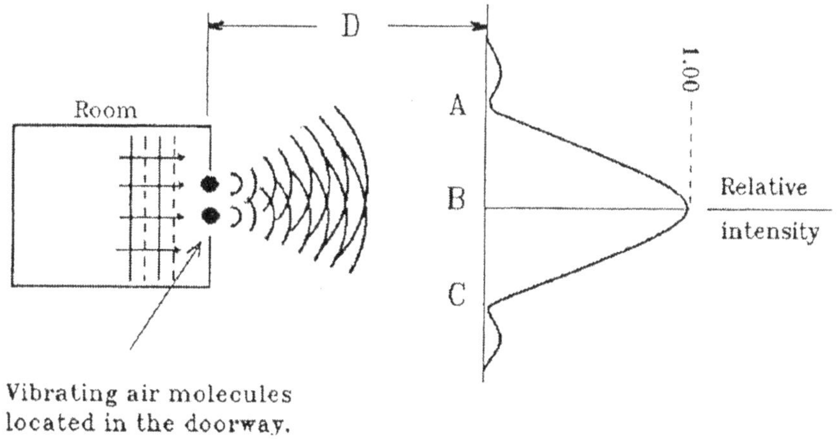

Figure 6.30 Diffraction of sound through an opening

Along the wall opposite the opening we would find the sound to be distributed in intensity in a bell-shaped pattern as shown. Simple application of trigonometry to the triangles in

Figure 6.31 shows how the spread of the bell-shaped pattern depends on the wavelength and, in turn, on the frequency of the sound wave.

Each point in the opening can be thought of as a sound radiator. If we take a ray (a line drawn perpendicular to the wave front) from the top of the opening and one from the middle, the path difference between these two rays would have to be $\lambda/2$ if the waves are to interfere destructively at point A. Now, take a ray just below the top and one coming from a point just below the middle and meeting at point A. Again the path difference would have to be $\lambda/2$. Continuing in this manner down through the opening, reveals that a ray from the middle of the opening and one from the bottom would also have to be $\lambda/2$ apart to give destructive interference at point A. This means that the path difference between a ray from the very top of the opening and one from the bottom would have to be λ as shown below in Figure 6.31.

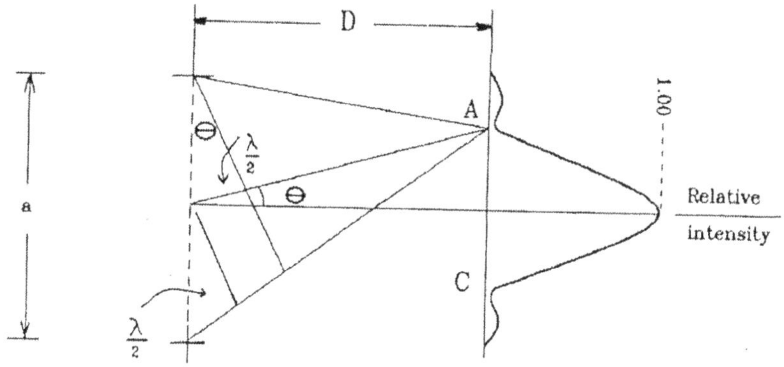

Figure 6.31 Diffraction

From *Computer Music: Science and Technology of a New Art,* Stanionis, Berberich. Copyright © 1989 by Research Foundation of State University of New York. Reprinted by permission.

Now two triangles can be constructed as shown and because the angle θ is small they can be termed similar triangles (Fig. 6.32).

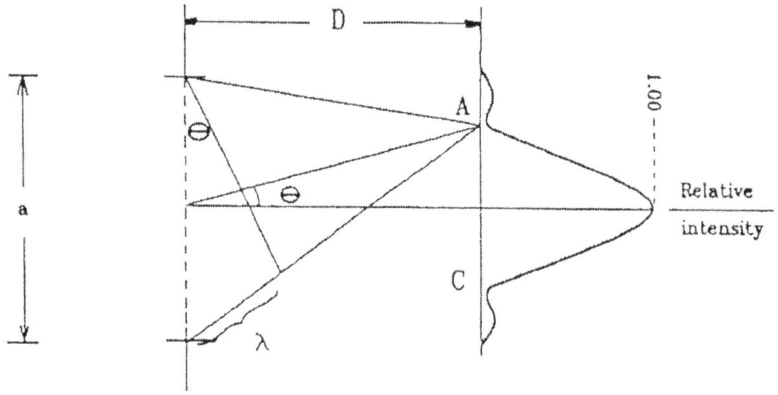

Figure 6.32 Diffraction

From *Computer Music: Science and Technology of a New Art,* Stanionis, Berberich. Copyright © 1989 by Research Foundation of State University of New York. Reprinted by permission.

See Trigonometry Review at the end of this chapter.

$$\sin \theta = \lambda/a$$

but $$\lambda = v/f$$

so $$\sin \theta = v/fa$$

As θ becomes larger, point A moves upward and the bell-shaped pattern spreads out. Notice that as f the frequency becomes larger and therefore θ smaller, the bell-shaped curve will contract. The minimum point A moves closer to the maximum. It is not as spread out. Just the opposite happens when f becomes small. Now the curve will spread out over a large distance and sound can be heard well to the side of the opening.

For a speaker creating sound in a room, the low frequencies spread out in all directions while the high frequencies are highly directional (Fig. 6.33). Different speakers spread the high frequencies at different angles with the axis of the speaker. This property of a speaker is referred to as its dispersion. The low frequencies below about 100 Hz have no directionality at all and make it impossible for a listener in a room to locate the source of such frequencies by listening. So to reproduce the extreme low frequencies, only a single speaker is necessary. In order to reproduce the spatial pattern of the original sound, the midrange and high frequencies must be reproduced stereophonically by two speaker systems on the left and right of the listener.

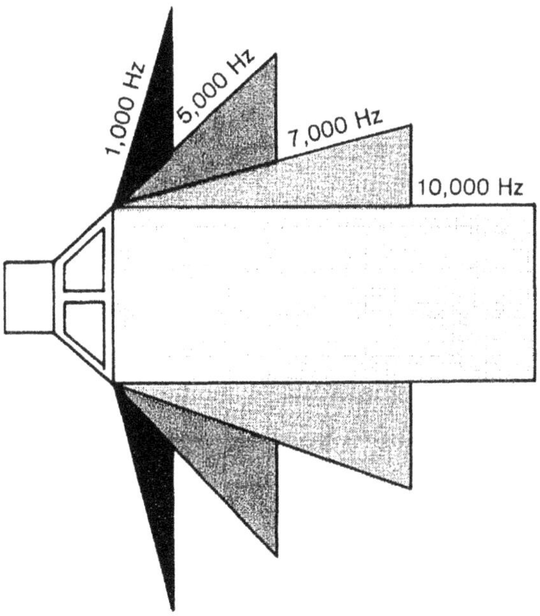

Figure 6.33 Speaker dispersion with frequency

From *Science of Hi-Fidelity* by Johnson, Walker, Cutnell. Reprinted by permission of the authors.

In a room a bulky subwoofer, which generates the low frequencies, can be placed anywhere and small satellites can be located inconspicuously on the left and right of the listener.

Trigonometry Review

Suppose you were given a triangle with one angle of 90°, a right triangle. Let the sides be labeled a, b, c. Consider all possible ratios that can be formed taking two sides of the triangle at a time. See Figure 6.34. They would be: b/c, a/c, b/a, c/b, c/a, a/b.

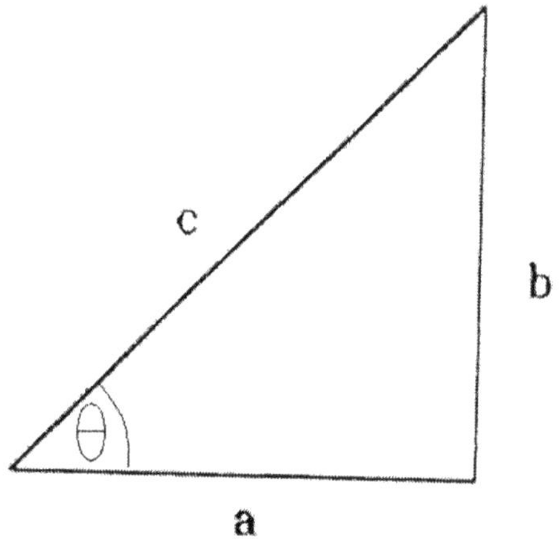

Figure 6.34 Triangle

In mathematics each of these ratios is given a name in terms of the angle θ that the side opposite the right angle makes with the base. The names are:

$$\sin\theta = b/c \qquad \csc\theta = c/b$$

$$\cos\theta = a/c \qquad \sec\theta = c/a$$

$$\tan\theta = b/a \qquad \text{cotan}\,\theta = a/b$$

The side opposite the right angle is called the hypotenuse. The side next to the angle is called the adjacent and the side opposite the base angle naturally is referred to as the opposite. So $\sin\theta$ = Opp/Hyp, $\cos\theta$ = Adj/Hyp, and $\tan\theta$ = Opp/Adj or in terms of an acronym, SOHCAHTOA. If you can pronounce this acronym, you may use it to help in remembering the meaning of the trigonometric ratios.

Review Questions

1. What is sound?
2. What is pitch?
3. How is the loudness of sounds measured?
4. Discuss the decibel scale and how it works.
5. What is meant by the intensity of a sound?
6. What do the Fletcher-Munson curves tell us?
7. What are the characteristics of sound?
8. What is the relationship between loudness and intensity of sound?
9. What is meant by presence when speaking about sound?
10. Discuss attack, sustain, and decay with reference to a sound wave.
11. What is reverberation time?
12. What role does interference play in terms of sound?
13. What is the effect of diffraction on the frequency spectrum of audible sound?

Multiple Choice Questions

1. Sound is a ___ wave motion.
a. transverse
b. circular
c. longitudinal
d. polarized

2. Air pressure varies by about _____ as a sound wave passes through it.
a. 14.7 psi
b. 32 psi
c. one millionth psi
d. 1 atmosphere

3. The human ear can detect vibrations between 20 and ______ Hz.
a. 200
b. 2000
c. 20000
d. 200000

4. The pitch of a sound is a characteristic of sound related to its _____.
a. loudness
b. quality
c. frequency
d. intensity

5. The loudness of a sound is measured in terms of its _____.
a. harmonics
b. quality
c. frequency
d. intensity

6. The intensity of a sound is measured in terms of the amount of _____ per unit area it contains.
a. energy
b. power
c. work
d. loudness

7. The smallest interval between pitches in Western music is _____.
a. full tone
b. full step
c. half step
d. octave

8. The amplitude squared is proportional to the _____ of a sound.
a. pitch
b. loudness
c. quality
d. frequency

9. The loudness of a sound is measured in terms of a logarithmic scale called the ______ scale.
a. log
b. trig
c. decibel
d. Bel

10. Increasing the intensity of a sound by 50 dB multiplies the perceived loudness of the sound by a factor of __________.
a. 2
b. 5
c. 10
d. 32

11. A ten decibel increase in the loudness of a sound is perceived as _____ the loudness of the sound.
a. doubling
b. quadrupling
c. increasing tenfold
d. increasing 100 fold

12. Compared to a twenty year old woman a 60 year old woman hears a low frequency sound to be_____.
a. twice as loud
b. half as loud
c. the same
d. a little lower

13. When a sixty year old man is compared to a twenty year old, the loudness of the 4000 Hz sound appears to be ______ of what the twenty year old observed.
a. ½
b. ¼
c. 1/8
d. 1/16

14. What characteristic of sound allows one to distinguish a piano sounding a middle C and a factory whistle sounding a middle C?
a. pitch
b. loudness
c. quality
d. vibrato

15. We can distinguish the sounds of different musical instruments sounding the same note because _____.
a. of the overtones present
b. of the relative amplitudes
c. of the number of frequencies present
d. of the relative amplitudes and number of harmonics present

16. Musical notes when sounded have transient characteristics called _____.
a. overtones
b. attack, sustain, decay
c. harmonics
d. duration

17. When speakers are moved outdoors, there is a definite decrease in volume due to a lack of_____.
a. refractions
b. interferences
c. reflections
d. diffractions

18. "Live" rooms for sound are characterized by having a large number of _____ surfaces.
a. rough
b. smooth
c. soft
d. upholstered

19. The reverberation time is the time required for the reverberated sound to decay to _____ of its original power level after the sound source has been turned off.
a. half
b. an eighth
c. a thousandth
d. a millionth

20. When learned men spoke in the Tower of Babel, their words were unintelligible because of _____.
a. multiple refractions
b. multiple reflections
c. multiple diffractions
d. multiple interferences

21. If one wave has a frequency of 300 Hz and is sounded along with one of 296 Hz, a beat frequency of _____ Hz is heard.
a. 597 Hz
b. 4 Hz
c. 298 Hz
d. 0 Hz

22. A standing wave has nodes and antinodes. Nodes are places of ____ vibration.
a. maximum
b. minimum
c. selective
d. periodic

23. Loops on a standing wave correspond to _____.
a. nodes
b. minima
c. antinodes
d. zero vibration

24. The phenomenon that occurs when sound passes through openings or bends around obstacles is called _____.
a. reflection
b. refraction
c. diffraction
d. interference

25. _____ frequencies suffer the greatest diffraction when passing through an opening.
a. low
b. midrange
c. high
d. ultrasound

26. The frequencies which are most directional are _____.
a. low
b. midrange
c. high
d. infra

27. Because of ____ effects a subwoofer can be placed at any location in a room.
a. reflection
b. refraction
c. interference
d. diffraction

28. _____ frequencies are not highly directional.
a. low
b. midrange
c. high
d. harmonic

29. Resonance is a phenomenon that occurs when two waves _____ one another.
a. reinforce
b. cancel
c. interfere destructively
d. eliminate

30. The Tacoma Narrows Bridge collapsed because of ____.
a. reflection
b. refraction
c. diffraction
d. resonance

Bibliography and Web Resources

Johnson, Kenneth, and Willard Walker and John Cutnell. *The Science of Hi-Fidelity.* Kendall-Hunt Publishing Company, Dubuque, Iowa, 1995.

Symko, Orest. *Physics of Hi-Fi.* Kendall-Hunt Publishing Company, Dubuque, Iowa, 1995.

Stanionis, Victor and Hugh Berberich. *Computer Music: Science and Technology of a New Art.* Research Foundation of State University of New York, Stony Brook, NY 1989.

Philharmonic Hall - **http://courses.ats.rochester.edu/ddouglass/PHY103/lecturenotes/32**

PhilharmonicHall - **http://charmian.sonoma.edu/~bryant/Fall2004/Physics%20300%20Problem%20Solutions%20Fall%202004/Auditoria%20WEB%20references/Bouncing%20off%20the%20Walls.htm**

Tacoma Narrows Bridge collapse **http://www.madsci.org/posts/archives/may98/892678504.Eg.r.html**

Chapter 7

Loudspeakers

"No one ever won a chess game by betting on each move. Sometimes you have to move backward to get a step forward."

Amar Gopal Bose, 1964

Founder, Bose Corporation

The most important components of the audio reproduction system are the speakers. These are the devices that turn the electrical signal into a sound wave that we hear, and yet the speakers are probably the weakest link in the hi-fidelity chain. There are more manufacturers of speakers than any other hi-fidelity component and all the manufacturers claim that their speakers are best.

Choosing a set of speakers is really a subjective judgment. There are many variables sound engineers use in designing a good set of speakers. The problem is that these variables interact and sometimes when optimizing one set, another is degraded. No present day speaker is free from sound distorting flaws and nearly all of them can't replicate the signal as recorded.

Goals

After studying this chapter, you should be able to:

- Describe the anatomy of the cone speaker.
- Describe the function of each part of the cone speaker.
- Understand the rationale for having several drivers in a speaker system.
- Understand the function of the crossover network.
- Understand the problem in producing low frequency sounds.
- Describe the function of a baffle.
- Understand the role that speaker enclosures play in a speaker system.
- Understand the operation of a horn speaker.
- Describe the operation of a waveguide in a speaker system.
- Understand the role dispersion plays in a speaker system.

Cone Speaker

The most common form of loudspeaker is the permanent magnet cone speaker. The anatomy of the cone speaker consists of six major parts: the diaphragm, flexible edge, spider, permanent magnet, voice coil, and basket (Fig. 7.1).

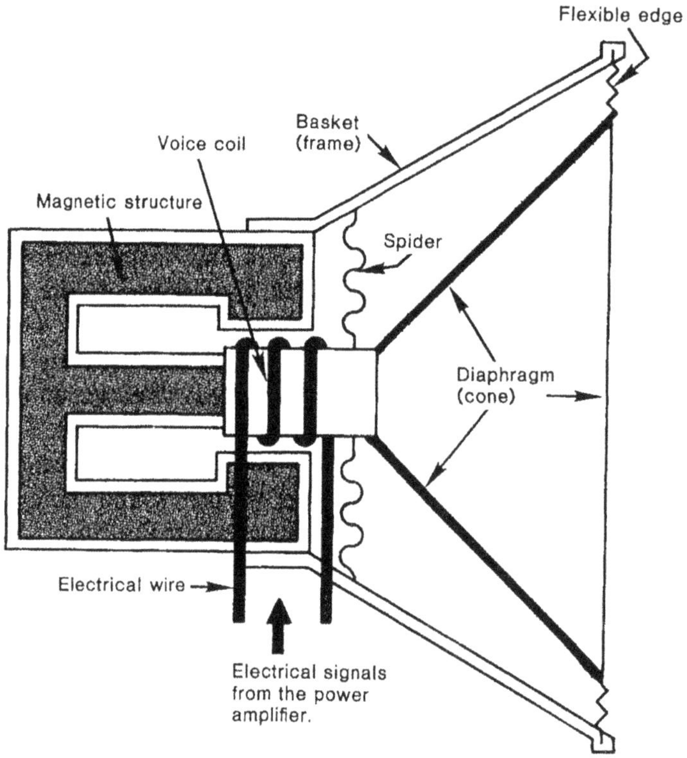

Figure 7.1 Building blocks of a cone speaker

From *Science of Hi-Fidelity* by Johnson, Walker, Cutnell. Reprinted by permission of the authors.

The basket is the metal frame that holds all the other components together. The diaphragm is the paper-like cone that pushes the air back and forth, creating the changes in pressure that we hear as sound. It is suspended from the front of the basket and held to it by the flexible edge. The voice coil is attached to the other end of the diaphragm and carries the electrical signal from the amplifier creating its own magnetic field, which then interacts with the magnetic field of the permanent magnet. The spider keeps the voice coil aligned so that it stays in a horizontal position, preventing vertical vibrations. This minimizes friction between the voice coil and the magnet where the clearances are measured in thousandths of a centimeter.

Flexible Edge (Surround)

The flexible edge, which holds the diaphragm at the top to the basket, is usually made of butyl rubber, santoprene, EPDM (ethylene-propylene-diene terpolymers) rubber, or compressed foam. Rubber and plastic based surrounds resist the elements better but compressed foam is lighter and maintains its physical characteristics better under variations in temperature. Some manufacturers will use the surrounds to limit the movement of the diaphragm and the voice coil attached to the other end, so as to keep the voice coil in the magnetic gap and lower wear and tear on the spider. However, limiting the movement usually results in higher distortion levels at higher volume. Oversized flexible edges are also used as a marketing tool (Fig. 7.2).

Figure 7.2 Cone speaker

Dust Cap

You may notice that at the center of the diaphragm, one side of the voice coil is covered by a dome shaped dust cap that prevents dust and other particles from entering the voice coil and magnetic gap. Many times you will find the logo of the speaker manufacturer imprinted on the dust cap. The dust cap, once it's bonded to the diaphragm, adds to its rigidity and extends its life.

Diaphragm (Cone)

Cones are made of polypropylene and paper. Some are made from composites such as carbon fiber, Kevlar, and fiberglass. It is important that a cone maintain its shape and rigidity and be strong, yet flexible enough to handle various applications (Fig. 7.4). The diaphragm creates sound by generating an oscillating pressure wave in the air that, upon interacting with the human ear, produces a hearing sensation. It can be modeled by a mass attached to a coil spring and hung from a support (Fig. 7.3),

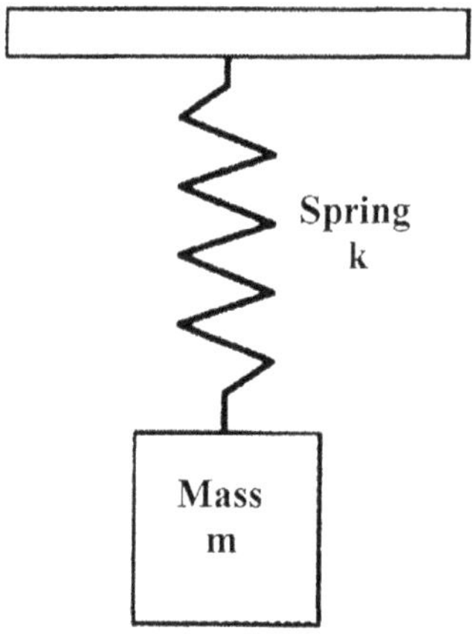

Figure 7.3 Mass and spring model

In order for the diaphragm (Fig. 7.4) to produce low bass sounds, it must be able to vibrate slowly to produce the low frequencies desired. In the spring model we learn from experimentation, that by hanging a heavy mass, m, on a given spring it will vibrate more slowly. Another way to make the mass vibrate more slowly would be to use a weaker more compliant spring as opposed to a stiffer one. Such a spring would have a spring constant, k (which measures the stiffness of the spring) with a small value. It can be shown that the period of oscillation of the mass is given by

$$T = 2\pi \sqrt{(m/k)}$$

and since $f = 1/T$,

$$f = (1/2\pi) \sqrt{(k/m)} .$$

The formula tells us that as m is increased f gets lower. Also, if k is smaller, f will be lower which confirms our experimental observations. If the spring is highly compliant, it will allow the mass to vibrate more slowly. Therefore to produce bass frequencies a speaker must have a highly compliant suspension and a massive diaphragm. The formula also allows us to compute the resonant frequency of the speaker, i.e. the natural frequency of oscillation. It has been found that frequencies that are lower than the resonant frequency of a speaker are attenuated.

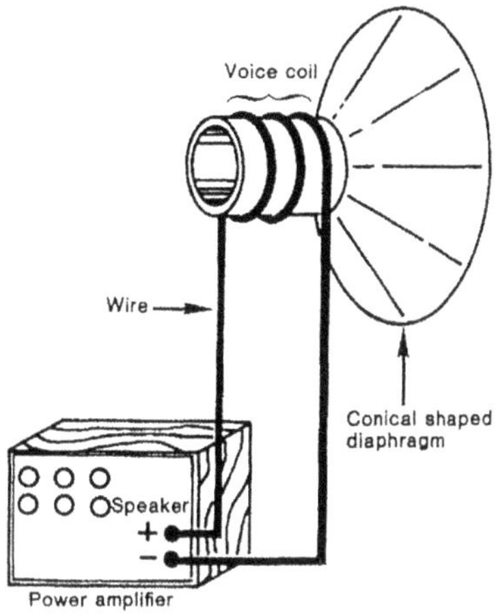

Figure 7.4 Diaphragm and voice coil

From *Science of Hi-Fidelity* by Johnson, Walker, Cutnell. Reprinted by permission of the authors.

Basket

Baskets are made from a variety of materials including steel, aluminum, and composites. It is the frame that holds all the other components together. The stronger the basket, the less likely resonance created by the vibrating diaphragm will occur among the other components.

Spider

Spiders, which keep the voice coil centered on the permanent magnet, come in two common forms: spider-like and corrugated (Fig. 7.5).

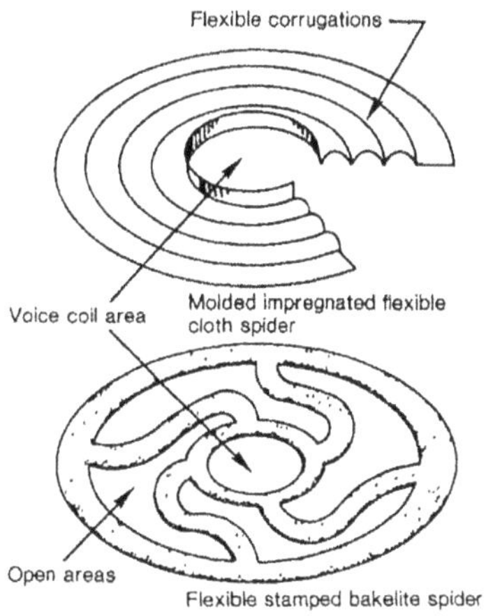

Figure 7.5 Spiders

From *Science of Hi-Fidelity* by Johnson, Walker, Cutnell. Reprinted by permission of the authors.

They are usually made of cotton, poly-cotton or conex. In the most common form, the corrugated spider, the differences are found in the heights of the hills and valleys. They affect the "throw" or horizontal distance the diaphragm is allowed to move. If they are not matched to the application properly, they will deteriorate rapidly and affect the sound from the speaker.

Voice Coil

The voice coil is essentially a hollow tube (like a cardboard toilet paper roll) with electrical wire wound around it (Fig. 7.4). It is physically attached to the diaphragm and spider and slides into a gap cut into the permanent magnet. As alternating electric current flows through it, a varying magnetic field is created around the coil, which then interacts with the magnetic field of the permanent magnet. The attractive and repulsive forces that occur cause the coil and the attached diaphragm to move. The cylinders around which the wire is wrapped in a coil are usually made of aluminum, Mylar or fiberglass. The wire used in the coil is usually copper, aluminum, or copper-clad aluminum which is bonded to the cylinder using special glue.

Magnet

The magnet (Fig. 7.6) is usually made of some iron, cobalt, or nickel compound. It is generally made from a ferrous material, which retains a permanent magnetic field around it. Some people think that the larger and heavier the magnet, the better the speaker will be. This is not true and we shall find out why later.

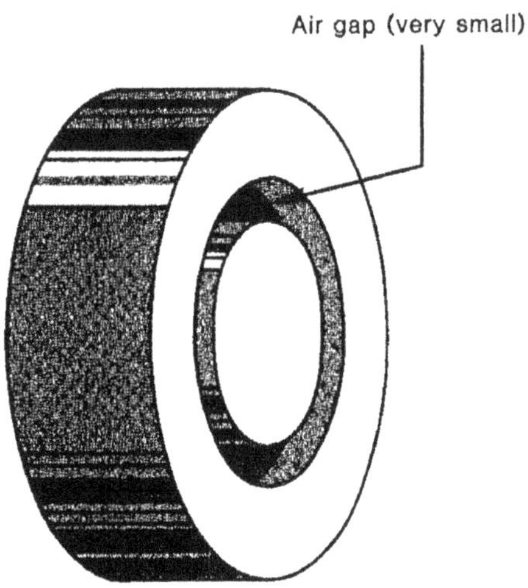

Figure 7.6 Magnet

From *Science of Hi-Fidelity* by Johnson, Walker, Cutnell. Reprinted by permission of the authors.

Speaker Systems

Like all systems, loudspeaker systems are made up of parts that work together to achieve a common goal. If you were to remove the front grill cloth cover on different speakers, you might be amazed at what you would see. Loudspeaker systems may contain different numbers of speakers or "drivers" as they are sometimes called. They may all differ in size. There may be open holes (ports) in the speaker enclosure or the speaker enclosure may be completely sealed. Why?

Many manufacturers of loudspeakers have found that they can improve the quality of sound generated by their loudspeakers using the strategy of "divide and conquer." They use different drivers to reproduce different segments of the audio frequency range (20 - 20,000 Hz) without distortion. To produce low frequencies (20 - 500 Hz) generally speaking a driver with large diaphragm called a woofer is used. Say the word woofer out loud: **WOOFER!** Note the low frequency sound generated.

To reproduce high frequency sounds (5000 - 20,000 Hz) in the audio spectrum a speaker with small diaphragm called a tweeter is used. Say **TWEETER** out loud. How does it sound in pitch? High or low? If a third driver, a midrange speaker, is added to reproduce frequencies between 500 – 5000 Hz, then the speaker would be called a three-way speaker system (Fig. 7.7). If a speaker enclosure contains only two drivers, we call it a two-way speaker system.

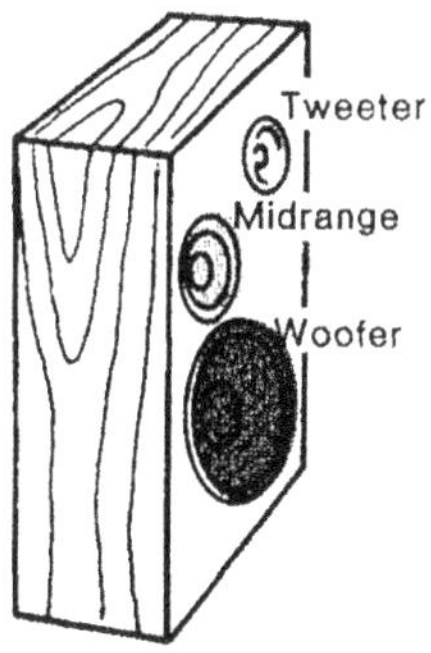

Figure 7.7 Three way speaker system
From *Science of Hi-Fidelity* by Johnson, Walker, Cutnell. Reprinted by permission of the authors.

When the electrical signal is sent from the amplifier containing all of the audio frequencies (treble, bass, midrange) mixed together to the speaker system, how does the speaker divide up the signal and send the appropriate audio frequencies to the appropriate driver? Inside the speaker enclosure is an electrical circuit called a crossover network that does the job. The electrical circuit (Fig. 7.8) is designed to switch, or crossover, at certain frequencies that are arbitrary but part of the manufacturer's design.

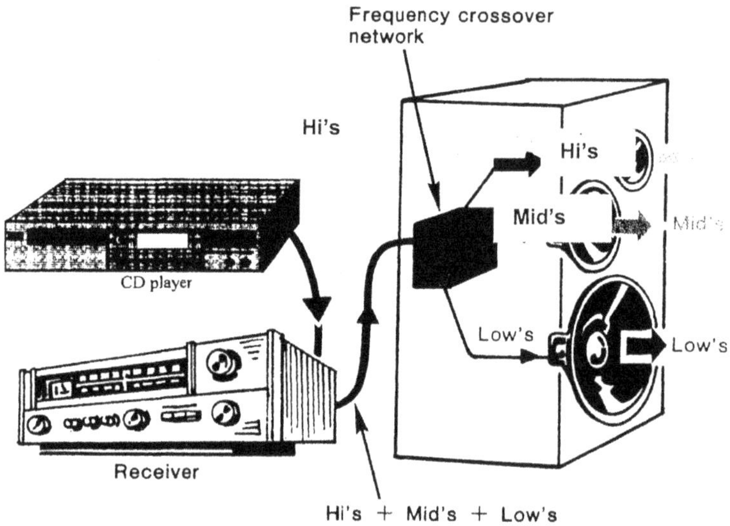

Figure 7.8 Three way speaker system with crossover network
From *Science of Hi-Fidelity* by Johnson, Walker, Cutnell. Reprinted by permission of the authors.

For example, in one design when the audio frequencies are over 500 Hz, the signal is then sent to the midrange speaker. It would handle the signal up to 5000 Hz, another crossover point. The high frequency signal would then be transferred to the tweeter (Fig. 7.9).

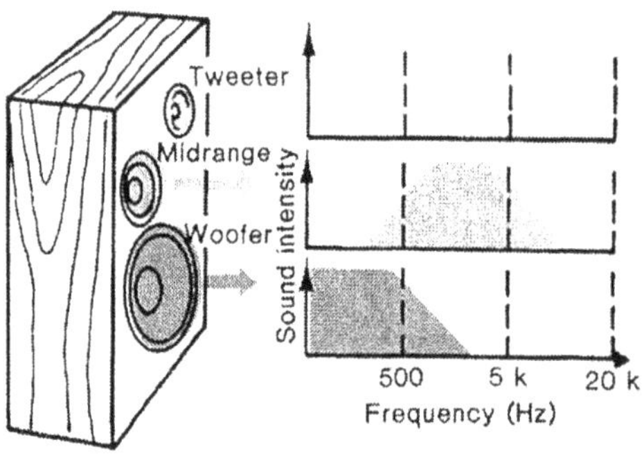

Figure 7.9 Crossover points
From *Science of Hi-Fidelity* by Johnson, Walker, Cutnell. Reprinted by permission of the authors.

Woofers

Audio engineers are faced with two major problems when designing woofers First, woofers are speakers with large diaphragms which are designed to move large amounts of air. They are the speakers that are designed to reproduce low frequency (20 - 500 Hz) sounds. To do this a large amount of energy is required to create the sounds at normal listening levels because the human ear is less sensitive to bass frequencies than those at the midrange and more sound intensity is required to hear the sound as compared to the midrange. See the Fletcher-Munson curves below in Figure 7.10.

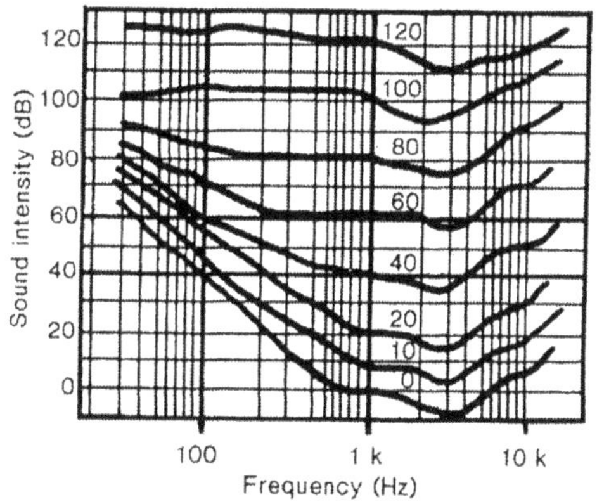

Figure 7.10 Fletcher-Munson curves
From *Science of Hi-Fidelity* by Johnson, Walker, Cutnell. Reprinted by permission of the authors.

Suppose the speaker was sounding a 1 kHz tone observed at 40 dB of sound intensity as seen on the graph. If we move to the left on the 40 dB curve (which means we are staying at constant loudness), we see that a 100 Hz tone would have to sound at 60 dB to be perceived as being at the same loudness level. The 20 dB difference tells us that the sound intensity of the speaker needed to produce the same loudness level as the 1 kHz tone would be 20 dB higher (100 times the intensity) than at 1 kHz (Fig.7.11). It is estimated that 40% of the sound energy contained in a musical composition falls in the low frequency range.

Power Ratio or Intensity Ratio	Difference in Decibels
1.00	0 dB
1.26	1 dB
1.58	2 dB
2.00	3 dB
2.50	4 dB
3.16	5 dB
4.00	6 dB
5.00	7 dB
6.30	8 dB
7.90	9 dB
10.0	10 dB
20.0	13 dB
31.6	15 dB
50.0	17 dB
$100 = 10^2$	20 dB

Figure 7.11 Power ratio or intensity ratio conversion to decibels and vice versa
From *Science of Hi-Fidelity* by Johnson, Walker, Cutnell. Reprinted by permission of the authors.

How do we solve the problem? We can make the diaphragm of the woofer larger so it moves a large amount of air, use more than one woofer, or increase the throw (displacement) of a smaller diameter speaker (Fig. 7.12). Because of the relatively inelastic flexible surrounds in the 1950s, diaphragms had to be large to move a large volume of air. However as more flexible materials were synthesized to replace the rubber used in flexible surrounds, the throw of the speaker could be increased and smaller diaphragm speakers were able to move the same volume of air as their older counterparts. Today we have small bookshelf speakers that can produce bass sounds by moving large volumes of air.

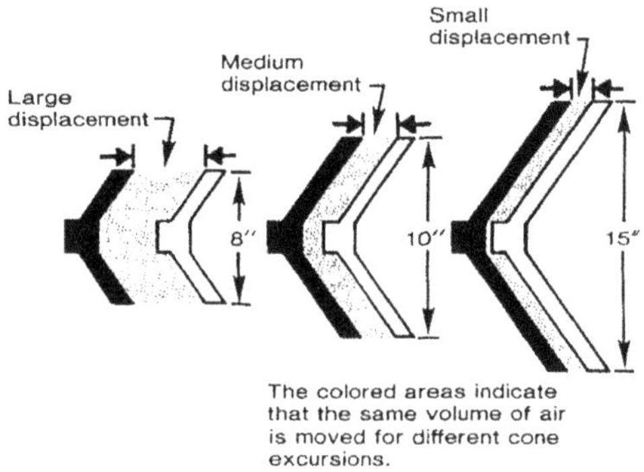

Figure 7.12 Displacement for different size diaphragms moving same volume of air
From *Science of Hi-Fidelity* by Johnson, Walker, Cutnell. Reprinted by permission of the authors.

Tweeters

Tweeters are designed to reproduce high frequencies and in order to do that they are small, with a diameter of 1 to 3 inches, and the diaphragm is made of lightweight material so they can vibrate extremely fast. A midrange speaker, ranging in size from 3-1/2" to 6-3/4", is dedicated to the reproduction of midrange frequencies. The diaphragm of a tweeter is usually dome shaped and dome, or "edge-driven", tweeters are the type found in most home speakers. They disperse their sound over a wider area because of their dome shape (Fig. 7.13).

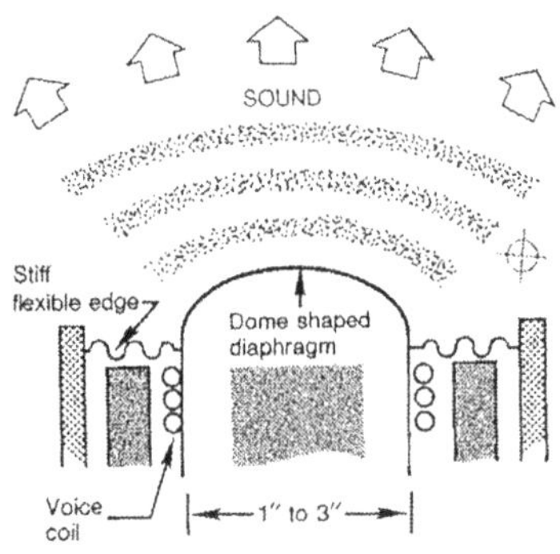

Figure 7.13 Tweeter speaker
From *Science of Hi-Fidelity* by Johnson, Walker, Cutnell. Reprinted by permission of the authors.

Horns

Many tweeters are horn tweeters in which the small driver is placed at the mouth of a horn or megaphone shaped enclosure. The horn or megaphone results in a louder sound with good dispersion. A horn effect is similar to cupping your hands around your mouth when yelling to someone at a distance. It makes it louder. Why? Well, a horn is like an acoustic transformer. A big horn or funnel effectively makes the source driver the entire area of air at the end of the horn. So, if the horn opening is ten times the size of the diaphragm of the source, then the driven area is ten times greater. The big black boxes at the bottom of a rock band's P.A. system are horns. The drivers inside the cabinets themselves are not very big compared with the box (Fig. 7.14).

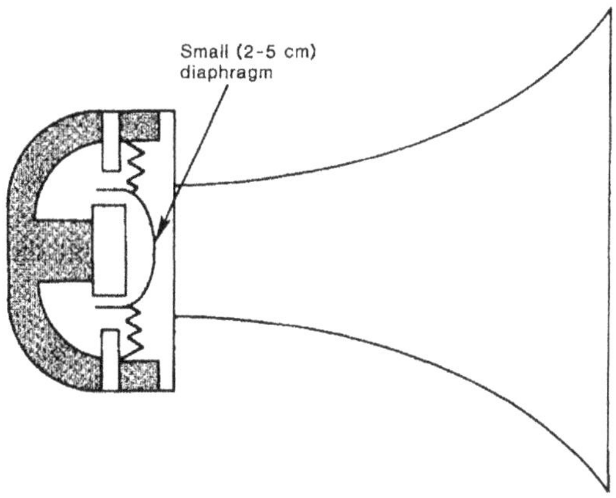

Figure 7.14 Horn speaker

From *Science of Hi-Fidelity* by Johnson, Walker, Cutnell. Reprinted by permission of the authors.

Horn speakers can be used to reproduce not only high frequencies but also midrange and bass frequencies. They are the most efficient loudspeakers in converting electrical energy into sound. All horn speakers have a cut-off frequency below which they will not generate any sound. The cut-off frequency is a function of the size of the horn. Horns that produce bass frequencies would have to be very large (Fig. 7.15).

We can calculate the cut-off frequency for a horn speaker by noting that the diameter of the opening of the horn speaker is approximately $\lambda/4$ of the sound wave associated with the cut-off frequency.

Figure 7.15 Horn speakers
From *Science of Hi-Fidelity* by Johnson, Walker, Cutnell. Reprinted by permission of the authors.

The formula tells us that a horn with an opening of 2 m in diameter can reproduce frequencies down to about 43 Hz.

Diameter = 2 m = $\lambda/4$

Therefore, $\lambda = 8$ m.

Since $v = f\lambda$,

$f = v/\lambda = (344 \text{ m/s})/8 \text{ m} = 43 \text{ Hz}$

where 344 m/s is the speed of sound at room temperature.

This diameter of 2 m (approximately 6 feet) is somewhat larger than your average vented cabinet. In fact, the speaker boxes at rock concerts are generally useless below about 50 Hz even though they seem to produce bone rattling sound waves. Therefore, most horn type speakers are used as midranges and tweeters.

Speaker Enclosures

The diaphragm on a speaker moves back and forth. When moving forward, it produces a compression at the front surface and a rarefaction at the back surface (Fig. 7.16).

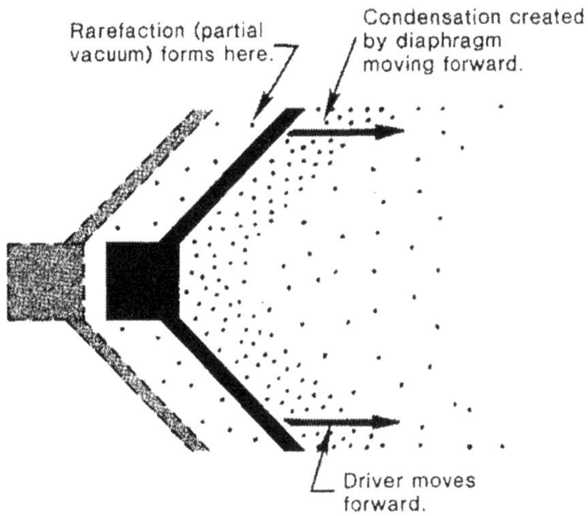

Figure 7.16 Moving diaphragm

From *Science of Hi-Fidelity* by Johnson, Walker, Cutnell. Reprinted by permission of the authors.

If the speaker is not mounted in a box or on a board the front and rear pressure waves will mix causing destructive interference, especially at low frequencies, greatly reducing the bass sounds. Why will the low frequencies be affected to a greater extent? Remember, the bass frequencies have greater dispersion characteristics, i.e. they spread out and bend more.

If we mount the speaker on a flat board with a hole cut in it for the diaphragm, we can prevent the front wave from canceling out the back wave. The board is then called a flat open baffle (Fig. 7.17).

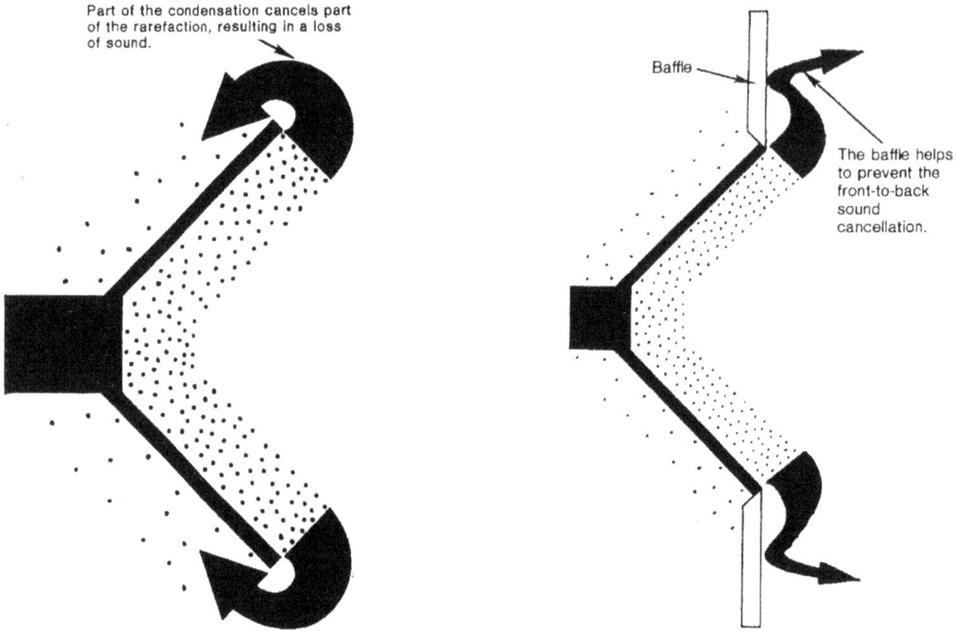

Figure 7.17 Flat open baffle

From *Science of Hi-Fidelity* by Johnson, Walker, Cutnell. Reprinted by permission of the authors.

If we add a top and sides, we then have a open back baffle. Adding the back side to form a fully enclosed box is called a closed baffle (Fig. 7.18).

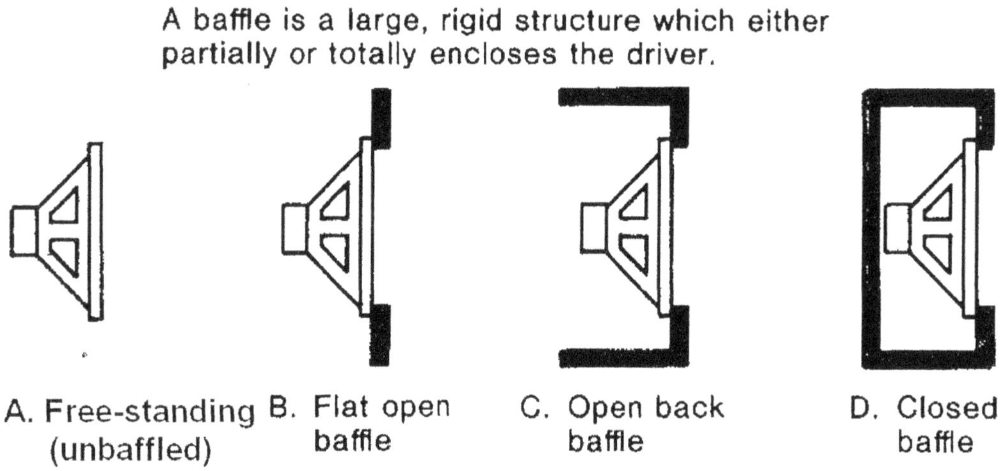

Figure 7.18 Types of baffles

From *Science of Hi-Fidelity* by Johnson, Walker, Cutnell. Reprinted by permission of the authors.

Sealed Enclosures

Sealed enclosures (Fig. 7.19) are made by mounting the driver in a big box (infinite baffle enclosure) or a small box (acoustic suspension enclosure).

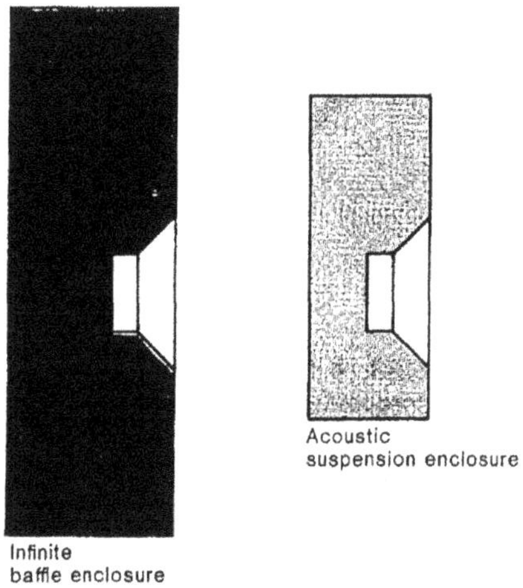

Figure 7.19 Sealed enclosures

From *Science of Hi-Fidelity* by Johnson, Walker, Cutnell. Reprinted by permission of the authors.

Acoustic Suspension

If we use a small box, or acoustic suspension enclosure, the amount of air in the box is limited and when the diaphragm on the driver moves the air in the box, the air will act as a cushion, limiting the throw of the diaphragm (Fig. 7.20). The small box with the entrapped air acts the same as if we attached a stiff spring to the back of the driver. How can we compensate for this so as to produce a small speaker with a good bass?

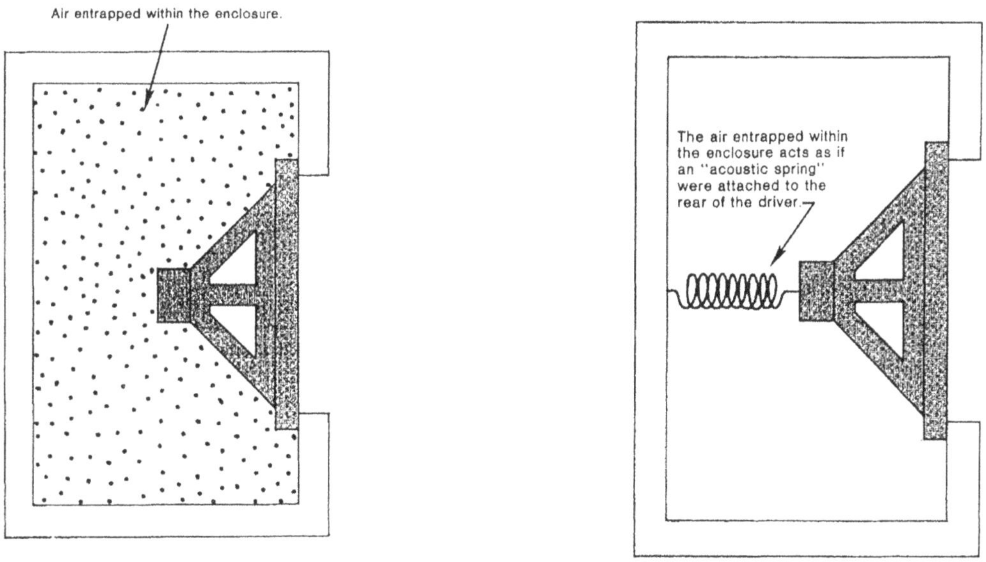

Figure 7.20 Acoustic suspension speaker

From *Science of Hi-Fidelity* by Johnson, Walker, Cutnell. Reprinted by permission of the authors.

Now every speaker has a frequency response (Fig. 7.21), which measures its ability to reproduce all frequencies in the range of hearing at the same sound intensity.

An ideal speaker would have a flat frequency response, that is it would reproduce the same sound intensity for all frequencies in the range of 20 Hz to 20,000 Hz

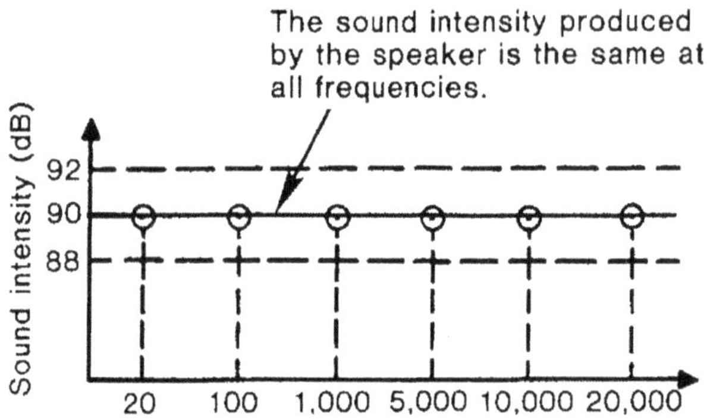

Figure 7.21 Frequency response curve for an ideal speaker

From *Science of Hi-Fidelity* by Johnson, Walker, Cutnell. Reprinted by permission of the authors.

However, real speakers usually have a resonance or natural frequency associated with them because of their mechanical construction and as a result tends to increase the sound intensity abnormally at their resonant frequency. Frequencies below the resonant frequency tend to diminish rapidly (Fig. 7.22).

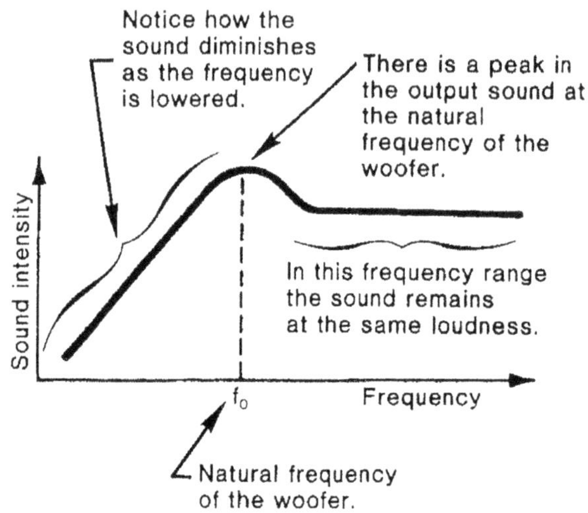

Figure 7.22 Resonant frequency in a speaker

From *Science of Hi-Fidelity* by Johnson, Walker, Cutnell. Reprinted by permission of the authors.

Placing a woofer in an enclosure increases the resonant frequency to an even higher value (Fig. 7.23).

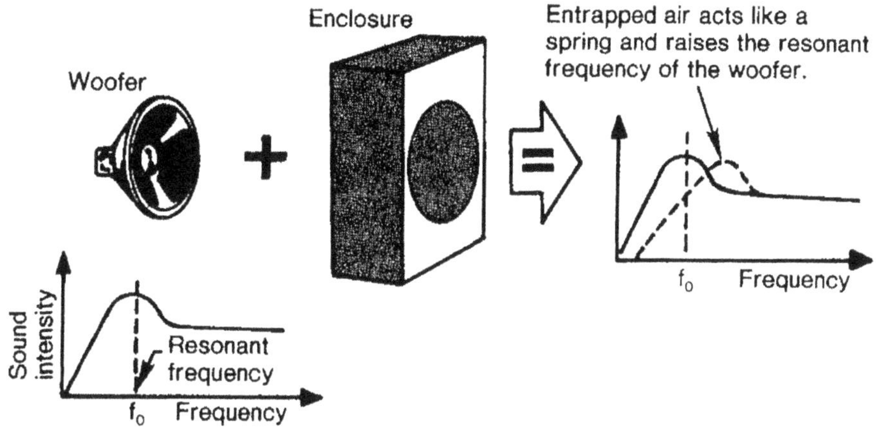

Figure 7.23 Woofer natural frequency in an enclosure
From *Science of Hi-Fidelity* by Johnson, Walker, Cutnell. Reprinted by permission of the authors.

We can compensate for the air cushion effect by increasing the throw of the speaker. This is done by using a loose (highly compliant) suspension. Usually the flexible edge is made of half a roll of very light and elastic material. Now the diaphragm can move through a longer distance, thus moving the same volume of air as a larger one.

In order to withstand the high speeds and accelerations associated with long throws the diaphragm must be massive and strong and the entire driver must be sturdily built. The more massive diaphragm usually gives the driver a lower resonant frequency. When it is placed in the speaker enclosure, the effect is to raise the resonant frequency.

Infinite Baffle

The big box, or infinite baffle, has to be large enough and contain enough air to render the air cushion effect negligible. Usually sound absorbing material is placed in the infinite baffle enclosure to absorb the rear wave, preventing reflections and interference with the motion of the diaphragm. This means that half of the sound power is absorbed.

Vented Enclosures

Vented enclosures such as the bass reflex, phase inverter and tuned port speakers are designed to increase the bass response. It does this by using the wave generated by the back surface of the diaphragm and reflected inside the enclosure. Usually a port or small opening appears on the front of the speaker enclosure through which the reflected wave can exit. It is designed to allow only low frequencies to pass through (Fig. 7.24).

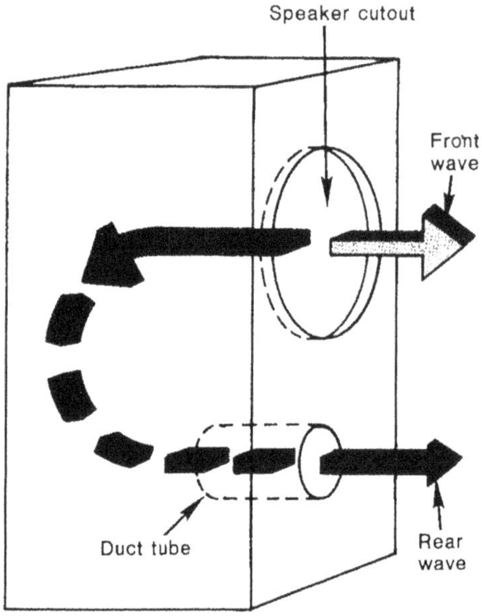

Figure 7.24 Vented enclosure

From *Science of Hi-Fidelity* by Johnson, Walker, Cutnell. Reprinted by permission of the authors.

If we are careful in designing the dimensions of the box, we can have the rear wave reflect from the rear of the enclosure and exit through the port in such a way that it is in phase with the wave being generated by the diaphragm. This results in constructive interference or a louder sound. The enclosure is designed to reinforce the bass sounds (Fig. 7.25).

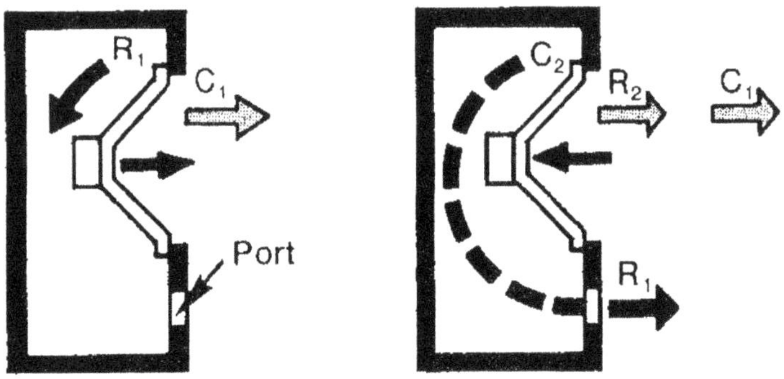

Figure 7.25 Reinforcement of sound by constructive interference

From *Science of Hi-Fidelity* by Johnson, Walker, Cutnell. Reprinted by permission of the authors.

As the cone moves forward, creating a condensation, a rarefaction moves backwards, reflecting off the back of the cabinet and emerging through the port as R_1. At this instant the diaphragm is forming a rarefaction R_2 as it moves backwards. The two interfere constructively resulting in a louder sound. Even though the two rarefactions are a half cycle apart, the difference in sound is negligible.

Bose Wave Guide System

Ordinarily, to reproduce a deep bass sound requires a large speaker diaphragm in a large cabinet driven by a large amplifier. Dr Amar Bose and one of his senior research engineers, Dr. William Short, have posed the question "is there a better way" of transferring sound from a speaker to the air in a room? Their method involves using the properties of an "acoustic wave-guide" to produce a wide range of frequencies from a small enclosure. Using an acoustic wave-guide a sound wave is confined to move over a fixed path. A wave-guide would be a device in which the sound is required to follow a certain path and is confined to that pathway. One example would be a stethoscope (Fig. 7.26).

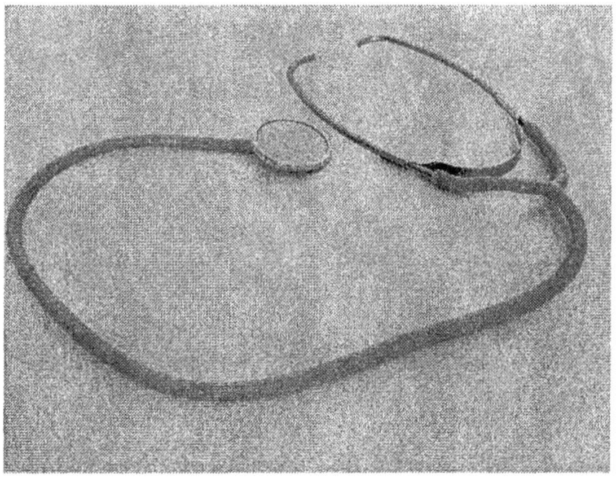

Figure 7.26 Stethoscope

Bose gives the simple example of a pipe organ, which uses a small amount of air to fill a cathedral with sound. The production of different notes requires a wave-guide of different lengths. In the case of the organ this would be a different length pipe. We would need many pipes to create different note sounds in the case of the organ.

Bose discovered that if a speaker was mounted in a tube, the vibration of the speaker could be coupled to the motion of air and it would act as a wave guide. Using signal processing techniques this wave guide was such that it could match the mechanical properties of the speaker over a wide range of frequencies allowing a small driver to produce clear sound without distortion at high volume levels and low frequencies.

They found that they could fold the wave-guide without loss in the quality of the sound. They were able to take one nearly seven feet (2 m) long and fold it into a foot high enclosure also folding a three foot (1 m) long wave-guide into a 14 inch wide cabinet. These form the basis for the Bose Acoustic Wave® music system and Wave® radio.

Loudspeaker Parameters

Manufacturers market loudspeakers by touting their various specifications. The manufacturers specifications may be some help in making a choice, as well as the tests performed by consumer organizations. Ultimately, the choice depends on you.

Remember the perfect speaker has a flat response over the hearing spectrum (Fig. 7.27).

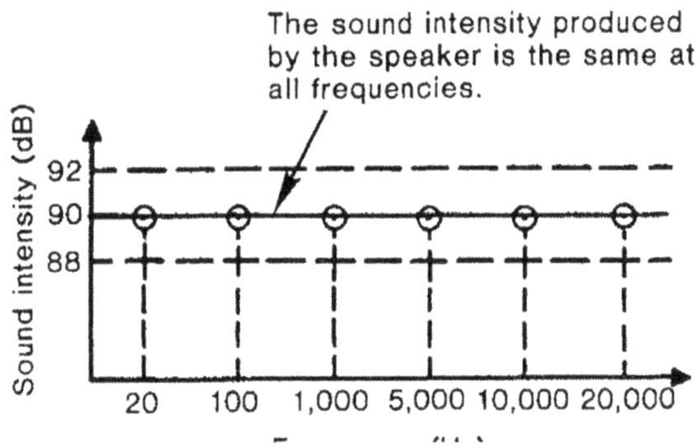

Figure 7.27 Perfect speaker

From *Science of Hi-Fidelity* by Johnson, Walker, Cutnell. Reprinted by permission of the authors.

In reality a speaker response curve is not flat (Fig. 7.28).

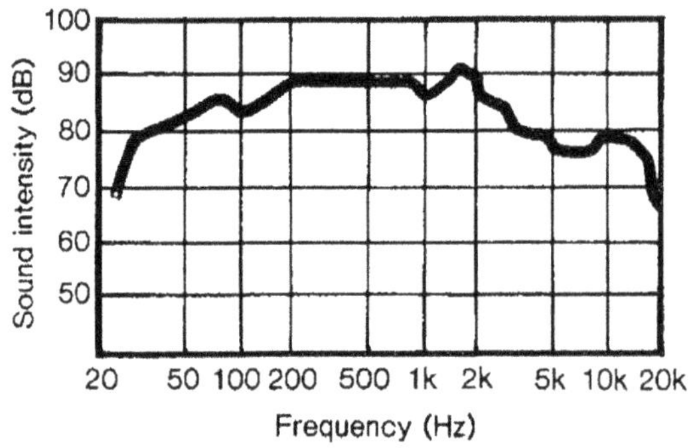

Figure 7.28 Ordinary speaker response curve

From *Science of Hi-Fidelity* by Johnson, Walker, Cutnell. Reprinted by permission of the authors.

Adjusting the treble and bass controls may help to get a more accurate response. Some consumer magazines rate speakers on accuracy. To do this, you have to specify not only the frequency range, but also the deviation from "flatness." For example a speaker could have a frequency range of 20 - 20,000 Hz but to specify its frequency response we must give the frequency range 20 - 20,000 Hz and the deviation ±3 dB (i.e. 20 - 20,000 Hz ± 3 dB).

Speaker Dispersion

We are also interested in how different sound frequencies spread out from the speaker as they are emitted. This data is usually not available at the typical consumer electronics outlet. High-end dealers can often provide this specification in terms of a polar coordinate graph of the dispersion characteristics of the speaker (Fig. 7.29).

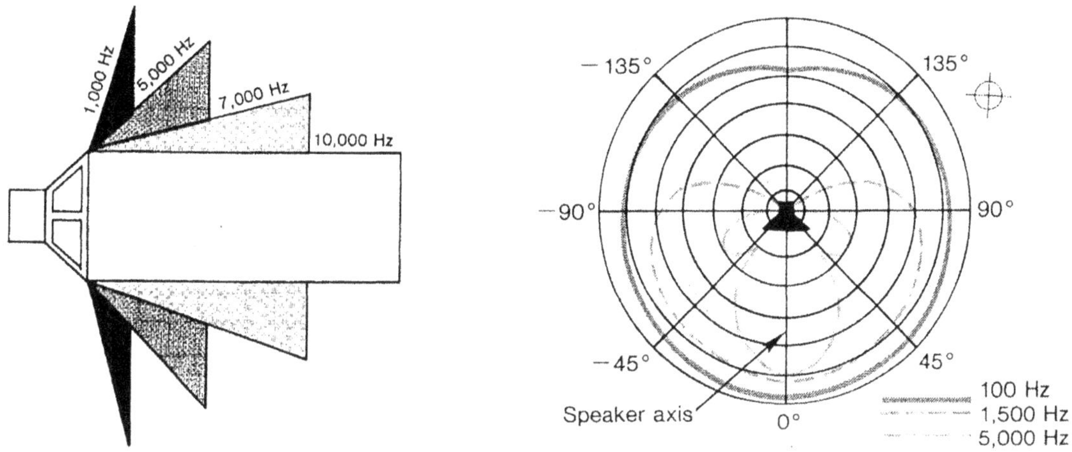

Figure 7.29 Polar coordinate dispersion graph for a speaker
From *Science of Hi-Fidelity* by Johnson, Walker, Cutnell. Reprinted by permission of the authors.

Notice the petal-shaped curve directly in front of the diaphragm of the speaker. This represents the dispersion of the high frequency sound. The heart shaped, or cardiod curve, corresponds to the way a 1500 Hz sound would spread out around the speaker. Finally, notice the almost circular curve surrounding the driver. This represents how the low frequency 100 Hz sound would spread out around the driver.

Satellite speaker systems are now used in hi-fidelity setups and depend on the dispersion characteristics of speakers. Great attention is paid to the placement of small satellite speakers because they emit the high frequency sound, which is very directional. They have to be placed at ear level and must also be spaced apart by 6 feet to 10 feet in order to experience the stereo effect. The subwoofer may be placed anywhere in the room since the bass (low frequencies) spread out evenly in all directions around the speaker.

Power Ratings

As a consumer, the fastest way to upgrade your audio system with the most noticeable improvement, is to add a new set of speakers. Usually the manufacturer will stamp the maximum and minimum power the driver can handle on the back of the driver. Attaching a speaker to a receiver with a large power output may cause the voice coil in the speaker to melt and evaporate if the maximum power rating of the speaker is exceeded. Someone may accidentally increase the volume and destroy the speaker.

The minimum power rating of the speaker is another important variable to consider when buying new speakers. If your receiver does not exceed the minimum power rating of your new speakers, the sound will be weak and distorted. The higher the minimum power rating of the speaker, the less efficient it is in converting electrical power to sound. Efficiency has nothing to do with the quality of sound it can produce.

Sensitivity Rating

The sensitivity rating of a loudspeaker has to do with a measure of the loudest sound that can be produced by the speaker. It is usually expressed in decibels but measured differently by different manufacturers. Although a speaker may have a high sensitivity rating, such as 100 dB, this has nothing to do with the quality of the sound it produces. It just indicates that the speaker will sound louder than one with a lower sensitivity rating.

There are other specifications that manufacturers list for their speakers such as impedance, number of drivers, crossover frequency points, magnet size, material, voice coil materials and sizes. For the most part, they are not directly related to the sound quality of the speaker.

Speaker Efficiency

The efficiency of a speaker is a measure of its ability to convert electrical energy into sound waves. Efficiency of a device is measured by dividing the work output by the work input. Multiplying by 100 gives us the efficiency as a percentage. The efficiency of a speaker is measured by dividing the acoustic power output by the electrical power input and multiplying the result by 100. If you had a speaker that was receiving 50 W of electrical power and producing 5 W of acoustic power, its efficiency would be 5 W/50 W x 100 = 10%.

Speaker efficiencies can vary from about 1 to 50 % depending on the type of speaker. There is no direct correlation between the efficiency of a speaker and its quality. A speaker can have a low efficiency but produce hi-fidelity sound.

Speaker Placement

Care must be taken in placing speakers at various places in a room. Reflections and dispersion effects play an important role. Remember that smooth, hard surfaces reflect more sound then rooms furnished with upholstered furniture and rugs. For best sound reproduction the characteristics of a "live" room and "dead" room have to be considered and addressed.

If speakers are placed flush against the wall of a room on one side, the sound they generate may reflect off the opposite wall and produce standing waves for certain frequencies. Standing waves, as you know, are characterized by nodes and antinodes and will lead to a distortion of sound (Fig. 7.30).

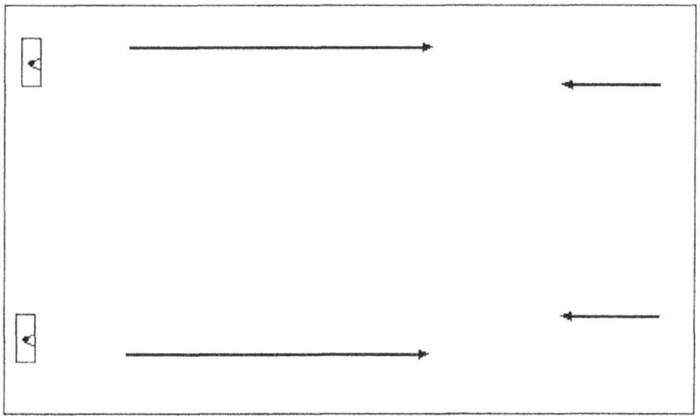

Figure 7.30 Reflection of sound from walls in a room

Placing the speakers at an angle of 10 degrees or so with the side wall will diminish the possibility of standing waves being set up. If the speakers are mounted high on a wall or placed high in a bookcase, the possibility exists that the high frequencies, which are most directional, will pass over the head of the listener. There are manufacturers that produce speaker stands which are adjustable in height so that the listener can raise the height of the tweeters to listening level whether one is reclining or seated when listening to music. If a speaker is placed in the corner of a room, reflection from the side wall and floor create two images of the speaker that also seem to be producing sound.

Care must be taken in not placing the speakers too close to one another or too far apart, otherwise the "stereo effect" will be lost. Since the high frequency notes are directional, as opposed to the low frequency sounds which spread out in all directions, it is the high frequency sounds that produce the stereo effect. So tweeters have to be placed at listening levels and no farther than 10 feet apart, or closer than 6 feet, for the best "stereo effect." Speaker placement requires some experimentation to achieve the best arrangement for listening (Fig. 7.31).

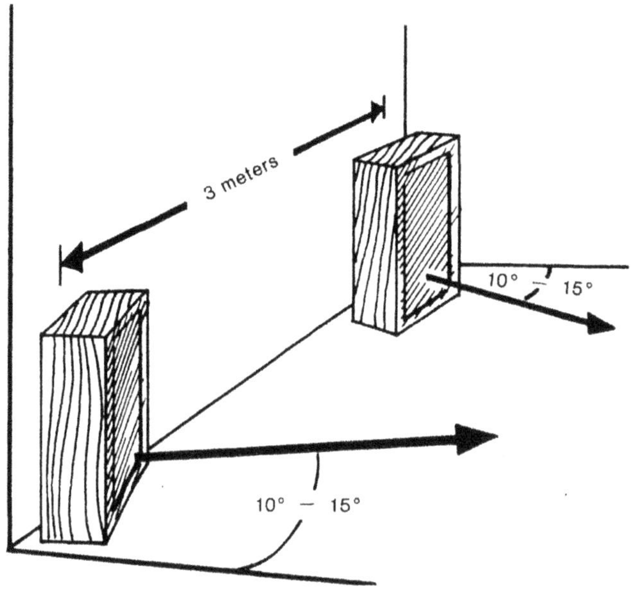

Figure 7.31 Guide for placement of speakers in a room
From *Science of Hi-Fidelity* by Johnson, Walker, Cutnell. Reprinted by permission of the authors.

Review Questions

1. Name the six parts of a cone speaker.
2. Explain the function of each of the six parts of the cone speaker.
3. Why do speaker systems have several drivers?
4. What is the purpose of a crossover network?
5. What is the function of a baffle?
6. What is the purpose of a speaker enclosure?
7. Discuss the two type of speakers systems that use a fully sealed enclosure to produce sound.
8. Why do some speaker enclosures have ports?
9. What role do the Fletcher-Munson curves play in speaker design?
10. How can a speaker with a small diaphragm produce a low bass sound?
11. How does a horn speaker differ from a cone speaker in producing sound?
12. How does the Bose waveguide speaker work?
13. What is speaker dispersion?
14. What role does speaker dispersion play in the placement of speakers in a room?

Multiple Choice Questions

1. The part of the speaker that pushes the air is called the_____.
a. flexible edge
b. basket
c. diaphragm
d. spider

2. The part of the cone speaker that carries the electricity is called the _____.
a. spider
b. diaphragm
c. voice coil
d. permanent magnet

3. The purpose of the permanent magnet in a cone speaker is to provide a magnetic field that the _____ can interact with.
a. flexible edge
b. voice coil
c. spider
d. diaphragm

4. The role played by the spider in a cone speaker is to keep the ______ aligned.
a. basket
b. flexible edge
c. voice coil
d. permanent magnet

5. The flexible edge attaches the _____ to the basket.
a. voice coil
b. spider
c. permanent magnet
d. diaphragm

6. One end of the voice coil is usually covered by a dome shaped _____.
a. dust cap
b. seal
c. name plate
d. push button

7 A frequency cross over network sends the bass signal to the _____.
a. tweeter
b. super tweeter
c. midrange
d. woofer

8. The driver in a speaker system that reproduces the low bass sound is called the _____.
a. tweeter
b. super tweeter
c. midrange
d. woofer

9. The speaker which reproduces the high frequencies is called the _____.
a. tweeter
b. subwoofer
c. midrange
d. woofer

10. Treble sounds are reproduced by the _____.
a. tweeter
b. subwoofer
c. midrange
d. woofer

11._____% of sound energy in a musical composition falls in the low frequency range.
a. 20
b. 40
c. 60
d. 80

12. A horn speaker acts like a _____ where the source drives the entire area of air at the end of the horn.
a. tunnel
b. funnel
c. black box
d. compressor

13. All horn speakers have a cut-off frequency that is a function of the ____ of the horn.
a. size
b. cross over network
c. treble frequency
d. diaphragm diameter

14. A rule of thumb tells us that a horn with a 3 m diameter is needed to produce frequencies down to about _____ Hz.
a. 10 Hz
b. 30 Hz
c. 100 Hz
d. 500 Hz

15. The function of a baffle, when used with a driver, is to _____.
a. separate the front and back sound waves
b. to support the driver
c. to allow the driver to be mounted in a box
d. to provide a tunnel for the sound

16. A large sealed enclosure containing large amount of air and a driver is called a(n) _____ speaker.
a. acoustic suspension
b. bass reflex
c. infinite baffle
d. tuned port

17. An enclosure with a port is called a _____ speaker.
a. acoustic suspension
b. bass reflex
c. infinite baffle
d. tunnel

18. The function of a port in a speaker enclosure is to allow the internal, reflected wave to exit the enclosure and interfere constructively with the external wave resulting in a louder _____ sound.
a. treble
b. midrange
c. bass
d. overall

19. An example of an acoustic wave guide is a(n) _____.
a. stethoscope
b. headphones
c. sonic amplifier
d. oscillator

20. Organs and flutes can generate loud sounds by setting up _____ for a single note.
a. traveling waves
b. standing waves
c. diffraction
d. dispersion

21. A typical frequency response for a speaker would be _____.
a. 20 - 20000 Hz
b. 10 - 30000 Hz
c. 10 dB
d. 20 - 20000 Hz ± 3 dB

22. A speaker's _____ characteristics tells us how different frequency sounds spread out.
a. diffraction
b. refraction
c. interference
d. compliance

23. The sounds that are least directional are _____.
a. high frequency
b. low frequency
c. treble sounds
d. midrange sounds

24. The most efficient speaker is the _____ speaker.
a. bass reflex
b. acoustic suspension
c. infinite baffle
d. horn

25. Sound radiating from a loudspeaker is highly directional for _____.
a. bass frequencies
b. mid range frequencies
c. treble sounds
d. midrange sounds

26. Since low frequency sound waves are least directional, a(n) _____ is used with the driver.
a. horn
b. baffle
c. stethoscope
d. acoustic reflex guide

27. A "live room" is characterized by _____.
a. good sound absorption
b. interference patterns
c. diffraction patterns
d. many reflections

28. For the best stereo effect speakers should be placed _____ feet apart.
a. greater than 10
b. less than 6
c. between 6 and 10
d. about 4

29. When a driver is placed in a small sealed enclosure, its resonant frequency is _____.
a. increased
b. decreased
c. unaffected
d. minimized

30. If the diaphragm in a speaker is made more massive, its natural frequency will be _____.
a. increased
b. decreased
c. unaffected
d. minimized

Bibliography and Web References

Johnson, Kenneth and Willard Walker and John Cutnell. *The Science of Hi-Fidelity*, Kendall-Hunt Publishing Company, Dubuque, Iowa, 1995.

Stanionis, Victor and Hugh Berberich. *Computer Music: Science and Technology of a New Art*. Research Foundation of State University of New York, Stony Brook, NY 1989.

Symko, Orest. *Physics of Hi-Fi*.. Kendall-Hunt Publishing Company, Dubuque, Iowa, 1995.

Bose - **http://www.bose.co.uk/technologies/acoustics/waveguide_intro.asp**

Bose – **http://qualitysound.bose.com/pg/learning/waveguide_solution.jsp**

Wave guides – **http://www.ieee-virtual-museum.org/collection/tech.php?id=2345890&lid=1**

Chapter 8

Electricity

"What hath God wrought!"

Samuel Morse, 1844
First electronic long distance message

To understand how a MIDI synthesizer works, a basic understanding of some fundamentals of electricity is needed. The ancient Greeks became aware of electricity when they first noticed that amber when rubbed with wool would attract small bits of straw, feathers, and dried grass. The word for amber in Greek is "electron". The Latin word, "electricus", means to "produce from amber by friction." It is from the Greek and Latin roots we have our English words for electron, electricity, electronic, etc. The study of electricity will help us understand the operation of all the components of a MIDI music system.

Goals

After studying this chapter, you should be able to :

- Describe a model of the atom.
- Understand how matter is made up of atoms.
- Know the difference between conductors and nonconductors.
- Know Coulomb's law as applied to electrostatic charges.
- Be able to work with current electricity.
- Know the meaning of current, voltage, and resistance.
- Understand what physical variables determine the resistance of a conductor.
- Work with simple circuits and combinations of resistors.
- Know what a crossover network is and how it works.
- Understand some of the dangers associated with electricity.

Electricity Fundamentals

Atoms are the fundamental building blocks of matter. According to current models of atomic structure, the particles which comprise atoms have electrical properties. Many sub-atomic particles are electrically charged. Scientists use models to enhance their understanding of physical phenomena and to make predictions. The model most commonly used to describe an atom is in the analogy with our solar system. Just as the planets circle the sun in their orbits, negatively charged particles, called electrons, circle and are attracted by a positively charged nucleus in the atom (Fig. 8.1).

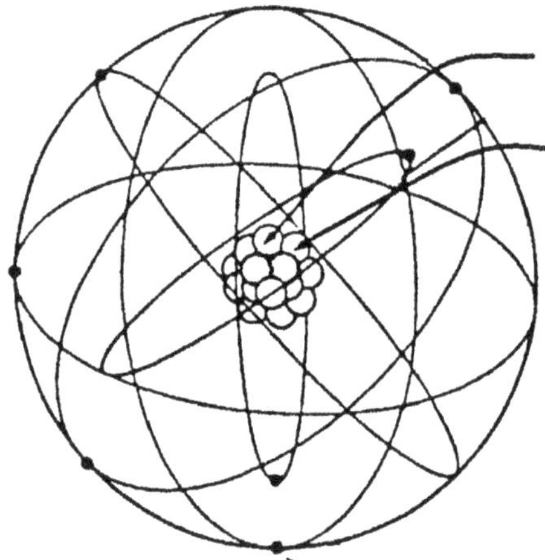

Figure 8.1 Solar system model of the atom

From *Computer Music: Science and Technology of a New Art,* Stanionis, Berberich. Copyright © 1989 by Research Foundation of State University of New York. Reprinted by permission.

Static Charges

The ancient Greeks first discovered that certain materials like amber, when rubbed, have the capability of attracting small bits of matter such as straw and feathers. Thales of Miletus is sometimes credited with being the first person to notice that amber will attract bits of dry grass or straw if it is rubbed with a wool cloth. What is amber? Where are most of the amber deposits found on the earth? Many people in the world will cut amber, which is honey-colored, into jewelry. A pendant with an insect trapped inside is highly valued. Do you remember the movie "Jurassic Park?" The key to reproducing the extinct creatures was contained in the amber which is petrified tree sap. How were the scientists able to reproduce the dinosaurs?

Rubbing amber or a rubber rod with fur charges it through friction so that it will attract small pieces of paper. An excess of electrons are deposited on the rubber rod through friction. Run a comb through your hair and hold it near some dry scraps of paper. What do you observe? Walk across a wool rug with your rubber-soled shoes on a cold, dry day and touch a light switch.

What happens? Why do you get an electric shock? If you do the same thing on a warm, humid summer day, why don't you get a shock?

Scientists discovered that there were two types of electric charges, named by Benjamin Franklin as positive and negative. In 1785, Coulomb studied the nature of the forces between charges and found that like charges repel and unlike charges attract. He also found that the magnitude of the attractive or repulsive force varies inversely as the square of the distance (r) and directly as the product of the charges (q).

Coulomb used a type of torsion balance in which he suspended one fixed sphere near another hanging like a dumbbell from a wire. See Figure 8.2.

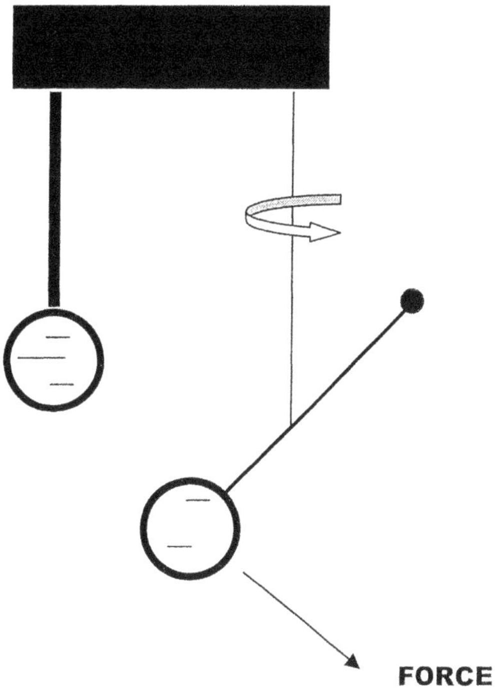

Figure 8.2 Coulomb torsion balance

Coulomb first gave each of the spheres a negative charge by conduction, by touching them with a negatively charged rod. He observed that the one hanging on the wire attempted to twist away from the one hanging on a fixed rod. When he tried the experiment with two positively charged spheres, the result again was repulsion between the two. However, a positive and negative sphere caused thc movable sphere to be attracted to the fixed one. So, Coulomb observed that like charges repel and unlike charges attract. Using this apparatus he was able to determine the mathematical relationship which determined the magnitude of the force of attraction or repulsion or Coulomb's Law:

$$F = k\, q_1\, q_2/r^2 ,$$

where k is a proportionality constant, the q's represent the magnitude of the charges, and r represents the distance between them.

In the atom the negatively charged particles are the electrons. An equal number of positively charged particles are contained in the nucleus of an electrically neutral atom and are called protons. The nucleus usually also contains neutral particles called neutrons. Both protons and neutrons are about 1,840 times as massive as the electron, and therefore, the nucleus contains most of the mass of the atom. Matter is made up of aggregates of such atoms. How the atoms are combined determines the electrical behavior of matter.

Models behave like real systems, but the number of variables used in them are usually much less than in a real system. By checking the model against the real system the scientist determines how good the model is. The true test for any model is experiment.

A useful model of solid matter is one in which the atoms of a material are connected to each other by coiled springs in a periodic arrangement (Fig. 8.3). At ordinary temperatures the atoms oscillate about their equilibrium positions. Most materials in nature have a crystalline form and behave like our mechanical model in many respects.

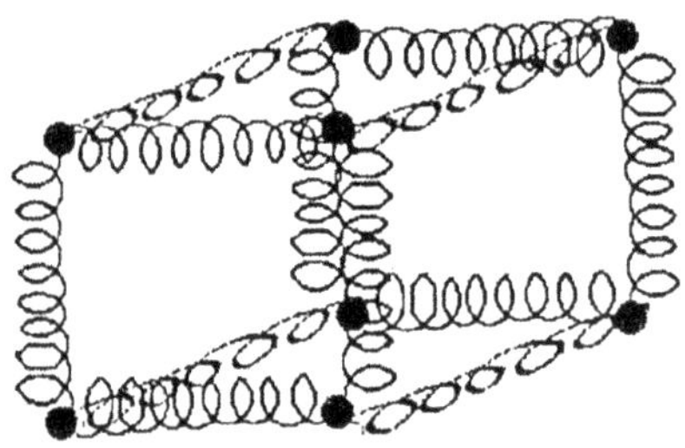

Figure 8.3 Model of matter

From *Computer Music: Science and Technology of a New Art,* Stanionis, Berberich. Copyright © 1989 by Research Foundation of State University of New York. Reprinted by permission.

In electrically conducting materials, some negatively charged electrons can move from atom to atom. These "conduction electrons" belong to the whole solid structure rather than to any particular atom. When an electric field is set up in the conductor, these charges experience a force and move through the conductor. This is what we call electric current. In any region of space where a charge experiences an electric force, we say an electric field exists. The electric fields themselves come about due to large distributions of positive and negative charges separated from one another (Fig. 8.4).

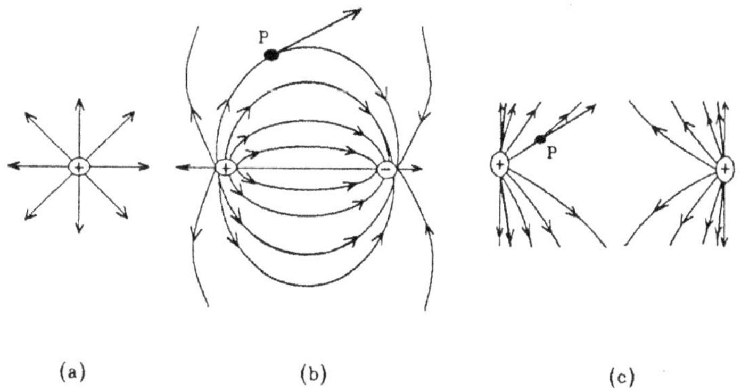

Figure 8.4 Electric fields between charges
From *Computer Music: Science and Technology of a New Art,* Stanionis, Berberich. Copyright © 1989 by Research Foundation of State University of New York. Reprinted by permission.

Materials which allow electrons to move freely from atom to atom are called conductors. Nonconductors are made up of atoms in which the electrons are tightly bound to each atom and cannot move freely, but can only shift from one side of the atom to the other. Between these two extreme classes of materials are semiconductors, a class of materials whose electrical behavior is intermediate between metals (conductors) and insulators (nonconductors). Conduction charges are less plentiful in semiconductors, but their availability can be dramatically increased by adding small amounts of impurities to the material or by increasing the temperature. On the other hand, in both conductors and semiconductors impurities and temperature hinder the movement of conduction charges and provide resistance to the flow of current.

What do you think is one of the best conducting materials? Gold. Why have people valued gold so much over the ages? Have you noticed that gold doesn't tarnish? Silver, although it also is a good conductor, tarnishes or oxidizes, meaning it combines with the oxygen in the air to form silver oxide. In the past automobile exteriors readily combined with the oxygen in the air to form rust. There are companies that sell connecting wires for stereo components that brag about the fact they use gold plated connectors which facilitates the flow of electrons and current electricity. If gold is such a great conductor and doesn't oxidize, why don't we use it for house wiring? How about silver? What about copper? Copper also oxidizes, forming a green patina as it combines with the oxygen in the air. This oxidized layer of copper protects the layers below, preventing them from oxidizing and giving the copper object a long life. Electrons are also able to pass through the oxidized layer easily. Copper wires seem to be the best and most economical materials for conducting electricity and are the metal of choice for wiring.

OPEC restricted the flow of gasoline into the United States in 1973 by raising the price of a barrel of oil from \$2.90 per barrel to \$11.60 per barrel and gasoline prices jumped from about 36 cents/gal to \$1/gal overnight generating sticker shock at the pump. The copper producers of the world also formed a consortium to boost the price of copper starting in the early 1960s. Manufacturers responded by making residential wiring out of aluminum and between 1962 and 1972 wired 2 million homes with aluminum wire. It turns out that when aluminum wire oxidizes, its resistance to the flow of current increases and when it is connected to a dissimilar metal like a copper switch. As a result aluminum wiring became a fire hazard. It expanded and contracted more than copper and thus when connected to a copper switch also loosened increasing the resistance at the connection. Warning signs became warm or hot switch plates, strange odors, switch failure, smoke and fire. The safest way to eliminate the fire hazard potential of aluminum wiring is to remove all of the aluminum wire and replace it with copper wiring. Today, we use copper for our electrical wiring applications.

The tripling of the price of copper overnight led many drug addicts seeking cash to support their habits to raid empty apartments in urban areas and steal the plumbing for resale to scrap dealers without bothering to shut off the water supply. This led to a cycle of empty apartments and urban housing deterioration. Even electrical wiring from overhead lines in rail yards was pilfered overnight, sometimes with deadly consequences for the thief.

Most matter we come in contact with is electrically neutral. The number of positive and negative charges in the matter is the same and distributed uniformly throughout. However, we can place two conductors in proximity and upset this balance by doing work by carrying negative charges from one plate to the other and setting up an electric field between the plates. See Figure 8.5. The total amount of work done in moving the charge is represented by the voltage, V, the work per unit charge measured in Joules/Coulomb. Voltage is the potential difference, or voltage difference, between the plates. Joules/Coulomb (J/C) are called volts.

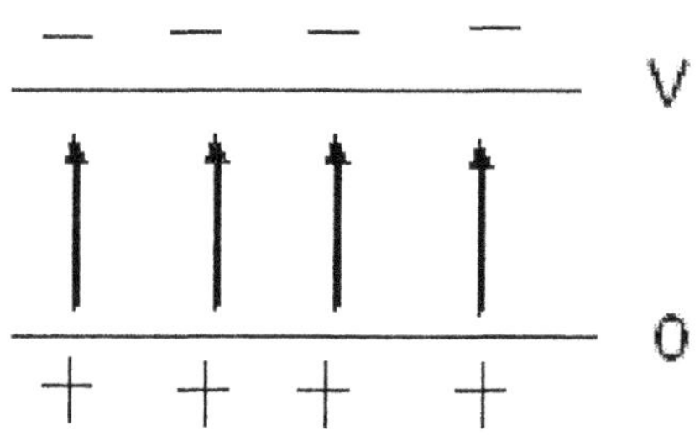

Figure 8.5 Two oppositely charged plates

We can think of the charges we moved to the top plate as having energy stored in the electric field between the plates.

Electric Current

The rate of flow of electrical charge through a wire is called electrical current. Historically, it was believed (incorrectly) that positive charge carried the electrical current in a circuit, and the direction of current in a circuit today is still labeled in many books as if the positive charges moved. In this book the direction of electric current is shown as the direction the negative charges move. Consider a simple electrical circuit with a switch and two charged plates. When the switch (K) is closed, the negative charges would be attracted to the positive ones and rush to meet them through the action of the electric forces acting between them (Fig. 8.6) in the wire. The amount of charge (which is measured in coulombs) that flows by any point in a circuit in one second is called electrical current I and is measured in coulombs/second (C/s) or amperes (A). Therefore $I = q/t$.

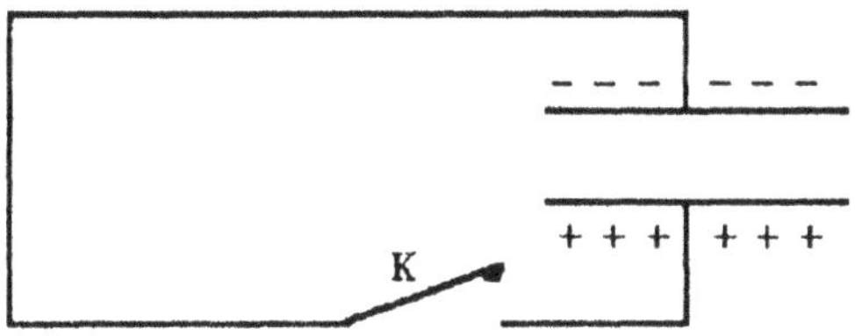

Figure 8.6 Simple electrical circuit

To separate the positive and negative charges onto the two separate plates requires work by an outside source. It is worth repeating that the work done in moving each positive charge to the top plate is measured in terms of the work per unit charge or joules per coulomb. Work per unit charge in physics is measured in terms of the electrical potential energy acquired and is called electrical potential which is measured in J/C or volts. Potential differences in a circuit measure differences in energy or voltage of the charges.

Keeping the charges circulating and creating a steady one-way (or direct) current requires the use of a charge pump, generator or some other energy conversion device, such as a battery, which converts chemical energy into electrical energy (Fig. 8.7). Such devices are called sources of electromotive force, or emf. This is a misnomer since they are sources of energy, not force. Many circuits have an alternating source of electromotive force which generates an alternating potential difference, causing current to alternately change its direction (AC current).

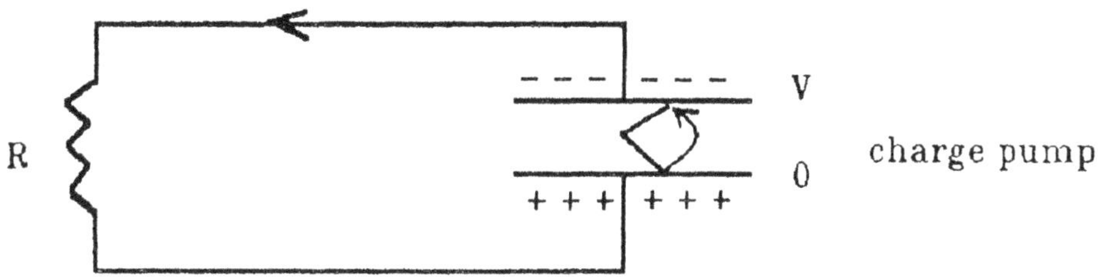

Figure 8.7 Electrical circuit with charge pump

In studying simple circuits, Ohm found that the energy the charges had initially was lost as they traveled through the conducting material and collided with atoms of the crystal structure of the conductor. Collisions between the atoms in the lattice structure of the conducting material and the electrons result in the electrons losing energy, which is apparent in the increased vibrations of the atoms in the crystal structure. We detect this increase in energy in the crystalline conductor as heat. So the energy gained by the electrons in establishing the original potential difference reappears in the form of heat. The conducting material is said to offer resistance to the flow of electrons.

In 1827, George Simon Ohm characterized this property of conducting materials as resistance. In particular, he found,

$$I \alpha V,$$

where I is the current through the conducting material and V the potential difference across the conducting material.

In mathematics, to go from the "proportionality sign" to the "equals sign", we insert a constant called the constant of proportionality. Ohm used 1/R so,

$$I = V/R \text{ which is Ohm's law.}$$

Solving for R, gives R = V/I or, in terms of units, volts/amp (V/A). This combination of units is called the ohm and is a measure of electrical resistance.

Suppose we have a speaker rated at 8Ω, where Ω is a symbol for the word ohm, and the voltage applied from the amplifier is 40 V. What is the current? See Figure 8.8.

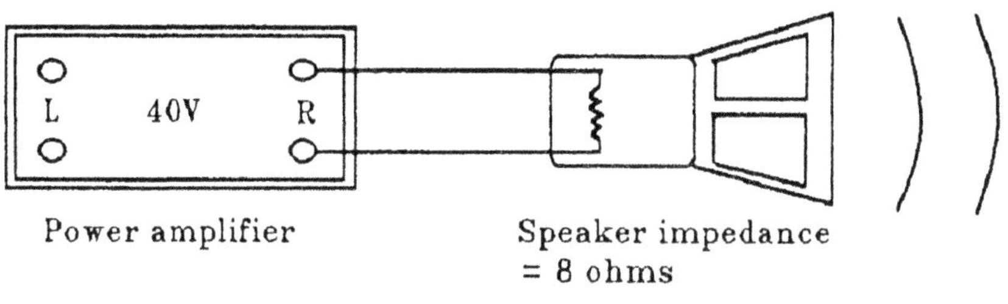

Figure 8.8 Simple speaker circuit

From *Computer Music: Science and Technology of a New Art,* Stanionis, Berberich. Copyright © 1989 by Research Foundation of State University of New York. Reprinted by permission.

$$I = V/R.$$

$$I = 40\ V/8\ \Omega.$$

$$I = 5\ A.$$

To summarize, we see that the flow of charge through a circuit is called current. A measure of the energy possessed by the charges is contained in a voltage measurement and the electrical resistance of the circuit is measured in terms of ohms. Voltage drives the flow of electrical current in a circuit.

Resistance

What do you think would happen to the resistance if the connecting wires were longer but the potential difference remained the same (Fig. 8.9)? More wire implies more crystal structure and, therefore, more resistance.

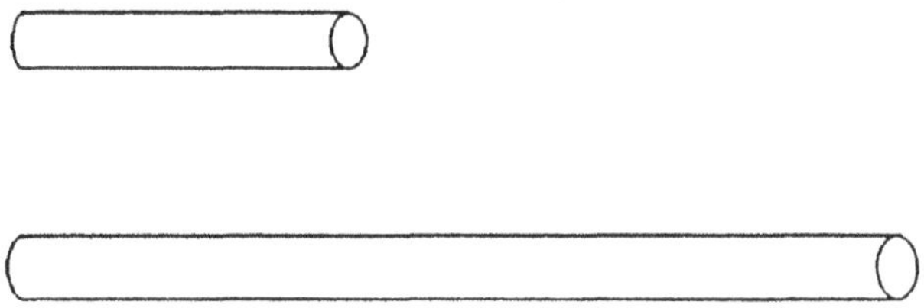

Figure 8.9 Resistance in different length wires
From *Computer Music: Science and Technology of a New Art,* Stanionis, Berberich. Copyright © 1989 by Research Foundation of State University of New York. Reprinted by permission.

If two wires of the same length but of different thicknesses were used for the same potential difference, which would offer greater resistance to the flow of charge (Fig. 8.10)? The thick wire offers many paths for the charges to take while the thin wire is limited. The charges can flow more easily through the thick wire which means it has less resistance.

Figure 8.10 Resistance of wires of different thickness
From *Computer Music: Science and Technology of a New Art,* Stanionis, Berberich. Copyright © 1989 by Research Foundation of State University of New York. Reprinted by permission.

Wire is sold in electronics and hardware stores according to its thickness or gauge number. Number 22 speaker wire is thin, inexpensive and cannot carry much current. Many household extension cords are number 18 wire and are also limited in the maximum current they can carry. Number 14 wire was once the standard thickness used for residential wiring and is limited to carrying a maximum of 15 A while newer houses use number 12 wire and can carry up to 20 A. "Boombox" cars have to add power amplifiers to generate the loud sounds they produce and wire them directly to the car battery using number 8 or number 10 wires. As the gauge number decreases, the thickness of the wire increases.

What happens if we heat up our wire with an external source of heat like a Bunsen burner? Well, the crystal structure with the atoms connected to each other would oscillate even more, impeding the flow of electrons. The hotter wire has a higher resistance.

So far we have been describing direct currents. These currents flow only in one direction. In our MIDI and hi-fidelity systems the electrical current, for the most part, is alternating current. As the name implies the current flows to and fro. The electrons flow for a time in one direction and then in the other because the voltage applied to the ends of the wire is also changing at a certain frequency. This property of AC is the reason that electrical current can carry the information contained in a sound wave. AC current is similar to the longitudinal waves that that travel down the length of a slinky spring. It is the motion of the wave that transfers energy between two points and because of this AC current is able to transmit audio information such as frequency and amplitude of a sound wave.

Since AC currents and voltages vary periodically, it is necessary to define what is meant by a numerical value of AC current or voltage. Since the heating effect produced by an electrical current when it passes through a resistor is independent of direction, it can be used as the basis for defining the effective value of the alternating current. In particular, the effective value of an alternating current is the amount of direct current that would produce the same amount of heat in a given resistor in the same time. This definition allows us to write that

$$I_{eff} = I_{max}/\sqrt{2} \quad \text{and} \quad V_{eff} = V_{max}/\sqrt{2}$$

Usually we omit the subscript "eff" for effective value when we denote AC currents and voltages.

Electric Power

The resistance of the circuit, due to the crystal structure of the matter, converts the energy of the electrons to heat no matter which way they flow. The amount of energy (W) converted into heat per unit time is called power (P) and is measured in watts.

$$P = W/t.$$

$$P = Vq/t \quad \text{but} \quad I = \text{charge}(q)/\text{time}(t).$$

$$P = IV \quad \text{or using Ohm's law} \quad P = I(IR) = I^2 R.$$

So, a light bulb using 1 amp at 120 V AC consumes power

$$P = IV = (1\ A)(120\ V) = 120 \text{ watts.}$$

Suppose 2 amps of current is sent to an 8 ohm speaker, how much power does it consume? See Figure 8.11.

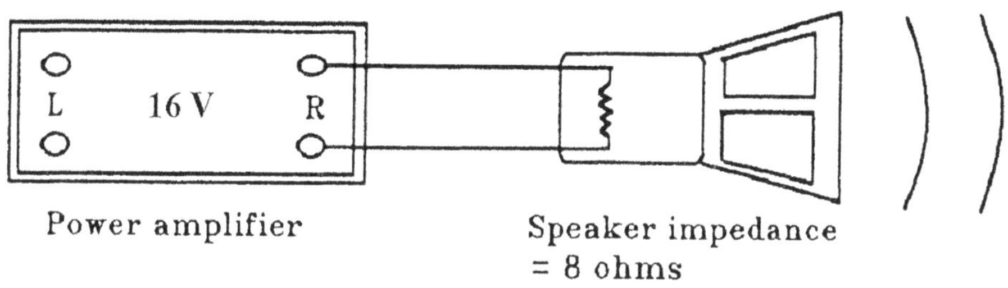

Figure 8.11 Speaker connected to an amplifier
From *Computer Music: Science and Technology of a New Art,* Stanionis, Berberich. Copyright © 1989 by Research Foundation of State University of New York. Reprinted by permission.

$$P = I^2R = (2\ A)^2 \times 8\ \Omega = 32 \text{ watts.}$$

All speakers should be fused in order to protect them from trying to convert too much electrical energy to heat and mechanical energy, destroying themselves in the process. Fuses usually consist of a soft metal strip connected in series with the speaker. When the current exceeds a certain prescribed value, the fuse will melt creating a break in the wiring of the circuit and stopping the flow of current (Fig. 8.12). Circuit breakers, which interrupt the flow of current also, are found on many speaker systems today. They can be reset if triggered and need not be replaced as in the case of a fuse.

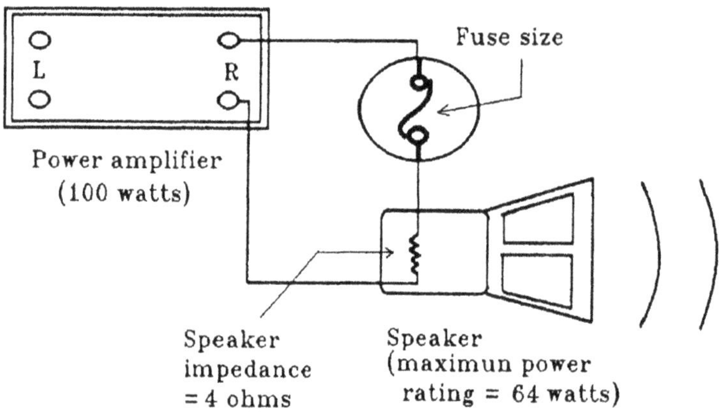

Figure 8.12 Speaker circuit with fuse

From *Computer Music: Science and Technology of a New Art,* Stanionis, Berberich. Copyright © 1989 by Research Foundation of State University of New York. Reprinted by permission.

Suppose you had a speaker rated at 64 W maximum power and an impedance or resistance of 4 ohms. What is the maximum current it could handle?

$$P = I^2 R.$$

$$64\ W = I^2 (4\ \Omega).$$

or

$$I^2 = 16\ A^2.$$

$$I = 4\ A.$$

Therefore, a fuse limited to slightly less than 4 A is needed to protect the speaker and should be connected to the speaker in series as shown in Figure 8.12.

Series and Parallel Resistances

When one end of a wire is twisted with another to form a longer wire, we effectively increase the path along which the electrons must flow. The wires are said to be connected in series. It is simple to see that when wires are connected in series, their resistances add (Fig. 8.13).

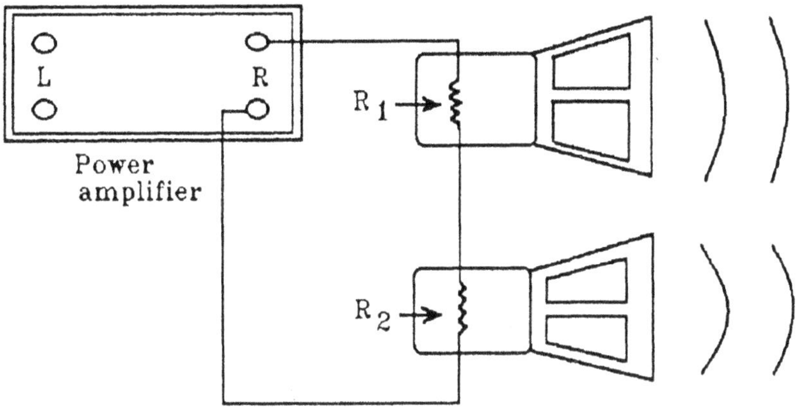

Figure 8.13 Two speakers connected in series
From *Computer Music: Science and Technology of a New Art,* Stanionis, Berberich. Copyright © 1989 by Research Foundation of State University of New York. Reprinted by permission.

In a series circuit the same current I passes from the amplifier through each resistance. However, the energy lost in each, measured by the voltage drop across each resistance will be different. So if the voltage drop across R_1 is V_1 and the voltage drop across R_2 is V_2 then by conservation of energy

$$V = V_1 + V_2$$

and using Ohm's law in the form V = IR and remembering I is the same for each resistance,

$$I\,R = I\,R_1 + I\,R_2.$$

Dividing both sides by I, $R = R_1 + R_2.$

For a series circuit: $R = R_1 + R_2$.

Suppose an 8 ohm and 4 ohm speaker are connected in series to a 24 volt power source. What is the current flowing through each?

$$R = R_1 + R_2 = 8 \text{ ohm} + 4 \text{ ohm} = 12 \text{ ohm}.$$

$$I = V/R.$$

$$I = 24\ V/12\ \Omega = 2\ A.$$

The power delivered to the first speaker is $P_1 = I^2 R_1 = (2\ A)^2 (8\ \Omega) = 32\ W$ and to the second $P_2 = (2\ A)^2 (4\ \Omega) = 16\ W$. So the amplifier is supplying a total of 48W.

In a series circuit, the current through each speaker is the same and the voltage drop and energy loss in each is different, but the total is what is available at the amplifier. If one of the speakers should burn out, the circuit will be broken and no sound will come from either. In addition, each speaker's sound intensity is reduced when connected in series since the current, and therefore, the power delivered to the speaker would not be as large as when one speaker alone was connected to the amplifier.

If each device is connected directly to the voltage source, we say they are wired in parallel (Fig. 8.14). The voltage across each is the same but the current in each is different.

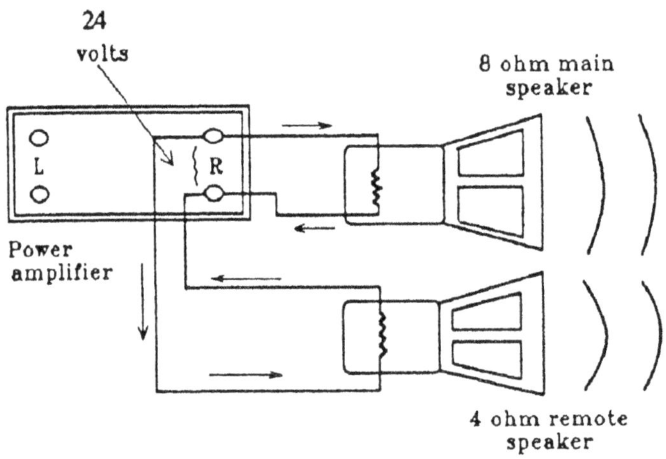

Figure 8.14 Two speakers connected in parallel

From *Computer Music: Science and Technology of a New Art,* Stanionis, Berberich. Copyright © 1989 by Research Foundation of State University of New York. Reprinted by permission.

So,

$$I = I_1 + I_2 = V/R_1 + V/R_2.$$

$$I = V/R = V/R_1 + V/R_2.$$

$$1/R = 1/R_1 + 1/R_2.$$

The total resistance is obtained by first finding the reciprocal of each individual resistor, then adding them all up and taking the reciprocal of the result.

$$1/R = 1/(8 \text{ ohm}) + 1/(4 \text{ ohm}) = 3/(8 \text{ ohm}).$$

$$R = (8/3) \text{ ohm} = 2\ 2/3 \text{ ohms}.$$

The total resistance is less than any of the individual resistances that were used. If we again have a 24V amplifier and the two speakers connected in parallel,

$$I = V/R_1 = 24 \text{ V}/(8\ \Omega) = 3 \text{ A}.$$

$$I = V/R_2 = 24 \text{ V}/(4\ \Omega) = 6 \text{ A}.$$

The total current is 9A. The power used by the first speaker is $P_1 = I^2R = (3 \text{ A})^2 (8\ \Omega) = 72$ W and for the second speaker $P_2 = I^2R = (6\text{A})^2 (4\ \Omega) = 144$ W or a total of 216 W. Which is much larger than the 48W supplied when the speakers are connected in series.

If one speaker burns out, the other continues playing. Since the currents taken by each speaker are additive, one has to be careful not to exceed the power rating of the amplifier by connecting multiple sets of speakers in parallel. In trying to provide the larger current called for, the amplifier might self-destruct.

There are many controls in an electronic audio system that depend on resistance for their operation. The volume control in a preamp is essentially a variable resistance connected in series with the speaker (Fig. 8.15). By increasing the resistance, the current is diminished and the sound from the speaker lessened accordingly.

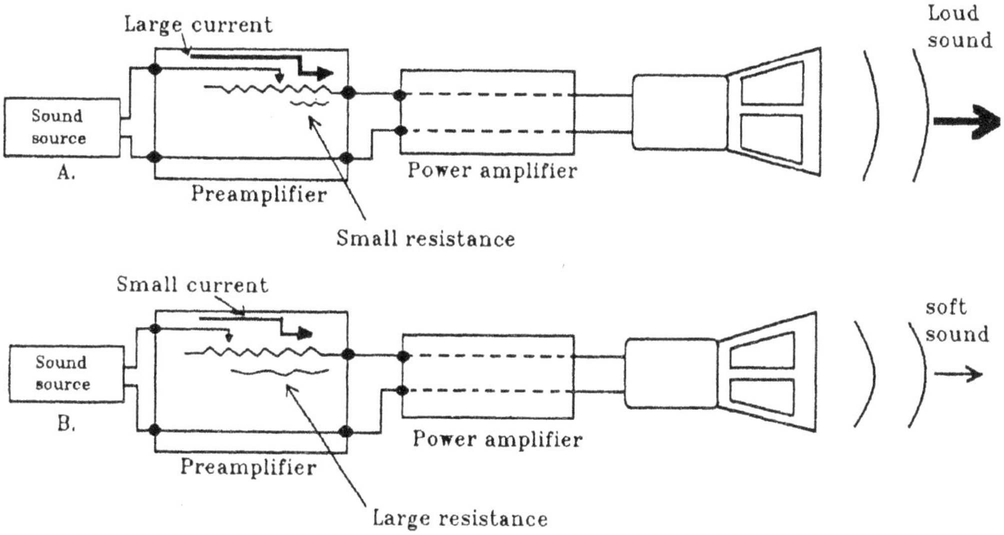

Figure 8.15 Volume control for speaker

From *Science of Hi-Fidelity* by Johnson, Walker, Cutnell. Reprinted by permission of the authors.

The balance control (Fig. 8.16) is also a variable resistance arranged in such a fashion that increasing the resistance for one speaker decreases the resistance for the other.

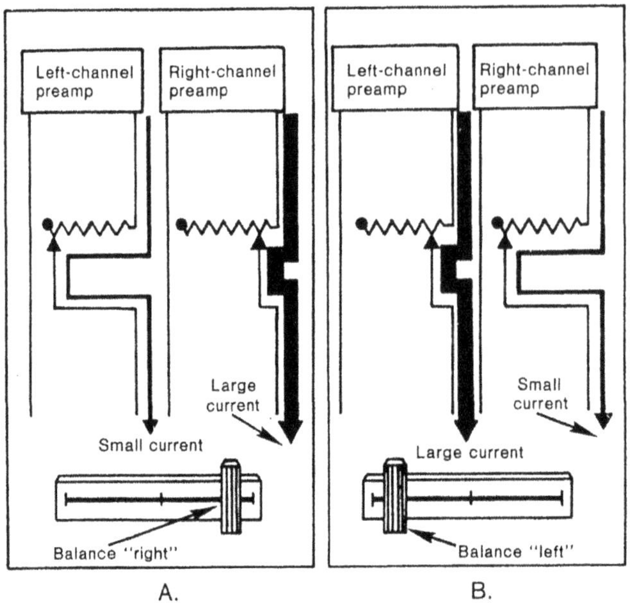

Figure 8.16 Balance control

From *Science of Hi-Fidelity* by Johnson, Walker, Cutnell. Reprinted by permission of the authors.

Crossover Networks

Electrical circuits may contain electrical components known as capacitors and inductors. The interesting thing about capacitors and inductors in a circuit is that their impedance, which is a property analogous to resistance, to the flow of alternating current (AC) depends on frequency of the AC current. This allows us to use them in filtering circuits which allow only certain frequencies to pass. In speaker networks, high or treble frequencies are directed to a tweeter and low frequencies to a woofer. In a purely resistive circuit the resistance is a constant over the entire range of frequencies 20 - 20,000 Hz as seen in Figure 8.17.

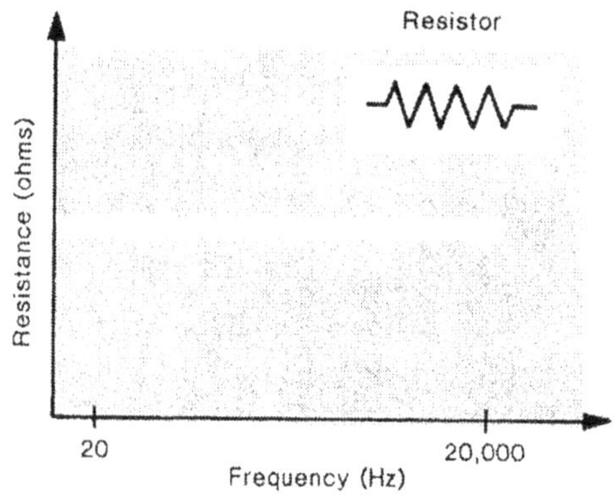

Figure 8.17 Resistance is constant over frequency range

From *Science of Hi-Fidelity* by Johnson, Walker, Cutnell. Reprinted by permission of the authors.

Any two conductors in proximity constitute a capacitor. In an electrical circuit, capacitors are used to store charge. Experiments with an almost purely capacitive circuit reveal that when low frequencies are applied to a capacitor, the resistance of the circuit in ohms turns out be high while for high frequencies it's low. In other words, it allows high frequencies to pass and block low frequencies (Fig. 8.18).

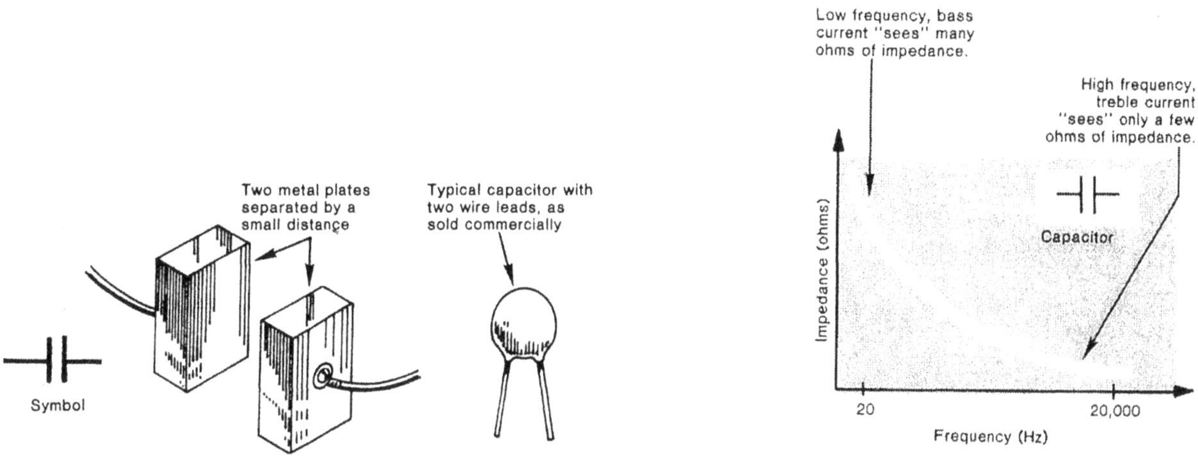

Figure 8.18 Capacitor's impedance decreases over with increase infrequency
From *Science of Hi-Fidelity* by Johnson, Walker, Cutnell. Reprinted by permission of the authors.

A capacitor, when connected in series with a speaker in a speaker system (Fig. 8.19), blocks bass signals and passes treble. High frequency treble current experiences low impedance. Thus the speaker becomes a tweeter.

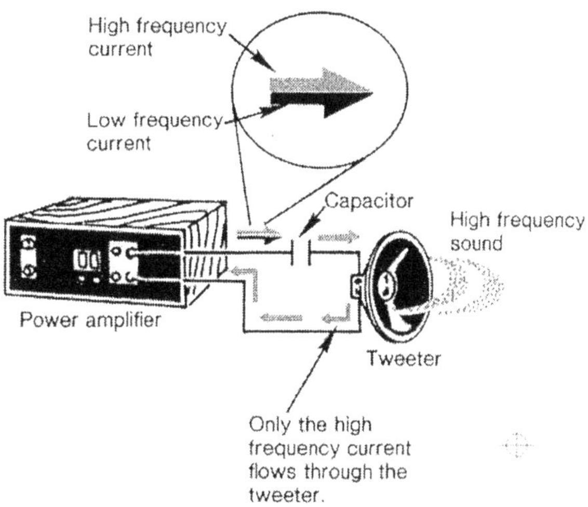

Figure 8.19 Capacitor connected in series with a speaker converts into a tweeter
From *Science of Hi-Fidelity* by Johnson, Walker, Cutnell. Reprinted by permission of the authors.

An inductor, which can be represented by a large coil of wire, derives its impedance from magnetic effects. Experiment reveals that the resistance is low for low frequencies and high for high frequencies. When an inductor is connected in series with a driver in a speaker system, it allows only low frequencies to pass (Fig. 8.20).

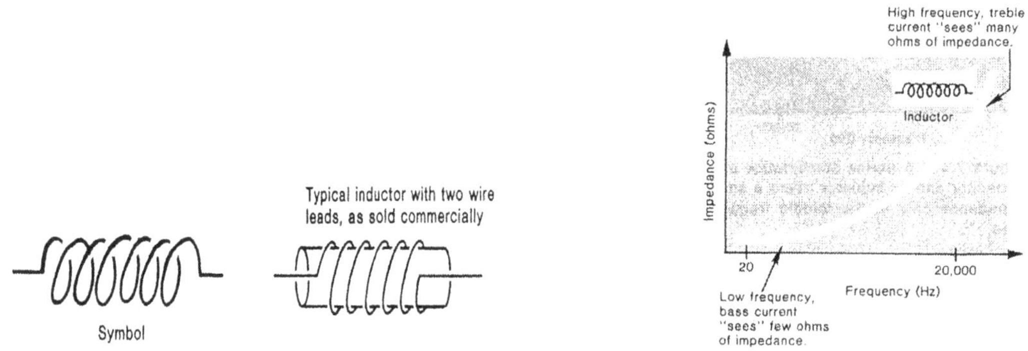

Figure 8.20 Inductor's impedance increases with increase in frequency
From *Science of Hi-Fidelity* by Johnson, Walker, Cutnell. Reprinted by permission of the authors.

High frequency treble current experiences high impedance and low frequency bass current experiences low impedance. Thus a speaker connected in series with an inductor allows only low frequencies to pass and becomes a woofer (Fig. 8.21).

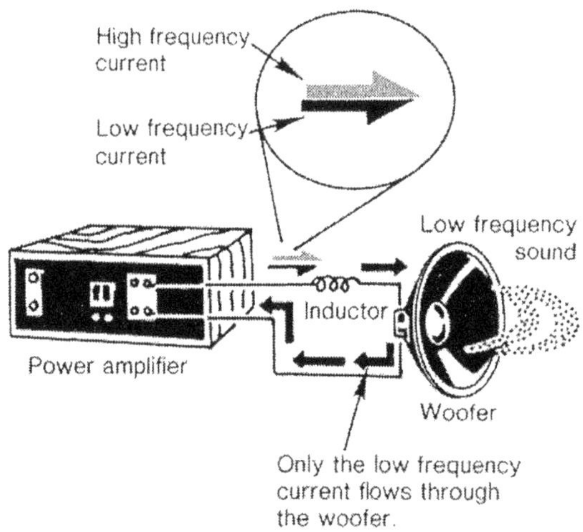

Figure 8.21 Inductor connected in series with a speaker converts it into a woofer
From *Science of Hi-Fidelity* by Johnson, Walker, Cutnell. Reprinted by permission of the authors.

When both the capacitor and inductor are connected in series with a driver, only intermediate frequencies will be allowed to pass (Fig. 8.22).

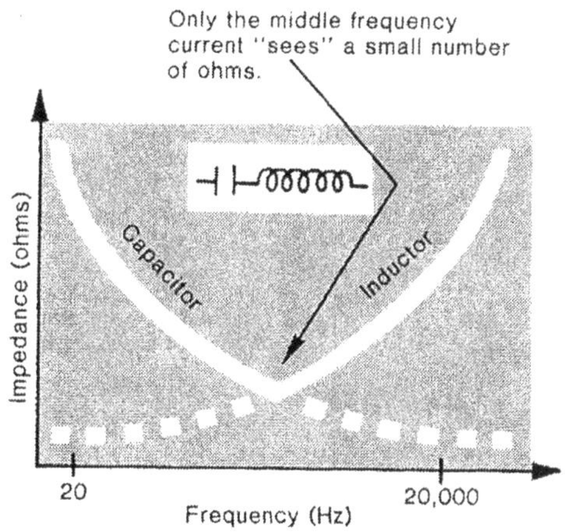

Figure 8.22 Inductor and capacitor in series with a speaker allow only midrange frequencies
From *Science of Hi-Fidelity* by Johnson, Walker, Cutnell. Reprinted by permission of the authors.

Crossover networks in speaker systems use capacitors and inductors to direct particular ranges of frequencies to the appropriate drivers (Fig. 8.23).

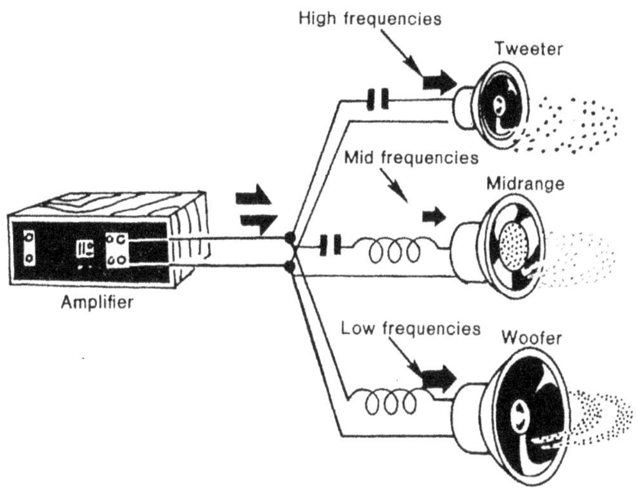

Figure 8.23 Cross over network with tweeter, midrange, woofer
From *Science of Hi-Fidelity* by Johnson, Walker, Cutnell. Reprinted by permission of the authors.

Household Wiring

Homes use alternating current for power. All electrical appliances in a home are connected in parallel to the power source. In a parallel wiring system one side is called the "line" and the other is the neutral side. The neutral side is always connected to a ground at the entrance panel. For a home, the ground may actually be an electrode that is a good conductor that is driven into the earth or it may be a connection to an incoming household water pipe (Fig. 8.24).

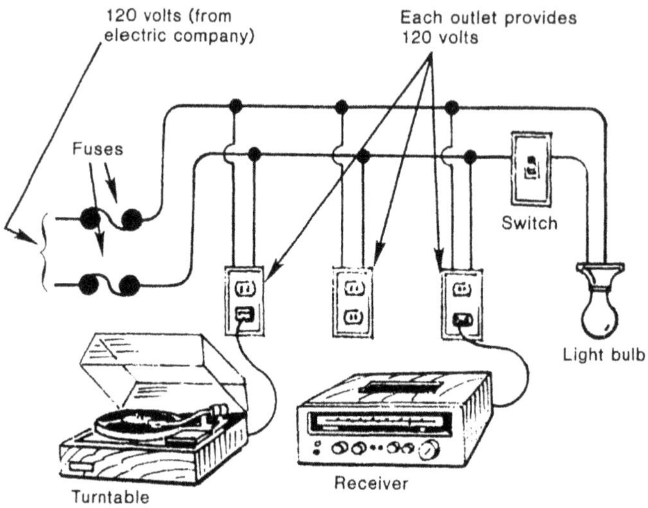

Figure 8.24 Home wiring

From *Science of Hi-Fidelity* by Johnson, Walker, Cutnell. Reprinted by permission of the authors.

Most modern home wiring systems have two "hot" lines coming into the home with opposite polarity with respect to the neutral. In the United States the voltage is approximately 120 volts, and 240 volts in Europe. The voltages are effective values which are $1/\sqrt{2}$ times the peak voltage supplied. The power company usually provides three conductors: the neutral, and two 120 volt lines at opposite polarity, giving a voltage of 240 between them. Lamps and appliances may be connected between either 120 volt line and the neutral. A device demanding high power such as an electric water heater, range, or clothes dryer can be connected between the two "hot" lines, which are usually coded with black or red insulation while the neutral side has white insulation. For grounding, a wire that is bare or has green insulation is used. In electronic devices and equipment the ground or neutral side of the line is usually black.

Suppose you had an electric drill with a two prong plug. If the internal wiring insulation should deteriorate, the outside of the metal drill case could conduct electricity through a person's hand, shocking them. If the drill had a three prong plug, the person would receive no shock since the current would be carried through the ground wire connected to the third prong into the ground rather than through the person.

The maximum current of each circuit is limited by the resistance of the wires and is protected by a fuse or circuit breaker from overloading or overheating. A fuse is usually made of a lead-tin alloy which melts and opens the circuit when the rated current is exceeded. A circuit breaker is an electromechanical device that opens the circuit when the rated current is exceeded. It can be reset after opening, while a fuse must be replaced.

Electric Shock

The severity of an electric shock you would experience depends primarily on the amount of current that passes through you. Ohm's law, $I = V/R$ tells us that the current depends on the voltage and resistance. If you were in contact with a 1 mA current for about a second, you would experience a tingling sensation. Remember, we have to have a complete circuit with a potential difference at the ends for current to flow. The three primary factors that affect the level of shock a person would receive when becoming part of an electric circuit depends on the amount of current flowing, its path thorough the body and the time of contact.

Physiological Effects of Electric Currents

Current level (in milliamperes)	Probable effect on human body
1 mA	Perception level. Slight tingling sensation. Still dangerous under certain conditions.
5 mA	Slight shock felt; not painful but disturbing. Average individual can let go. However, strong involuntary reactions to shocks in this range may lead to injuries.
6-30 mA	Painful shock, muscular control is lost. This is called the freezing current or "let-go" range.
50-150 mA	Extreme pain, respiratory arrest, severe muscular contractions. Individual cannot let go. Death is possible.
1000-4300 mA	Ventricular fibrillation (the rhythmic pumping action of the heart ceases.) Muscular contraction and nerve damage occur. Death is most likely.
10,000 mA	Cardiac arrest, severe burns and probable death.

Figure 8.25 Physiological effects for different electric currents
http://www.osha.gov/SLTC/etools/construction/electrical_incidents/eleccurrent.html

A bird can perch on a 50,000 volt power line and experience no ill effects since both feet are at the same potential, or voltage. The bird is not grounded. If you were to come in contact with a refrigerator in which the insulation of the internal wiring has broken down over time, the outside metal case could be at a high voltage. Just touching the case would not give you a shock as long as your body was not grounded. However. turning the water on in the sink while opening the refrigerator with the other hand completes a circuit and the effects could be lethal. Some electricians will put one hand in their pocket while working around a live wire so as to not accidentally ground themselves.

A current of 6 to 30 mA (Fig. 8.25) causes muscles to contract and seize and you won't be able to let go of the live conductor until the circuit is broken. In the meantime you dissipate electrical energy in your body. You burn. At currents in the range of 100 to 300 mA your heart loses it rhythmic beat and starts to fibrillate instead of pumping blood to your brain. In a very short time you die if the heart's rhythm is not restored.

More and more we see defibrillators that can be used by ordinary citizens becoming available at sporting events, on aircraft, in schools and with local police and firemen. The defibrillators basically use currents on the order of 6A to "jump start" the pumping rhythm. The heart muscle contracts and, when released, starts to beat. Tasers or stun guns work in a similar fashion in that the victim is momentarily shocked with a high voltage pulse between two electrodes causing the victim to lose control over their muscles for a few minutes even when the current is stopped. Other factors such as the voltage, moisture, heart cycle, and general health of the person are factors in determining the severity of the shock.

Body fluids are good conductors because of their ion concentrations. However, the resistance of dry skin can vary from 500,000 ohms to 1000 ohms for wet skin. If it were not for the high resistance of skin, it is possible that a 1.5 volt battery could produce a lethal shock. Suppose you dampen your fingers and place them across the terminals of a 1.5 volt D cell battery. The current that passes through your fingers, assuming your wet fingers have lowered your resistance to 20,000 ohms, will be:

$$I = V/R$$

$$I = 1.5 \text{ volts}/20{,}000 \text{ ohms} = 0.0001 \text{ amps}$$

or 0.1 mA corresponding to a power of $P = IV$ or 0.00015 W and no noticeable sensation.

For example, suppose your body's resistance is 100,000 ohms on a particular day and you touch a device at 120 V connecting yourself to ground. The current will be:

$$I = V/R$$

$$I = 120 \text{ volts}/100000 \text{ ohms}$$

$$I = 0.0012 \text{ A or } 1.2 \text{ mA.}$$

You will probably feel a tingling sensation. On the other hand, suppose you are down in the damp basement with a clothes washer which has overflowed, wetting the floor. Now stepping in the water in your socks your resistance may be lowered to 1000 ohms. Touching the live washing machine and completing the circuit allows a current to flow.

$$I = 120 \text{ V}/1000 \text{ ohms}$$

$$I = 120 \text{ mA}.$$

This current may give you a lethal shock and cause your heart to seize or to go into ventricular fibrillation. If the extensor muscles are excited by the shock, the sudden reaction the person experiences may cause you to be thrown back and fall suffering severe bodily trauma from the fall.

Low voltage can be hazardous as well as high voltage, Suppose a person experiences a muscular contraction that causes their muscles to seize so they can't let go. Contact with a high voltage immediately reduces the body's resistance by breaking down human skin and increasing current flow. For example, Ohm's law, $I=V/R$ tells us that 1000V applied to a lowered bodily resistance of 500 ohms results in a current of 2 A which can cause cardiac arrest and damage internal organs. A difference of less than 100 mA can produce a killer current.

Review Questions

1. Discuss the solar system model of the atom.
2. Discuss Coulomb's discovery for electrostatic charges.
3. What makes a material a conductor or a nonconductor?
4. What is electric current?
5. What did George Ohm discover regarding simple electric circuits?
6. How does the resistance of a wire vary with respect to length, cross sectional area, and temperature?
7. What is the difference between direct current electricity and alternating current electricity? Explain.
8. What is power in an electric circuit?
9. Discuss the differences in voltage, current and power dissipation in resistors for series circuits and parallel circuits.
10. Describe how the volume control on an audio system works.
11. Describe the operation of a balance control in an audio system.
12. What happens to the audio frequencies when a capacitor is connected in series with a speaker?
13. Suppose an inductor were connected in series with a speaker. What sort of frequencies would be produced by the speaker? Explain.
14. Explain the operation of a cross over network in a speaker system.
15. Discuss the operation of a home wiring system.
16. Discuss the physiological effects of electricity.

Multiple Choice Questions

1. In the atomic model of the atom, the small central nucleus is thought of as containing _____.
a. electrons
b. protons
c. protons and electrons
d. protons and neutrons

2. The properties of an atom are determined by its configuration of _______ in the outer orbits.
a. electrons
b. protons
c. protons and electrons
d. protons and neutrons

3. In a conductor the outer electrons in orbit in an atom are ______.
a. tightly bound
b. bound to each other
c. free to wander from atom to atom
d. volatile

4. Amber is a type of _____.
a. rubber
b. lodes stone
c. petrified tree sap
d. secretion of the lac insect

5. When amber is rubbed with wool, it acquires a ______.
a. honey color
b. negative charge
c. soft patina
d. positive charge

6. How many different types of charge are there?
a. 1
b. 2
c. 3
d. 4

7. Coulomb discovered that _____.
a. unlike charges attract
b. opposite charges repel
c. positive and negative charges repel
d. the force between charges depended inversely on the distance between them

8. The best conducting material is_____.
a. copper
b. aluminum
c. silver
d. gold

9. The rate of flow of charge in a conductor is called electric_____.
a. voltage
b. resistance
c. impedance
d. current

10. The unit for electric current is the _____.
a. coulomb
b. volt
c. ampere
d. ohm

11. Ohm's law states _____.
a. I = VR
b. V = I/R
c. R = V/I
d. V = R/I

12. If the current in a circuit is 4 A and the voltage is 12 V, the resistance is _____.
a. 3 ohms
b. 1/3 ohms
c. 48 ohms
d. 8 ohms

13. Applying a voltage of 100 V across a 50 ohm resistor will cause a current of _____ amps to flow through it.
a. 2
b. ½
c. 50
d. 150

14. A speaker has an impedance of 8 ohms and a voltage of 40 volts across it. The current through it is ____ amps.
a. 48
b. 1/5
c. 5
d. 320

15. Doubling the length of a wire increases its resistance by a factor of _____.
a. 2
b. 4
c. 8
d. 16

16. The resistance of #22 gauge wire is _______ the resistance of #18 wire.
a. the same as
b. smaller than
c. greater than
d. ½

17. The power dissipated in a circuit may be obtained from P = _____.
a. IR
b. IR^2
c. IV
d. I/R

18. Electrical power is measured in terms of _____.
a. coulombs
b. ampères
c. volts
d. watts

19. A driver can be converted to a tweeter by connecting a __ in series.
a. inductor
b. coil
c. impedance
d. capacitor

20. If a speaker has an impedance of 8 ohms and a current of 2 A, the power it is receiving is _____ W.
a. 16
b. 32
c. 64
d. 4

21. If a speaker has a 4 ohm impedance and a limit of 64W, then a fuse lower than ____ amps may be connected in series to protect it from burnout.
a. 2
b. 4
c. 6
d. 16

22. If three 6 ohm resistors are connected in series, the total resistance will be _____ ohms.
a. 6
b. 18
c. 2
d. ½

23. If the three 6 ohm speakers are connected in parallel, the total resistance will be _____.
a. 6
b. 18
c. 2
d. ½

24. When resistors are connected in parallel, the total resistance will be ______ any of the individual resistances.
a. larger than
b. less than
c. the same as
d. parallel to

25. When two speakers are connected in series and one of them burns out, the sound from the other _____.
a. becomes louder
b. becomes softer
c. remains the same
d. stops

26. When a driver in a speaker system is connected to an inductor, it plays ________frequencies.
a. all
b. low
c. high
d. midrange

27. When a capacitor is connected in series with a speaker, it allows only ______ frequencies to pass.
a. low
b. high
c. midrange
d. bass

28. Two speakers with impedances of 4 Ω and 8 Ω are connected in parallel to a 32 V amplifier. The power dissipated by the 4 Ω speaker is ______ W.
a. 256
b. 128
c. 64
d. 8

29. If a 4 Ω and a 16 Ω speaker are connected in series to a 32 V amplifier, the power dissipated by the 4 Ω speaker is _____ W.
a. 25.6
b. 10.1
c. 64
d. 0.4

30. If the 16 Ω and a 8 Ω speaker are connected in series to a 48 V amplifier, the current in the 8 Ω speaker will be _____ A.
a. 2
b. 0.5
c. 384
d. 6

Bibliography and Web Resources

Young, Hugh and Roger Freedman. *University Physics*. Addison-Wesley Publishing Co., San Francisco, CA, 2004.

Stanionis, Victor and Hugh Berberich. *Computer Music: Science and Technology of a New Art*. Research Foundation of State University of New York, Stony Brook, NY 1989.

Johnson, Kenneth and Willard Walker and John Cutnell. *The Science of Hi-Fidelity*. Kendall-Hunt Publishing Company, Dubuque, Iowa, 1995.

Wiring -
http://www.pathnet.org/si.asp?id=314

Shock Table –
http://www.osha.gov/SLTC/etools/construction/electrical_incidents/eleccurrent.html

Shock Table –
http://hyperphysics.phy-astr.gsu.edu/hbase/electric/shock.html#c3

Electrical shock -
http://acept.la.asu.edu/courses/phs110/ds/appendixC.html

Tasers
http://www.allaboutcircuits.com/vol_1/chpt_3/2.html

Chapter 9

Synthesizers

"Computer performance of music was born in 1957 when an IBM 704 in NYC played a 17 second composition on the Music I program which I wrote."

Max Mathews, 1957
Father of Computer Music

A synthesizer is an electronic musical instrument. You may wonder what the role of a synthesizer is. Some may think of a synthesizer as a device that will imitate the sound of acoustical musical instruments. Why bother when we can listen to the actual instruments themselves? We should think of the synthesizer as a musical instrument itself that can produce new and unique sounds. In our attempts to synthesize the sounds of acoustical instruments, we learn a great amount of information about how they work, allowing us to improve them and produce replicas of outstanding instruments at an affordable price.

Synthesizers can be divided basically into two subgroups: analog and digital. Analog synthesizers use circuits that have varying voltages. Adjusting variable resistors and capacitors using knobs and switches changes sounds in an analog synthesizer. A digital synthesizer is basically a specialized computer that manipulates binary code and solves mathematical equations to produce a given sound.

Goals

After studying this chapter, you should be able to:

- Understand how an analog synthesizer works.
- Know what subtractive synthesis is.
- Understand the role of the gate signal and envelope generator in shaping the sound.
- Know what digital synthesis is.
- Understand the difference between analog and digital synthesizers.

- Discuss the differences between AM and FM broadcasting.
- Describe how each of the following synthesis methods work: additive, granular, wavetable, frequency modulation, phase distortion synthesis, and physical modeling synthesis.
- Understand how a sampling synthesizer works.
- Explain the roles of sampling frequency and resolution in a sampling synthesizer.

Analog Synthesis

An analog synthesizer is a system that produces a complex electrical voltage by generating component AC voltages, then adding, subtracting, and otherwise shaping them until they represent a sound signal with the desired qualities (pitch, timbre, etc.). It does not make any audible sounds by itself. Voltages can be used to produce a current in a loudspeaker system to generate the sounds desired. All analog synthesizers, in general, have sound generators, controls, memory and perhaps, a keyboard (Fig. 9.1).

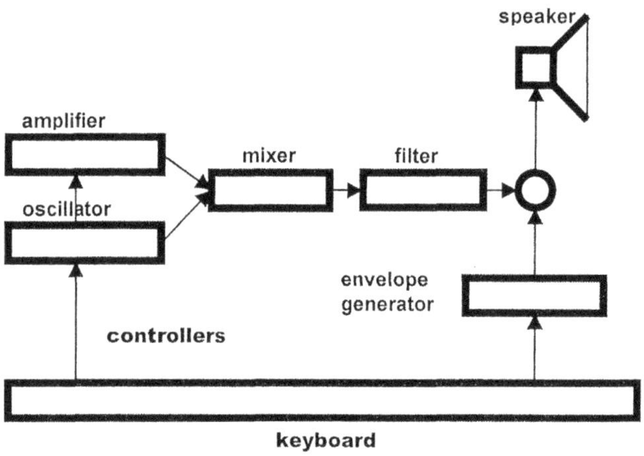

Figure 9.1 Analog synthesizer

Many synthesizers have keyboards that act as input devices but there are other types of controllers, such as wind and string devices, which may be used. The heart of a synthesizer is made up of the sound generators it contains. It is these sound generators that produce waveforms which, when modified, imitate the complex waveforms of real musical instruments or create new and complex sounds. The synthesized signals, when amplified and passed on to the speaker, produce the sounds we hear. Synthesizers can have anywhere from four to over one hundred sound generators.

Oscillator electrical circuits convert DC voltages into AC. They create the oscillating voltage by resonating electronically when current flows into them. The frequency of the oscillations can be changed from a few Hertz to millions of Hertz. The output of an oscillator is constant in amplitude and frequency. Oscillators, when combined with electrical filters and other electronic components, can be made to produce sine waves, triangle waves, square waves, and sawtooth waves. The sine wave produces a fundamental tone while the square, rectangular, and triangular waveforms contain only odd harmonics. The sawtooth wave contains all the harmonics (Fig. 9.2). The three main types of sound synthesizers are: analog, digital, and sampling.

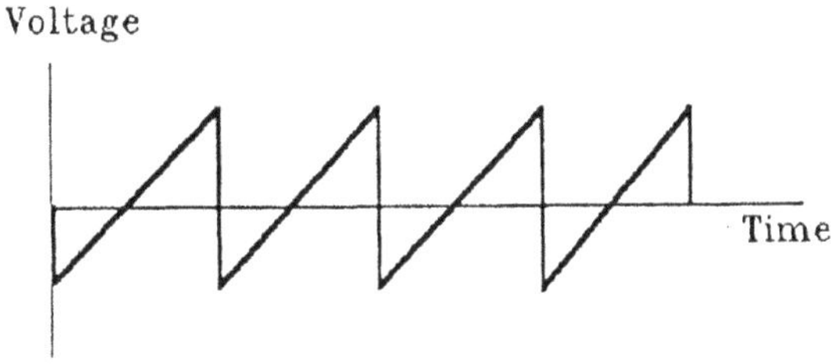

Figure 9.2 Sawtooth wave

From *Computer Music: Science and Technology of a New Art,* Stanionis, Berberich. Copyright © 1989 by Research Foundation of State University of New York. Reprinted by permission.

The controls on a synthesizer include an on/off switch and a volume control in the form of a knob or slider. In addition keyboard synthesizers will normally have a pitch bend wheel and a modulation wheel. A pitch bend wheel can be used to vary the pitch of a sound anywhere from a single half-step up to an octave. A modulator wheel is used to set the amount of modulation in the synthesized sound, usually in terms of its vibrato (a pulsating effect added to a tone).

Some synthesizers have the ability to offer a keyboard split which allows a player to play different sound designs on each half of the keyboard. Each half acts as if it is a different keyboard tied into the MIDI system. A foot control can be used to vary the volume continuously while a foot pedal has only two settings, on or off, and is usually used to turn the sustain part of a note on and off. Sustain keeps the notes playing even after the keyboard keys are released. Other controls may generate control voltages or envelopes that shape the frequency, amplitude or timbre of each note a synthesizer plays, or to select stored waveforms.

Subtractive Synthesis

Remember that in 1958, RCA developed the Mark II analog Music Synthesizer and Robert Moog displayed his synthesizer at the Audio Engineering Society Convention in 1964. The Hammond electric organ of the 1930s used additive, or Fourier, synthesis. Almost all traditional analog synthesizers use what is called subtractive synthesis. They start out by using waveforms having large numbers of harmonics like a triangle wave, or a square wave. For the most part, they cannot create sine waves with the accuracy needed for additive synthesis. In analog synthesis, the oscillators themselves can produce basic waveforms (such as: sine wave, triangle wave, sawtooth wave, square wave) quite easily with their harmonics. Using one of these basic waveforms the harmonics can be mixed together and modified to generate a complex waveform voltage to send through the amplifier and on to the speakers. Filters can alter the quality or timbre of a musical sound by removing selected frequencies, which is a form of subtractive synthesis. High pass filters absorb middle and low frequencies and pass high frequencies. Band pass filters pass middle frequencies while low pass filters pass only low frequencies. Filters can also be regulated by a control voltage which would fix the cut-off frequencies of the filter. The Moog is an analog synthesizer.

When a key is pressed on a synthesizer, a voltage is generated in the voltage-controlled oscillator (VCO). It is a tone generator whose pitch or frequency is determined by a control voltage which is generated at the keyboard. The greater the amount of voltage developed at its input, the higher the frequency of the VCO. When different keys are pressed, different control voltages are sent to the VCO, generating different frequencies. The synthesizer can play monophonically, when all the sound generators are playing the same note at the same time, and polyphonically, when each sound generator plays a different note.

Conventionally, the control voltage for VCO's have a one volt per octave difference. Therefore, the pitch doubles each time the voltage increases by one volt. In a single octave each key generates a voltage 1/12 volt higher than the previous (Fig. 9.3). Remember there are twelve notes to an octave. Control voltages are used to transmit pitch information. It is possible to generate a fuller sound by using the control voltage from a key to control more than one VCO at a time.

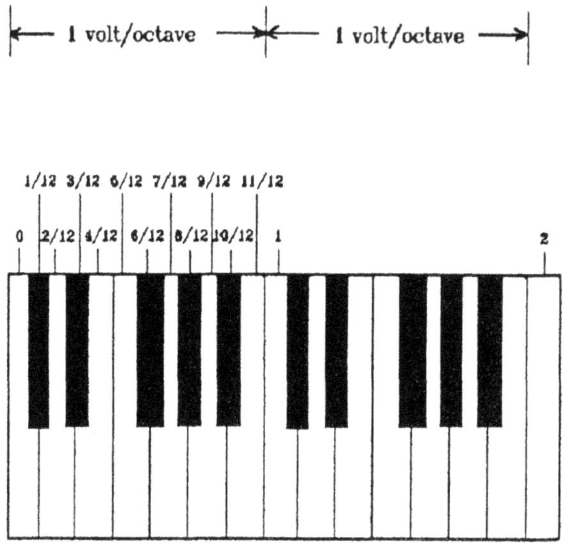

Figure 9.3 Control voltages

From *Computer Music: Science and Technology of a New Art,* Stanionis, Berberich. Copyright © 1989 by Research Foundation of State University of New York. Reprinted by permission.

The signal from the VCO next goes to the voltage-controlled amplifier. A VCA (voltage-controlled amplifier) controls the amplitude of the audio signal. In addition to sending out a control voltage to the oscillator, the keyboard generates another signal called a gate. The gate voltage is zero when the key is not pressed. When pressed, it jumps from zero to some constant fixed value (Fig. 9.4). This turns the VCA on and off. If there is no control voltage, the sound is turned off. The VCA level varies directly with the control voltage at its input. The higher the control voltage is the greater the amplification and, therefore, the louder the sound. The source of this voltage is usually the keyboard.

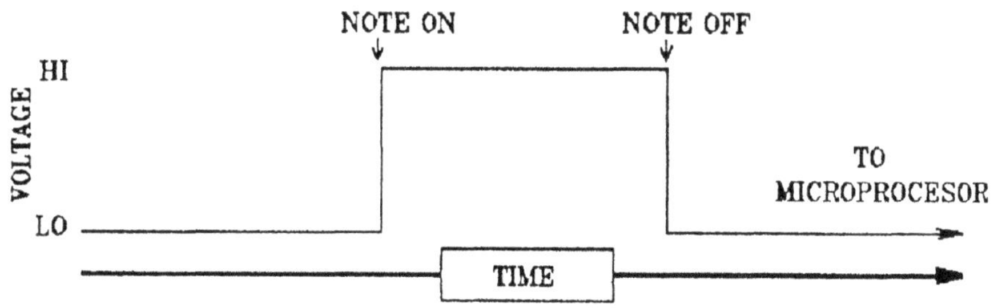

Figure 9.4 Gate signal
From *Computer Music: Science and Technology of a New Art,* Stanionis, Berberich. Copyright © 1989 by Research Foundation of State University of New York. Reprinted by permission.

A circuit called an envelope generator (EG) can be set into operation by a gate. An envelope generator governs the rate at which a sound builds up to its maximum value, how long it lasts, and how it fades away. An EG creates a control voltage that can change a great deal as a note is played. For example, if such a control voltage is used on a VCA (Fig. 9.5), the sound will change in terms of its dynamics (attack, decay, sustain and release).

The attack phase of a tone is how the tone first sounds and can range from the sudden blast of a trumpet to the slow build-up of an organ note. The voltage then drops to a lower value in the decay phase. The sustain phase corresponds to the length of time the musical tone sounds between the decay and release points. The release time is perceived in how the tone ends and fades out as the voltage drops to zero. The same process will take place if the EG controls the frequency of a VCO or other electronic component, such as a filter. The envelope generator can create a control voltage that can vary over the length of time the note is sounded and can be used to modify the output from a VCO, a VCA, or filters.

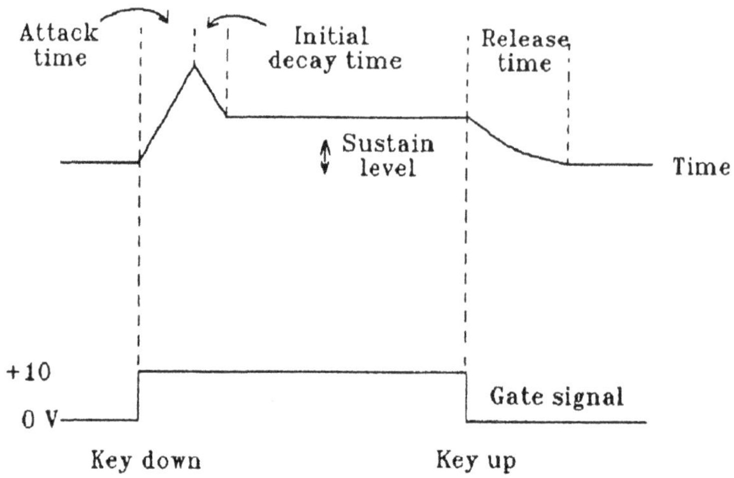

Figure 9.5 Gate activated envelope generator
From *Computer Music: Science and Technology of a New Art,* Stanionis, Berberich. Copyright © 1989 by Research Foundation of State University of New York. Reprinted by permission.

Pressing a key generates a voltage that starts the attack phase, which can range from a millisecond to seconds. It then drops over an adjustable period of time which lowers the VCA signal. The sustain level corresponds to the length of time the key is held down. Releasing the key turns the gate off. In the release phase the VCA fades to 0. This kind of an envelope is called an ASDR–envelope (Attack, Sustain, Decay, Release). The shape of this envelope can be changed using controls on the synthesizer so that the levels of each voltage can be adjusted, as well as their duration. There can be more than one envelope generator on the synthesizer and it changes not only the volume but the timbre and pitch.

As synthesizers evolved in the 1970s and 1980s, for a time there were hybrid combinations using subtractive synthesis. The electronic circuitry that produced the sound waves in the hybrids did so using subtractive synthesis. The earlier analog synthesizers with oscillators that responded to voltages or VCOs were converted to DCOs in that they responded to digital control signals. In the analog synthesizer the oscillator was an electronic circuit that generated a varying electronic signal, while in the digital case the oscillator was a table of numbers stored in memory and read out. The envelope generator in the hybrid was also digitally controlled shaping the amplitude of the wave and and passing the signal onto the DAC.

LFO

Synthesizers usually have a low frequency oscillator to generate vibrato, or modulation of the oscillator. Vibrato is an important sound component of most musical instruments and the human voice. Another change in sound associated with periodic modulation is tremolo which affects the amplitude of the sound rather than the pitch. The LFO (low frequency oscillator) is able to accomplish both these effects in a synthesizer. However, the synthesized vibrato is almost always too machine-like and the human brain is quick to recognize this fact. It doesn't have the variation and emotional component that a natural vibrato has.

Digital Synthesis

Music was first synthesized on mainframe digital computers under software control. In 1957 at Bell Labs Max Mathews, who is referred to by many as the Father of Computer Music, synthesized music on a mainframe digital computer with his software program called Music I. In digital synthesis a microprocessor controls digital circuits. The complex waveform voltages, which are fed to the speakers to produce the sound we hear, are generated from mathematical equations that are built into the synthesizer. They are not formed by manipulating control voltages and filters to shape a complex wave from a VCO, as in the analog case. However, physically, the digital synthesizers resemble their analog counterparts in that they have controls to generate envelopes, modify loudness, pitch, and timbre. In reality, they are parts of a mathematical system which uses mathematical equations and computer subroutines to create the waveform. The digital synthesis of sound tends to be "thin", because of difficulty in duplicating fully the complex waveforms associated with sound from traditional musical instruments.

A complex wave form is created, or digitally synthesized, by using the numbers calculated by the microprocessor and stored in memory in the digital synthesizer. The numbers are used to create a waveform table which, when plotted graphically, might look like the black rectangles shown in Figure 9.6. If the time interval between waveform readings for the height of each rectangle is small, then the rectangles will be very narrow and the waveform will be smoother. The numbers are sent to a digital to analog converter (DAC) which converts them into a voltage which, after amplification and passing through a speaker system, produces the sound we hear. For the sound to be realistic, the rate numbers are sent to the DAC by the sound generators must be around 50,000 times per second.

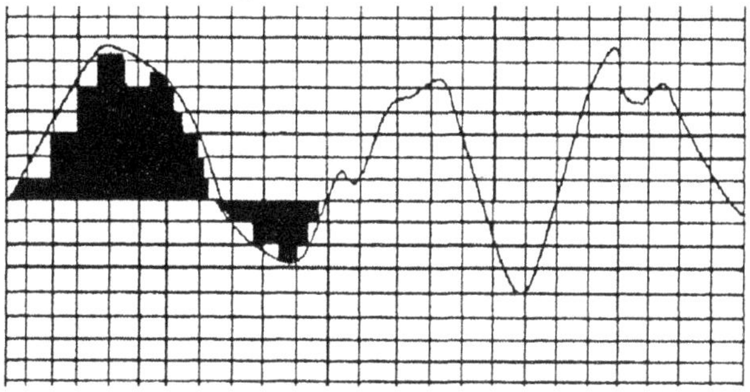

Figure 9.6 Digital synthesis

From *Computer Music: Science and Technology of a New Art,* Stanionis, Berberich. Copyright © 1989 by Research Foundation of State University of New York. Reprinted by permission.

When a person identifies the acoustical instrument playing a particular note, it is the attack phase, the initial sound that is heard, that is the most important in identifying the instrument. When we analyze the waveform of the initial sound, the waveform is so complex that it is closer in form to noise rather than some periodic waveform. The very complicated waveform associated with noise makes it very difficult to duplicate. Synthesizers try to duplicate such complex sounds by summing simple sine waves. Each one of the sine waves is said to be a partial or a harmonic. To simulate an acoustical instrument sound, 12 to 24 harmonics are required, each at the appropriate intensity. Digital synthesis gives us the ability to do this. Since the intensity of a note varies over the time it is sounded, an envelope must be defined for each of the 12 to 24 harmonics. Remember, it is the envelope that controls the volume generated by the waveform and the harmonics for a note. What makes it difficult to simulate a musical instrument is the complex envelopes that must be created for each partial.

Fourier Synthesis or Additive Synthesis

The main types of digital synthesis used today are: subtractive, additive, granular, wavetable, frequency modulation, phase distortion synthesis, and physical modeling synthesis. Fourier synthesis uses additive synthesis of sine waves to create complex waveforms. To understand additive synthesis and how it works, the concept of interference and law of superposition (as treated in the chapter on waves) should be reviewed at this time. Fourier discovered that square waves, saw tooth waves and triangle waves could be created by adding together sine and cosine waves and their multiples (Fig. 9.7). He discovered a way of representing certain mathematical functions as a sum of sine and cosine waves.

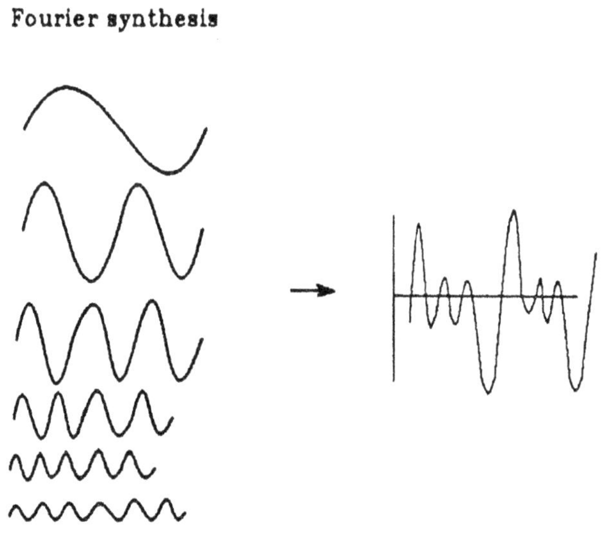

Figure 9.7 Fourier or additive synthesis
From *Science of Hi-Fidelity* by Johnson, Walker, Cutnell. Reprinted by permission of the authors.

Multiples of the fundamental frequency of a sound source are called harmonics. When a musical instrument is generating sound, the fundamental is excited along with various harmonics of differing intensities and this is what gives the musical instrument its characteristic sound. In theory once the harmonic spectrum of a sound is known (Fig. 9.8), that sound can be simulated by adding the various harmonics to the fundamental. This process is known as Fourier synthesis or additive synthesis. At least 12 to 24 harmonics are needed to recreate the sound in a realistic fashion. Organs are an example of basic additive synthesis.

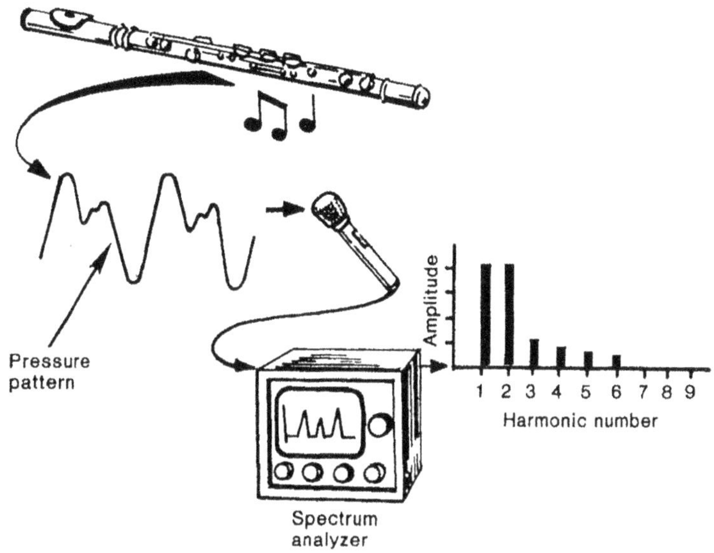

Figure 9.8 Spectrum analyzer for decomposing a sound into it components
From *Science of Hi-Fidelity* by Johnson, Walker, Cutnell. Reprinted by permission of the authors.

Using a device known as a spectrum analyzer we can play any musical instrument through a microphone and analyze the frequencies and amplitudes of the various components of the complex wave it creates. Knowing the frequencies and the associated amplitude associated with a flute, for example, allows us to generate the flute sound using our synthesizer. It is like a recipe for the sound in that it tells us which fundamental and harmonics are to be used and how much of each.

Digital synthesizers can use the recipe to create the complex waveforms that are needed. The sine waves are generated from a sine wave table stored in ROM. The amplitude of each harmonic is inserted according to the intensity of that harmonic. These waves are all summed together digitally, using the principle of superposition and interference, to create a complex wave. Finally, the wave is fed to a digital-to-analog converter, filtered and amplified to be heard as sound produced by the speakers (Fig. 9.9).

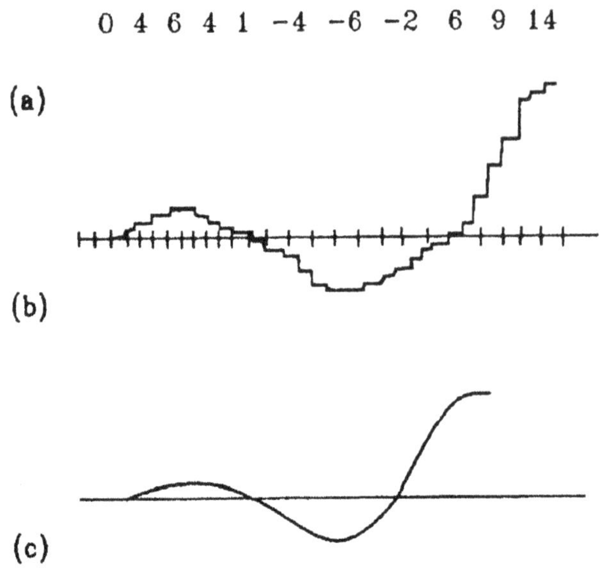

Figure 9.9 Digitally synthesized signal
From *Computer Music: Science and Technology of a New Art,* Stanionis, Berberich. Copyright © 1989 by Research Foundation of State University of New York. Reprinted by permission.

Remember, if several waves of differing phases and differing frequencies are added, the graph would be (Fig. 9.10):

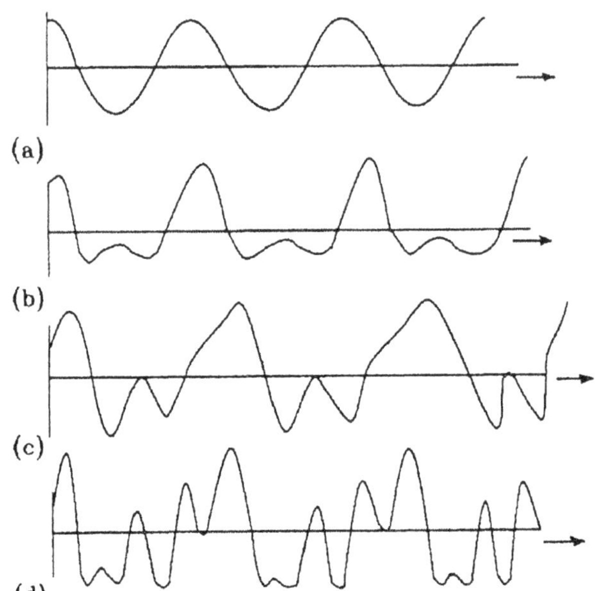

Figure 9.10 Fundamental with harmonics
From *Computer Music: Science and Technology of a New Art,* Stanionis, Berberich. Copyright © 1989 by Research Foundation of State University of New York. Reprinted by permission.

This graph corresponds to the fundamental with several harmonics added to it. Note how the waveform has become more complex in relation to a fuller sound.

FM Synthesis.

Before we analyze FM synthesis, let's review some of things we already know about AM and FM radio broadcasting. As you know AM is short for amplitude modulation and FM short for frequency modulation. To modulate something is to change it. We can consider a drawing straight line which can be thought of as unmodulated. However, when we write a word, we change it (Fig. 9.11).

Figure 9.11 Unmodulated line and modulated line

From *Computer Music: Science and Technology of a New Art,* Stanionis, Berberich. Copyright © 1989 by Research Foundation of State University of New York. Reprinted by permission.

AM broadcasting depends on modulating the amplitude of the carrier wave to convey information. It has a nominal range of several hundred miles but its range may be increased at night. The earth has a large number of charged particles trapped in its atmosphere that act like a reflective shield for AM waves. After sundown, this shield (the ionosphere), moves farther out allowing longer reflections to take place, increasing the range of the AM broadcast (Fig. 9.12).

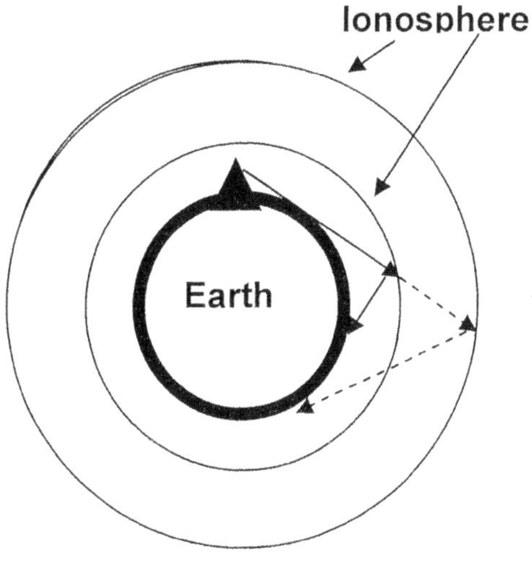

Figure 9.12 Reflection from ionized atmospheric layers

AM broadcasters have been assigned the frequencies between 535 - 1605 kHz for broadcasting. This means that 1010 on your AM dial corresponds to a frequency of 1010 kHz or 1,010,000 Hz. AM transmits audio frequencies in the range of 200 - 5000 Hz which is suitable for voice broadcasting but not hi-fidelity. The wavelength of AM waves is about three football fields in length. Compare this with FM broadcasts, which have a wavelength of about ten feet. FM broadcasters have been assigned frequencies between 88 - 108 MHz and they can broadcast audio frequencies from 30 - 15,000 Hz which is suitable for broadcasting music at hi-fidelity. Both methods of broadcasting use a carrier wave assigned to their radio station that they can modulate (Fig. 9.13).

BAND	ALLOCATION
30 – 535 Hz	Maritime aeronautical navigation
535 – 1605 kHz	AM radio
1605 kHz – 30 MHz	Amateur radio government radio, international shortwave, fixed and mobile communications
30 – 50 MHz	Government communication: fire, police, etc.
50 – 54 MHz	Amateur
54 – 72 MHz	Television channels 2 to 4
72 – 76 MHz	Government and nongovernmental services.
76 - 88 MHz	Television channels 5 and 6
88 – 108 MHz	FM broadcast
108 – 122 MHz	Aeronautical navigation
122 – 174 MHz	Government and non-governmental
174 – 216 MHz	Television channels 7 to 13
216 – 470 Mhz	Amateur, government, aeronautical, citizen's radio
470 – 890 MHz	UHF television channels 14 to 83
890 – 3000 MHz	Aeronautical, amateur, government and non-governmental , radar
3,000 – 30,000 MHz	Government and non-government, fixed and mobile, amateur, radio navigation
30,000 – 300,000 MH	Experimental, government, amateur

Figure 9.13 FCC allocations

In AM broadcasting the amplitude of the carrier wave (Fig. 914) is modified by the signal generated by the broadcaster's microphone and it is this modified wave that is transmitted (Fig. 9.15). When it is received by an AM tuner, its circuits detect the amplitude changes and produce an output audio signal.

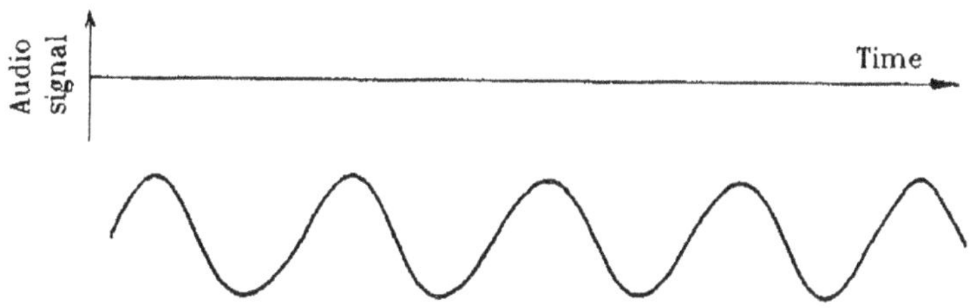

Figure 9.14 Unmodulated carrier signal
From *Computer Music: Science and Technology of a New Art,* Stanionis, Berberich. Copyright © 1989 by Research Foundation of State University of New York. Reprinted by permission.

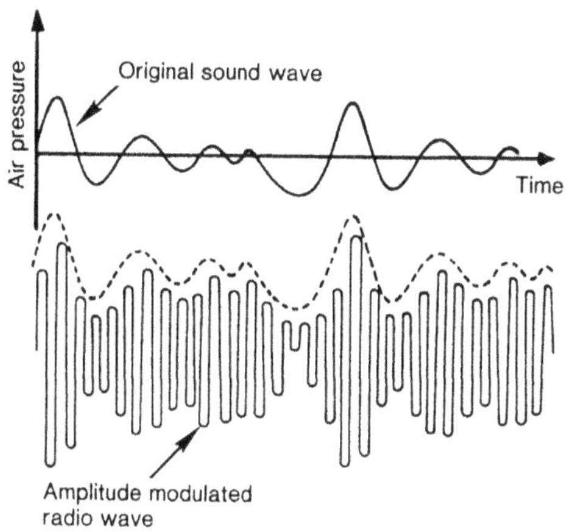

Figure 9.15 Amplitude modulated carrier signal
From *Science of Hi-Fidelity* by Johnson, Walker, Cutnell. Reprinted by permission of the authors.

In FM broadcasting it is the frequency of the carrier wave that is changed within certain limits of about 75 kHz from the station's assigned frequency. The changes in frequency correspond to the input signal generated by the music and are broadcast to the public. The closest any two stations can possibly be is 200 kHz. The station is allowed a variation of 75 kHz to either side of its assigned frequency for a total of 150 kHz, 50 kHz is left over providing a so-called guard band of 25 kHz on either side.

Again, it is now the FM tuner that detects the change in frequency and converts it to a audio signal. It is interesting to note that FM stations are assigned frequencies by the government which are 400 kHz apart which is more than the 200 kHz deemed necessary. This creates an unused station between two operating stations. For example, you may have a station broadcasting at 99.9 Mhz on your dial. The closest stations to this one would be 100.1 and 100.3. There would be no station broadcasting at 100.1. Check it out for yourself. What do you notice about the numbers designating broadcasting stations? Are they all odd numbers? The reason for leaving the blank stations in between is to prevent any possible overlap, even if it is accidental. See Figure 9.16.

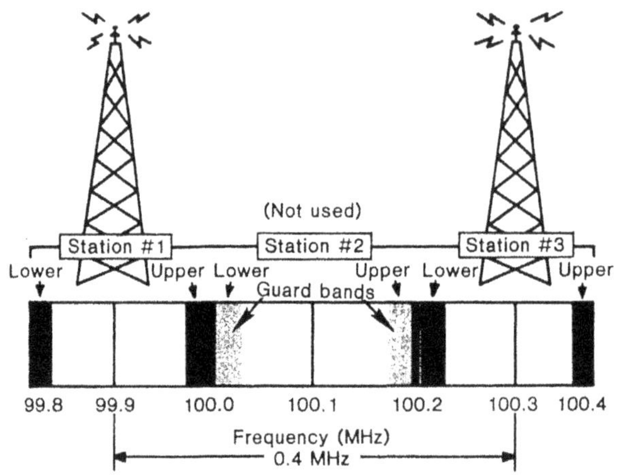

Figure 9.16 FM stations

From *Science of Hi-Fidelity* by Johnson, Walker, Cutnell. Reprinted by permission of the authors.

FM (frequency modulation) synthesis generates a complex waveform from two sine waves. One of the digitally generated sine waves is called the modulator and the other the carrier. The modulator has the ability to modify or change the frequency of the second waveform. FM modulation works by changing the frequency of a wave and keeping its amplitude constant. The frequency is changed to reflect the information it is carrying.

Suppose we have a 1000 Hz signal that is unmodulated. The carrier wave would look like Figure 9.14. The intensity of the carrier wave determines the overall volume of the tone. Now. if we modulate the frequency of the wave slightly using another signal, there will be a region where the wave is compressed and others where it expands in step with the signal being used to modulate the 1000 Hz carrier signal (Fig. 9.17).

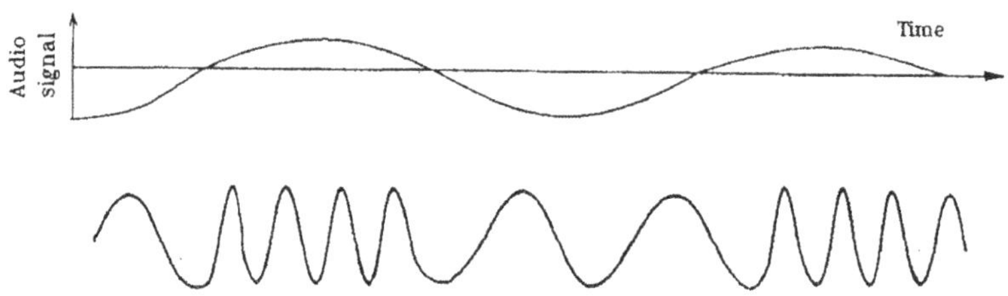

Figure 9.17 Modulated carrier wave

From *Computer Music: Science and Technology of a New Art,* Stanionis, Berberich. Copyright © 1989 by Research Foundation of State University of New York. Reprinted by permission.

A complex musical signal would look like (Fig, 9.18):

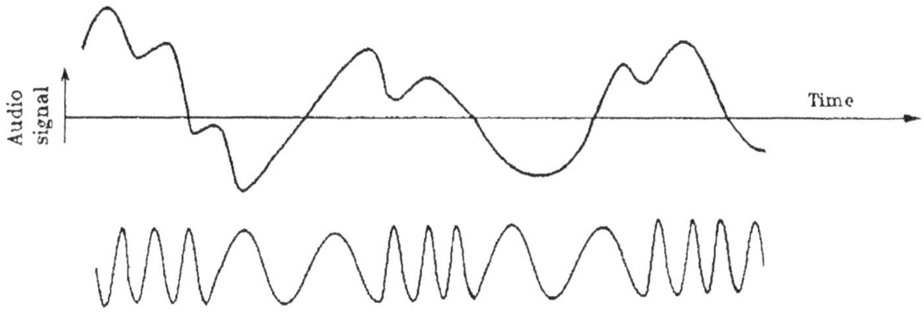

Figure 9.18 Information carried by carrier wave

From *Computer Music: Science and Technology of a New Art,* Stanionis, Berberich. Copyright © 1989 by Research Foundation of State University of New York. Reprinted by permission.

The modulator's intensity has the ability to modify the relative amplitudes of the harmonics. As it is varied while a note is sounded, it changes the quality or timbre of the note smoothly. Sophisticated FM synthesizers use multiple carriers and modulators to create more complex waveforms. Each sine wave in a more complex synthesizer is called an operator.

Phase Distortion Synthesis

In phase distortion synthesis a different method is used to produce the complex waveforms. A sine wave is distorted by playing different sections at different speeds. To fully understand how this method works, we have to plot a simple sine wave mathematically. Consider y = sin x. Let's construct a table of values choosing values of x between 0 and 2π radians and computing values of y or sin x using a calculator (Fig. 9.19).

x	0	$\frac{\pi}{6}$	$\frac{\pi}{3}$	$\frac{\pi}{2}$	$\frac{2\pi}{3}$	$\frac{5\pi}{6}$		$\frac{7\pi}{6}$	$\frac{4\pi}{3}$	$\frac{3\pi}{2}$	$\frac{5\pi}{3}$	$\frac{11\pi}{6}$	2π
y	0	0.5	0.87	1	0.87	0.5	0	−0.5	−0.87	−1	−0.87	−0.5	0

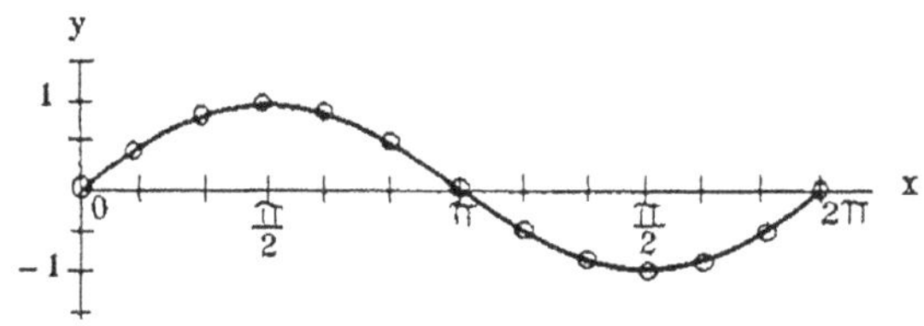

Figure 9.19 Graph of y = sin x

From *Computer Music: Science and Technology of a New Art,* Stanionis, Berberich. Copyright © 1989 by Research Foundation of State University of New York. Reprinted by permission.

If we chose to draw a graph of a slightly more complicated sine wave such as y = A sin (bx + c), we would use basically the same method. In our previous example the amplitude or A =

1, b = 1 and c, the phase angle was zero. Consider the plot of y = sin (x + π/4). The phase angle is π/4 or 45 degrees (Fig. 9.20).

Remember:
2π radians = 360 degrees. So, 45 (1 degree) = 45 (2π/360) radians = (π/4) radians.

Figure 9.20 Graph of y = sin (x + π/4)

From *Computer Music: Science and Technology of a New Art,* Stanionis, Berberich. Copyright © 1989 by Research Foundation of State University of New York. Reprinted by permission.

Note how the graph has shifted. When y = 0,

$$\sin(x + \pi/4) = 0$$

and therefore

$$x + \pi/4 = 0$$

or

$$x = -\pi/4.$$

So the graph begins at x = -π/4. The whole graph has shifted a distance of -π/4 to the left. As we compute values for y the phase angle, π/4 will always remain constant.

If we let the phase angle vary in the course of creating the table of y values, the sine wave will be distorted. In a PD type synthesizer, if the sine wave form is read from the ROM at a constant rate, we see a regular sine curve. When the phase angle speed is increased or decreased somewhere in the middle of the reading the wave changes in form, producing the complex forms needed to simulate musical instruments (Fig. 9.21).

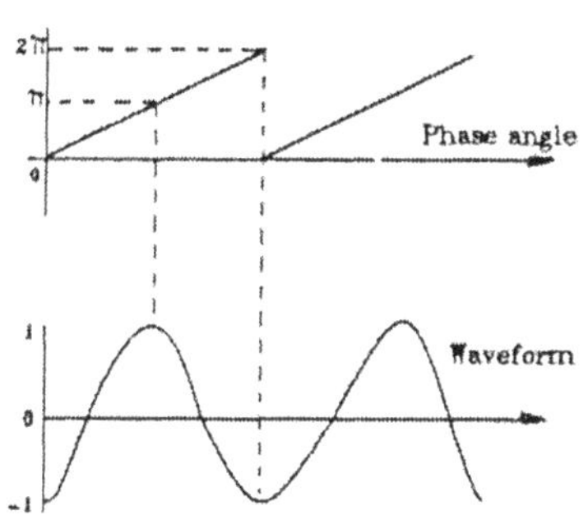

Phase angle read linearly produces a sine wave.

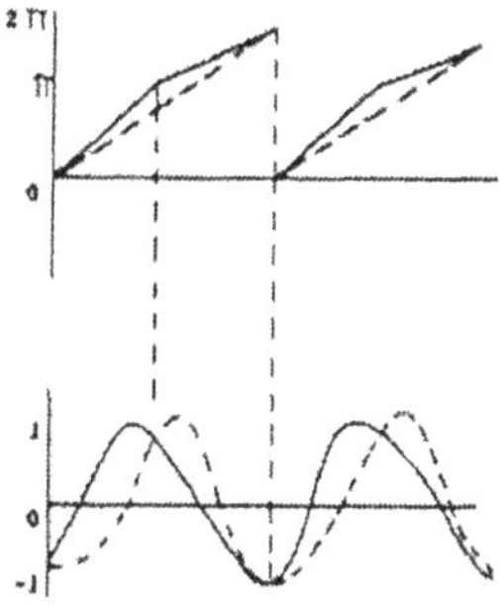

Phase angle read nonlinearly produces another wave form.

Figure 9.21 Phase distortion

From *Computer Music: Science and Technology of a New Art,* Stanionis, Berberich. Copyright © 1989 by Research Foundation of State University of New York. Reprinted by permission.

Physical Modeling (PhM) Synthesis

In this method for synthesis computer software has been developed to recreate the actual behavior of an instrument. A set of mathematical equations are used to simulate the behavior of a musical instrument. The equations define exciters and resonators. The exciters are simulated from the input, which may be the nuances of pressing a key on the keyboard, and resonators respond to the exciter by simulating the physical attributes responding to the exciter. The simulated sound is generated based on the initial set of parameters being used.

To model the sound of a guitar there would be a formula describing how plucking the guitar string imparts energy to it and how it generates resonance and the damping behavior of the strings. It would describe the transfer of string vibrations through the bridge, and finally, the resonance of the soundboard in response to those vibrations. A wind instrument would be modeled by having the computer solve equations that describe breath pressure, inflow angle in the mouth piece and back pressure in the instrument and their relationships to resonances in the instrument. The computer would process all the information recreating the actual operation of the instrument.

Sampling Synthesis

Sampling involves first recording a complex real sound and converting it to digital form using an analog-to-digital converter. Theoretically, a computer can use mathematical equations to digitally synthesize the sound of acoustic instruments by generating the waveforms associated with the musical instrument. However, in most instances the sound will be lacking in harmonic content. A sampling synthesizer records the acoustic waveform of a musical instrument in memory and then plays it back to produce different notes. Actually, in the strict sense of the term it is not synthesizing the sound.

The sampling synthesizer takes the complex waveform, divides it up into small pieces using an analog to digital converter, and stores the information in a waveform table. A sample of the incoming waveform is recorded periodically, depending on how fast the clock is ticking. For a voice recording it would take about 11,000 samples per second for recording and for a CD it would take 44,100 samples per second. The breaking of the sound into discrete samples means that there is some loss of information between samples. Fortunately, the human ear is limited in its resolution. So if the resolution we use is high enough, the ear will believe all the sound is present when played back (Fig. 9.22). If the intervals between samples are too long, the sound would not sound accurate on playback. A mathematical theorem published by Harry Nyquist, an engineer at the Bell Telephone Laboratories, tells us that to sample a signal properly, the sampling frequency must be at least twice as high as the highest frequency component of the signal. Note that when we double the sampling frequency how much better the analog curve is approximated by the digital steps.

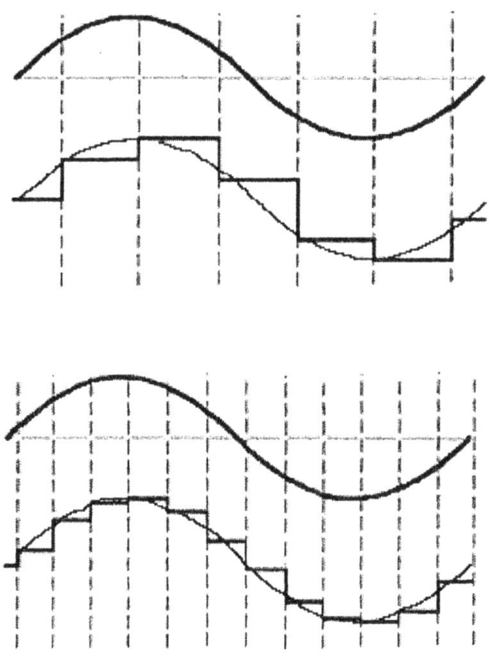

Figure 9.22 Two sampling frequencies

If a sound is sampled properly, it can then be reproduced, creating new sounds and notes. The synthesizer can truncate the waveform, repeat it, or loop it and process it using filters and amplifiers to produce a "fat" sound. The success of the process depends on the resolution of the sampler and its sampling frequency. It is like taking a photograph. For best resolution a 16 bit linear sampler and a sampling frequency of 44 - 50 kHz is desirable (Fig. 9.23) so that the rough edges that occur when the wave is converted into numbers is minimized (Fig. 9.22 above). One

Hertz corresponds to one sample per second, and one kilohertz (kHz) corresponds to one thousand samples per second.

Samples per second	Description
11kHz	8 bits, mono, for voice
22kHz	8-bit mono, good for voice and music
44.1kHz	16-bit standard CD sampling rate

Figure 9.23 Sampling rates for recording

Sampling frequency

Basically, sampling frequency corresponds to the number of samples taken per second. The waveform is closely approximated when the vertical bars shown in Figure 9.24 are closer together or the sampling frequency or number of measurements per second is high.

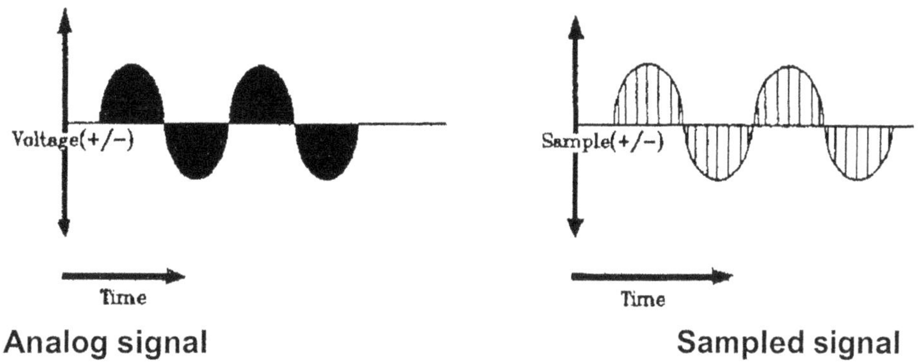

Figure 9.24 Digitized waveform

From *Computer Music: Science and Technology of a New Art,* Stanionis, Berberich. Copyright © 1989 by Research Foundation of State University of New York. Reprinted by permission.

The sound generator, when it synthesizes a sound, calculates the appropriate numbers needed to create the waveform and sends them to the digital-to-analog converter. In order for the sound to be realistic, the sound generator must send the numbers at a speed close to 50,000 times per second and it is then processed by the EG filters and amplifier to smooth out the rough edges. If the sampling frequency is not at least twice as high as the highest frequency being recorded, then aliasing occurs. Aliasing results in a distortion of the sound being played back. To remedy this problem the sampling rate can be increased or anti-aliasing filters can be used to restrict the bandwidth (or frequency content) of the signal.

For example, suppose we are using a sampling rate of 44.1 kHz on a tone that is slowly increasing in frequency. When we sample the tone when its frequency is low, we have many sample points per cycle. However, as its frequency increases the cycles are shorter, the waves are closer together, and fewer points are sampled than needed to reproduce it accurately. Once it reaches 22 kHz we reach the Nyquist limit. It is the highest frequency that can be recorded without aliasing. It is one-half the sampling frequency. As the frequency exceeds the limit, the number of sample points needed to reproduce the waveform is not available and instead the tone is recorded at a lower frequency. These recorded tones combine with the real frequencies to produce distorted sound on playback due to aliasing. At the lower frequency 13 points are sampled for one oscillation while at the higher frequency only 5 points are sampled for one oscillation (Fig. 9.25).

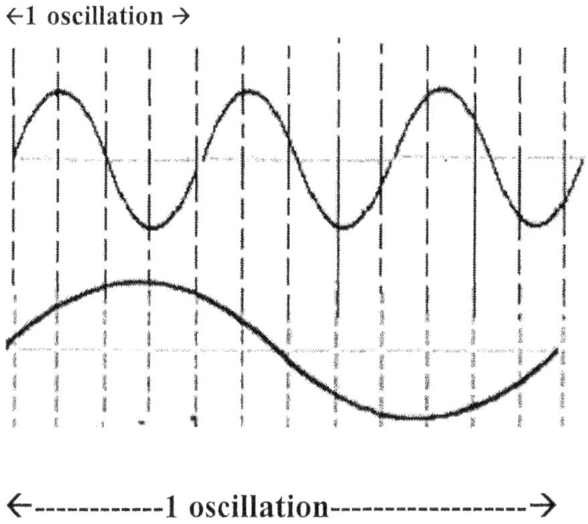

Figure 9.25 Sampling a high frequency tone and a low frequency tone

Sending the numbers from the table to the DAC for play back at different speeds, allows the synthesizer to generate different notes. Filters and envelope generator computations can be used to further modify the sound.

Using a microphone you can record the sound of someone saying the word green. The sampled sound is digitized and stored internally in the memory of the synthesizer in a waveform table. You can now use the synthesizer's sound generators controlled by a keyboard to play scales and chords using the recorded sound of the word green.

Sampling allows one to play any sound that can be heard. A sampler allows one to create a library of multisounds that are stored in memory. The multisounds consist of the sampled sounds or synthesized waveforms that have been modified by truncating them, looping them, and tuning them. They are then assigned to locations on the keyboard to create a musical composition. If you were take video on your camera at 15 frames per second, the movie on play back would seem herky-jerky. At 30 frames per second the movie action is smooth and continuous. In the case of sound if the sampling rate is not high enough, a similar situation arises and all the frequencies in the original sound are not captured.

Sampling rates can usually be set by using software but are limited by the hardware available. We see in the Figure 9.23 for CD sound we need a sampling rate of 44.1 kHz for high quality sound. Microsoft specifies that the sound hardware be capable of sampling at 11.025 kHz and 22.050 kHz, at a minimum. On playback the sound will be accurately reproduced if we use the exact same sampling frequency as when recording. If the sampling rate is increased, the pitch will be higher and if it decreased, the pitch will be lower. It's similar to the effect of speeding up a recording to get a sound similar to the Chipmunks or if lower, similar to the Church monks. We can use this technique involving transposition to play the sound over different octaves. However, as we move farther away from the original, we lose harmonics. To compensate for this effect usually five or ten samples are taken over the keyboard range in different octaves.

Resolution

Resolution tells us the amount of memory available to store each sample. The number of bits assigned to each sample defines the resolution of the sample. The more bits used in storing the information about the sample, the more accurately the sound on playback. If a sampler has 8 bit resolution, it can store $2^8 = 256$ different volume or dynamic levels. If the amplitude of a sample lies between two volume levels, it is rounded-up to the nearest one. With 16 bit resolution, 2^{16} (= 65,536) or thousands of distinct volume levels are stored. The more bits that are used, the closer the sample will represent its analog counterpart which can be thought of has having infinite resolution.

Each bit of resolution in which the amplitude of the wave is recorded corresponds to about 6 dB of sound. So if we have 8 bit resolution, the range of the sound would be about 6 x 8 or 48 dB, which is about the range of the sounds stored on an audio cassette. For 16 bit resolution, the range would be 6 x 16, or 96 dB, which corresponds to the range you have on a CD.

Sixteen bits is the lowest resolution for recording and playback of high fidelity sound, with twenty-four bits used by professionals. The higher resolution naturally requires more memory for recording. The resolution of a sampler is usually determined by its hardware.

Wave Table Synthesis

Wave table synthesis uses pulse code modulation (PCM) in which a digital sample of an acoustical instrument is sampled, stored in the wave table memory, and a series of wave tables is created in which one cycle of the sampled sound is placed in each table. Actually, in wave table synthesis sound is not being synthesized but played back from the wave table. The one note that is recorded is modified and played back imitating the actual sound of the instrument over its entire range. The note is recorded at a fixed sampling frequency, which is twice the highest frequency contained in the note. Lower and higher notes are then created by decoding the note at lower and higher sampling frequencies, respectively. The harmonic content of the note varies with pitch and several samples may have to be taken over a range of octaves to faithfully reproduce the sound of the acoustical instrument.

Granular Synthesis

Granular synthesis operates in much the same way as digital sampling. The difference is that entire samples are not used but instead split into small pieces of different time intervals in milliseconds. These small grains are then combined to produce unusual sounds which can be used for sound effects.

Multitimbral and Polyphonic

Early synthesizers had only a single voice and were monophonic. Pressing several keys on a keyboard at once would result in a single tone. By adding additional oscillators, synthesizers were built to generate four and eight voices (or tones or notes) at once. Today's polyphonic synthesizers are available with 32, 64 and even 128 voices or notes that can be played at once. In addition, synthesizers today are multitimbral in that they can play several instrumental sounds at once: piano, trumpet, guitar, drums etc. The Roland MT-32 synthesizer (or Roland multitimbral, 32 note polyphonic) can play 32 notes simultaneously, distributed over 8 musical instrumental sounds. It can sound like an 8 piece band but can't play more than 32 notes at once.

One of the major limitations common to all synthesizers has been the machine-like rendition of sound. When a poet recites a poem, there is feeling, emotion, expressiveness. These characteristics have been difficult to incorporate into a musical performance using a synthesizer.

Review Questions

1. Describe the operation of an analog synthesizer.
2. What role do oscillators play in an analog synthesizer?
3. What is subtractive synthesis?
4. What roles do the VCO and VCA play in an analog synthesizer?
5. What is a gate signal?
6. What is an envelope generator?
7. What is the role of the LFO?
8. What is the difference between an analog synthesizer and a digital synthesizer?
9. What is Fourier synthesis?
10. What is frequency modulation?
11. What are the differences between amplitude modulation and frequency modulation?
12. What is FM synthesis?
13. What is PD synthesis?
14. Discuss sampling synthesis.
15. What is the sampling frequency?
16. What role does resolution play in sampling?
17. What is PCM synthesis?
18. What is granular synthesis and physical modeling synthesis?

Multiple Choice Questions

1. Analog synthesizers synthesize sound using _________.
a. mathematical equations
b. varying voltages
c. sampling methods
d. a cpu

2. The wave form that can generate all the harmonics is the _____.
a. sine
b. triangle
c. square
d. saw tooth

3. A pitch bend wheel changes the _____ of a sound from a half step to an octave.
a. amplitude
b. vibrato
c. pitch
d. loudness

4. Almost all analog synthesizers synthesize sound using _____ synthesis.
a. additive
b. subtractive
c. Fourier
d. sampling

5. When a key is pressed on a synthesizer, a voltage is generated in the _____.
a. VCO
b. VCA
c. DAC
d. ADC

6. Conventionally, the control voltage over one octave is _____ volt(s).
a. 1/12
b. 12
c. 1
d. 8

7. In addition to generating a voltage in the VCO, when a key is pressed on the keyboard a _______ signal is activated.
a. VCA
b. terminal
c. voltage
d. oscillator

8. A(n) _____ called an envelope generator is set into operation by a gate voltage.
a. device
b. circuit
c. oscillator
d. key

9. The EG generates a control voltage that governs the dynamics of a sound in terms of its _____, sustain, decay and release.
a. loudness
b. amplitude
c. attack
d. vibrato

10. A LFO is used to generate _______.
a. vibrato
b. sustain
c. release
d. simpatico

11. Digital synthesizers produce sounds that are generated from ______.
a. VCO
b. mathematical equations
c. wave guides
d. continuous variable voltages

12. A digital synthesizer uses a microprocessor to calculate numbers for a waveform stored in a table. The numbers are sent to the ________ where they are converted into an analog wave.
a. DAC
b. VCO
c. ADC
d. VCA

13. The portion of the waveform that allows a person to identify a musical instrument immediately occurs in the ______ phase of a sound.
a. sustain
b. attack
c. release
d. decay

14. To simulate an acoustical instrument _____ harmonics are required.
a. 4 to 8
b. 8 to 12
c. 12 to 24
d. 1000

15. Fourier synthesis uses _____ synthesis of waves.
a. additive
b. subtractive
c. granular
d. frequency modulation

16. When a wave signal is changed in some way, we say it is _____.
a. modified
b. modulated
c. manipulated
d. monetized

17. FM stands for _______ modulation.
a. Fourier
b. frequency
c. forced
d. fortified

18. In phase distortion synthesis a complicated wave is created by letting the phase angle _____ as the wave is generated.
a. equal π
b. be a constant
c. vary
d. be periodic

19. Sampling synthesis, technically speaking, does not ______ the sound of instruments.
a. synthesize
b. record
c. modulate
d. resolve

20. In sampling synthesis the analog wave created by an acoustical instrument is ____ as its being recorded.
a. chopped up
b. modulated
c. recreated
d. synthesized

21. In order to reproduce a sound accurately the sampling frequency must be _______ the highest frequency in the sound being recorded.
a. twice
b. equal to
c. ½
d. less than

22. A synthesizer that can generate several instrumental sounds at once is said to be _____.
a. polyphonic
b. multisonic
c. multitimbral
d. monophonic

23. A synthesizer that can play 32 notes at once is said to be ______.
a. polyphonic
b. multisonic
c. multitimbral
d. monophonic

24. _________ synthesis uses a set of mathematical equations to simulate the behavior of a musical instrument.
a. PhM
b. Ph. D
c. PD
d. FM

25. _______ synthesis uses small pieces that are sampled and then are combined to produce unusual sounds which can be used for sound effects.
a. FM
b. PD
c. microbit
d. granular

26. For 16 bit resolution the amplitude of the wave recorded corresponds to _____ dB of sound.
a. 16
b. 32
c. 48
d. 96

27. Each bit of resolution in which the amplitude of the wave is recorded corresponds to about _____ dB of sound.
a. 2
b. 4
c. 6
d. 8

28. The number of samples taken per second by a sampler is called _____.
a. sampling frequency
b. resolution
c. aliasing
d. the Nyquist rate

29. A sampling rate for recording onto a CD would be _____.
a. 11 kHz
b. 22 kHz
c. 44.1 kHz
d. 48 kHz

30. The Nyquist limit is equal to _____ times the sampling frequency.
a. one
b. two
c. one eighth
d. one half

Bibliography and Web Resources

Johnson, Kenneth, and Willard Walker and John Cutnell. *The Science of Hi-Fidelity*. Kendall-Hunt Publishing Company, Dubuque, Iowa, 1995.

Stanionis, Victor and Hugh Berberich. *Computer Music: Science and Technology of a New Art*. Research Foundation of State University of New York, Stony Brook, NY 1989.

Sampling -
De Furia, Steve and Joe Scacciaferro. *The Sampling Book* . Hal Leonard Books, 1987.

Digital audio and sampling
Pohlman, Ken. *Principles of Digital Audio*, 3rd edition McGraw Hill Text, 1995.

Max Mathews -
http://www.csounds.com/mathews/

Sound waves adding harmonics -
http://www.phy.ntnu.edu.tw/java/sound/sound.html

Aliasing -
http://www.earlevel.com/Digital%20Audio/Aliasing.html

PCM or Wave Table -
http://www.ias.uwe.ac.uk/~a-winfie/teach2003/st_pcm.htm

Sampling -
http://www.turtlebeach.com/site/kb_ftp/8956010.asp

Sampling -
http://ist.uwaterloo.ca/ec/courses/Audio_Web/quality.html

Sampling –
http://ahds.ac.uk/creating/guides/audio-resources/GGP_Audio_3.2.htm

Chapter 10

Amplifiers

"An inexpensive instrument, not bigger than a watch, will enable its bearer to hear anywhere, on sea or land, music or song, the speech of a political leader, the address of an eminent man of science, or the sermon of an eloquent clergyman, delivered in some other place, however distant."

Nikola Tesla, 1908
Electrical Inventor and Master of Lightning

The basic function of an amplifier be it a source amplifier, a preamplifier, or a power amplifier is to faithfully and accurately enlarge the electrical signal it receives. Each source has built into it a preamplifier of its own which processes weak electrical signals generated by the electronics of a microphone, tuner, CD player, synthesizer, tape deck, etc., and enlarges it to about 0.5 volts before it sends it to the system pre-amp. Each of the source pre-amps also help to reduce noise in the signal. This process is called equalization. The 0.5 volt signal is passed on to the system pre-amp where it is processed so as to produce certain tonal qualities such as volume, bass, treble, etc. Finally, once these controls have been exercised the signal is enlarged once more by the power amplifier to the point where it is large enough to drive the speakers and produce sound.

Goals

After studying this chapter, you should be able to:

- Understand the function of sound source pre-amps.
- Understand the control functions of the system pre-amp.
- Know the function of the volume control and the loudness control.
- Know the effects of the bass and treble controls.
- Understand the function of the high and low filters.
- Understand the role of the power amplifier and how it operates.
- Know the definitions of total harmonic distortion and intermodulation distortion and how they differ.
- Understand what is meant by continuous power output.

- Know the definitions of and roles that dynamic headroom, stereo separation, input sensitivity, maximum input signal, signal to noise ratio, frequency response, and damping factor play in an amplifier.

Source Pre-amps

Each of our sound sources generates a very small electrical signal. As sound moves the diaphragm of a microphone, a tiny electrical signal is produced by an electromagnetic induction effect. The microphone pre-amp raises the voltage from approximately 0.0005 volts to about 0.5 volts while also equalizing the signal.

An AM-FM tuner also has its own pre-amp, which takes the weak electrical signal received by the antenna and boosts it to about 0.5 volts. It also reduces the amount of noise in the signal.

At the phono input, the signal received from the magnetic cartridge holding the stylus is on the order of 0.005 volts and must be enlarged. Again the size of the signal must be boosted to 0.5 volts and the phono pre-amp must compensate for the equalization curve applied to the audio that was cut into the vinyl record. The equalization is accomplished by the RIAA equalization amplifier, which adheres to the standards set by RIAA (Record Industry Association of America). The other input jacks on a receiver such as Aux, Tape, etc. do not provide the necessary amplification and equalization that the phono input does. If one were to attach a CD player to the phono input, the sound would be distorted. Because of the demise of vinyl recordings, many receivers do not have phono input jacks any longer and a separate phono pre-amp may have to be purchased in order to listen to a record.

A tape deck actually contains more than one pre-amp. On playback the signal from the tape is raised to 0.5 volts and is strong enough to be heard through a pair of headphones. On recording, the tape's pre-amp boosts the signal whether it comes from a microphone or another input.

System Preamplifier

The system preamplifier is the control center for the hi-fidelity system. It is here where the volume, bass, and treble controls are located, as well as filters such as the high and low filters. The loudness switch is also found in the pre-amp section. All of these control the amount of power reaching the speakers. They act almost like faucets which control the amount of water flowing into a sink (Fig. 10.1).

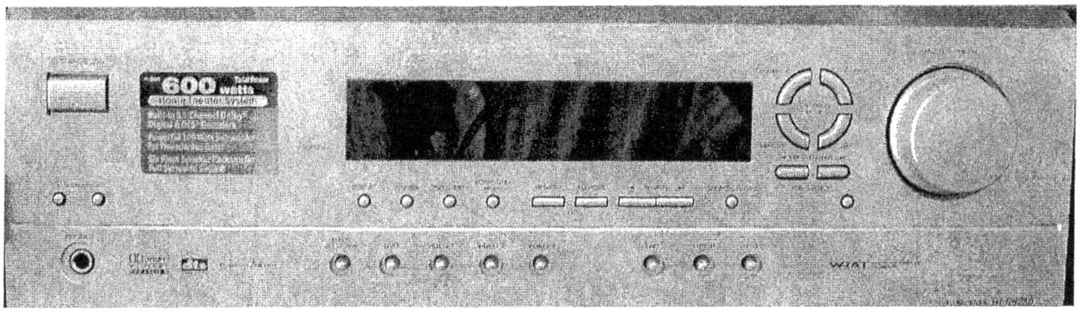

Figure 10.1 Receiver

One consumer publication, when testing speakers for accuracy, prints the variation in intensity of each speaker over the audible frequency range, showing how each speaker deviates from a flat response line. They then use the controls on the receiver to try to adjust the tonal qualities and response of the speakers and print a second curve revealing the speakers ability to more accurately reproduce all frequencies over the hearing range.

Normally the tonal controls (volume, filters and loudness switches) are used to compensate for the effect of the environment in which the receiver is being used. Is the room heavily furnished with sound absorbing furniture, carpets, and wall hangings? Is the room furnished sparsely with a large number of smooth and hard exposed surfaces? Is it very small or very large? How old are you? Do you have any hearing deficiencies? The preamplifier control center allows one to compensate for these environmental factors.

When we discussed the application of resistors in an electrical circuit, we found that we could regulate the amount of current and therefore the amount of power reaching the speakers with the volume control which was actually a variable resistance. The bass and treble controls are able to boost or cut the sound power over a selected frequency range. The action of the bass and treble controls is revealed in studying a typical response curve plotting intensity output versus frequency (Fig. 10.2).

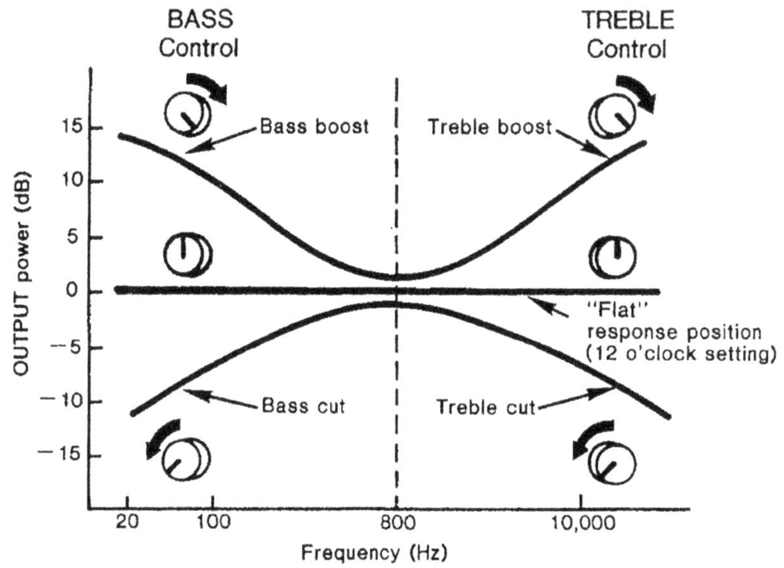

Figure 10.2 Bass and treble controls

From *Science of Hi-Fidelity* by Johnson, Walker, Cutnell. Reprinted by permission of the authors.

Notice that when the bass control is set at 12 o'clock the output power gives a flat response but when the bass control is moved to 5 o'clock all frequencies up to 800 Hz are boosted especially those in the low frequency range. There the boost is about 10 dB, which is almost twice the loudness it had when set at 12 o'clock. Similarly, we see that when the treble control is turned to 7 o'clock, the power being sent to the high frequencies is reduced and is increased when set to the 5 o'clock position.

The high and the low filters can also reduce the power sent to the high frequencies and the low frequencies, but they do it abruptly as compared to the gradual attenuation of the treble and bass controls. Once the sound reaches a definite cut-off frequency with the high filter on, the signal is reduced rapidly. Most of the static noise and hiss we hear on a hi-fi system is in the high frequency range. Therefore, by using the high filter we eliminate this noise while hearing more of the high frequency music than if we had used the treble control to eliminate the noise. The treble control would remove the high frequency noise and at the same time a good portion of the music at high frequency (Fig. 10.3).

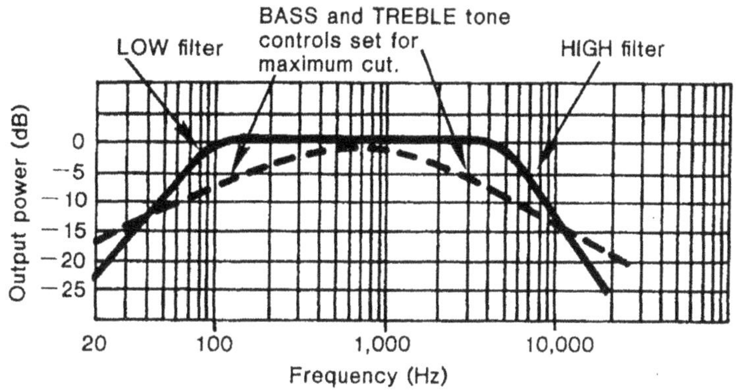

Figure 10.3 Bass and Treble control compared to high and low filter switches
From *Science of Hi-Fidelity* by Johnson, Walker, Cutnell. Reprinted by permission of the authors.

We have already spoken about the loudness control, which controls a circuit that sends power to boost the low and high frequencies at low volume levels. In the chapter on electricity we discussed the balance control, which controls the sound coming from the left and right speakers. We saw that both are connected in tandem to variable resistors, so that when one has low resistance the other has high resistance and little sound is heard. Moving the control to the opposite setting reverses the process. By adjusting the control we can balance the sound coming from each speaker so that their intensities seem the same.

Power Amplifier

A power amplifier receives the signal after it has been processed by the preamplifier and enlarges it sufficiently to drive the speakers to produce sound. It uses a power supply as its source of power and a device that behaves as a variable resistor. The variable resistor controls the power coming from the receiver's power supply to enlarge the signal it is receiving from the preamplifier. When signal from the preamplifier is large, the variable resistor device will allow a large current from the power supply to enlarge it even further before sending it on to the speakers. When the signal is small, very little reinforcement current will be allowed to flow from the power supply (Fig. 10.4).

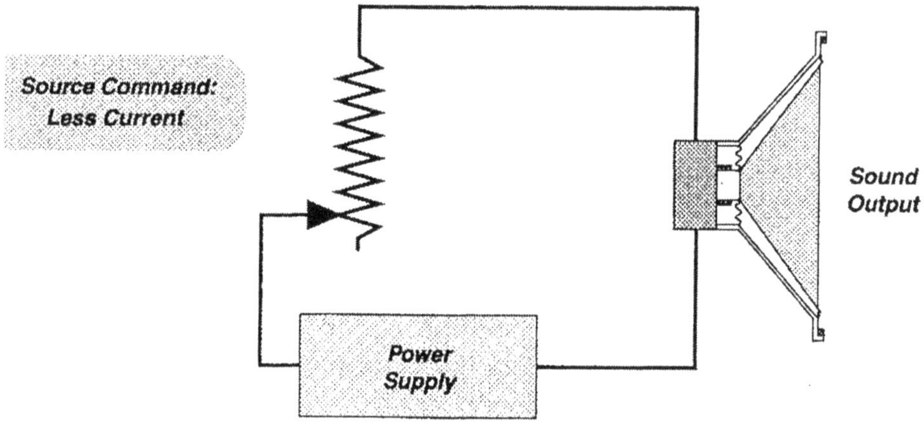

Figure 10.4 Simple amplifier
From *Physics of Hi-Fi* by Symko. Reprinted by permission of the author.

The variable resistance not only controls the amplitude of the output signal but also has to be able to respond to frequencies from 20 - 20,000 Hz. The device that has this capability is called a transistor (Fig. 10.5).

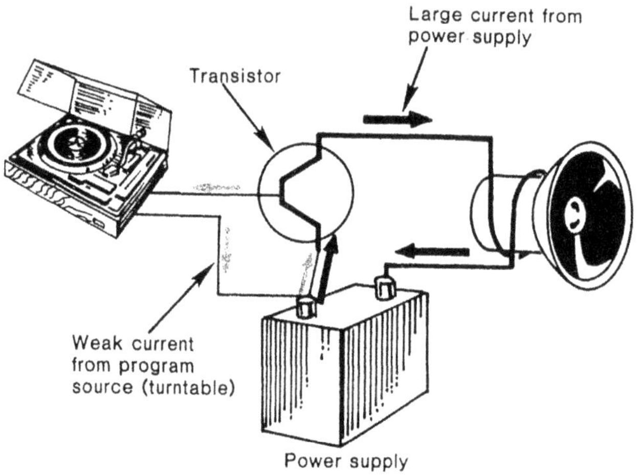

Figure 10.5 Transistor in an amplifier
From *Science of Hi-Fidelity* by Johnson, Walker, Cutnell. Reprinted by permission of the authors.

Power Amplifier Properties

All amplifiers are made from materials which may cause defects in their performance. In fact, all audio components are guilty of distorting the output signal in some way because of their material construction. The total distortions contributed by all the components of a hi-fidelity system are additive and when they reach a total of about 4%, most people will be conscious of them. The human ear is capable of detecting distortions on the order of 1%. Speakers are the major contributors to distortion.

Therefore, our goal would be to minimize distortion from the electronic components to as small a value as possible.

The basic function of the power amplifier is to reproduce the signal it is receiving from the preamplifier and faithfully and accurately enlarging it without adding or subtracting anything from the signal (Fig. 10.6).

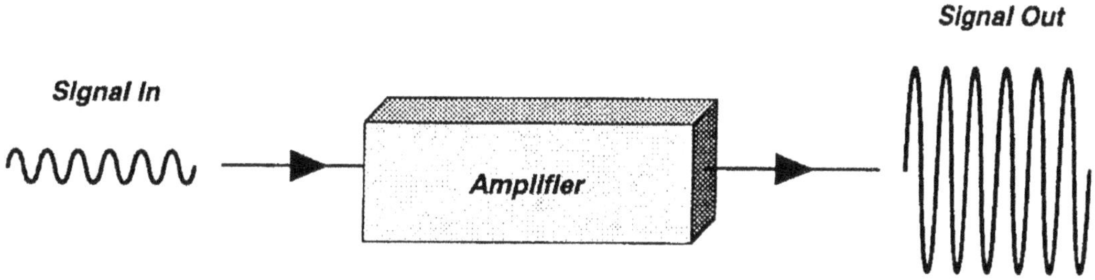

Figure 10.6 Basic function of an amplifier
From *Physics of Hi-Fi* by Symko. Reprinted by permission of the author.

In designing amplifiers electrical engineers usually employ feedback, which consists of returning some of the output to a system as input so as to modify the net output of the system. An audio amplifier employs negative feedback in trying to enlarge and reproduce a signal accurately while minimizing the effects of distortion produced by the imperfect electronic components. Negative feedback always changes the direction of the system. It is controlled in such a way to reduce the error between the desired output and actual output to zero. If the system were to continue in the same direction or manner as a result of the feedback, we call it positive feedback (Fig. 10.7).

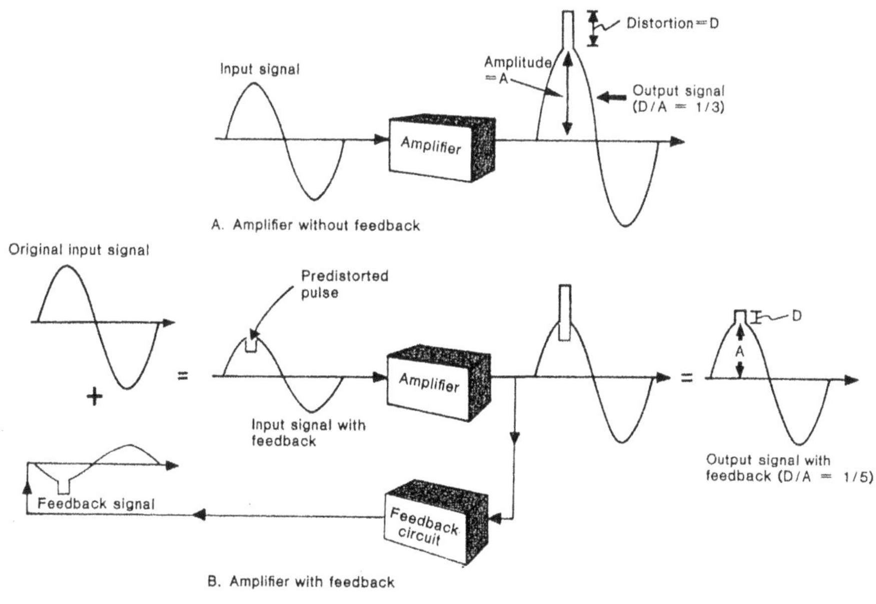

Figure 10.7 Negative feedback in an amplifier
From *Science of Hi-Fidelity* by Johnson, Walker, Cutnell. Reprinted by permission of the authors.

Amplifier Specifications

The important specifications for an amplifier are:

1. Total Harmonic Distortion
2. Intermodulation Distortion
3. Continuous Power Output
4. Dynamic Headroom
5. Frequency Response
6. Signal-to-Noise Ratio
7. Input Sensitivity
8. Maximum Input Signal
9. Stereo Separation
10. Damping Factor

Total Harmonic Distortion

Total harmonic distortion (THD) occurs when the output signal is distorted because the harmonics of the fundamental frequency produced by the amplifier internally are added by the amplifier to the fundamental. Musical instruments automatically produce harmonics along with the fundamental, but amplifiers should not. It may occur if the amplifier is limited and works only up to some frequency less than 20,000 Hz or because of non-linear gain in that it doesn't amplify all frequencies by a constant factor (Fig. 10.8). We can specify the THD by the following formula: THD = (voltage of harmonics/voltage of fundamental frequencies) x 100

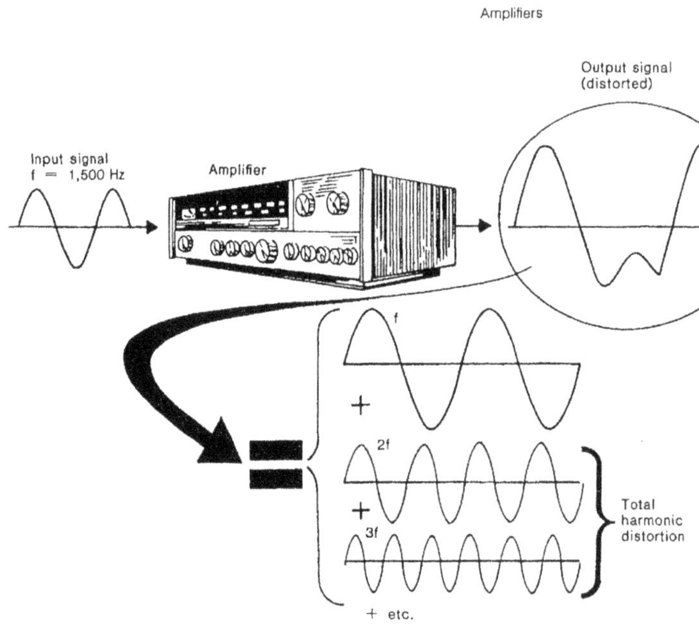

Figure 10.8 Amplifier producing THD

From *Science of Hi-Fidelity* by Johnson, Walker, Cutnell. Reprinted by permission of the authors.

THD increases significantly when the maximum rated power of the amplifier is exceeded. Normally we expect an amplifier to produce less than 0.5% THD to be considered as a hi-fidelity amplifier.

Intermodulation Distortion

Intermodulation distortion or IM distortion occurs when two or more signals are being amplified and the output contains frequencies made up of the sums and differences of the two signals. It also occurs because of the non-linear response of the amplifier.

IM = amplitude of new frequencies/amplitude of original frequencies x 100.

As with THD, it increases significantly as the maximum rated power of the amplifier is exceeded. It is also expected that the IM distortion be less than 0.5 % for it to be considered as a hi-fidelity device (Fig. 10.9).

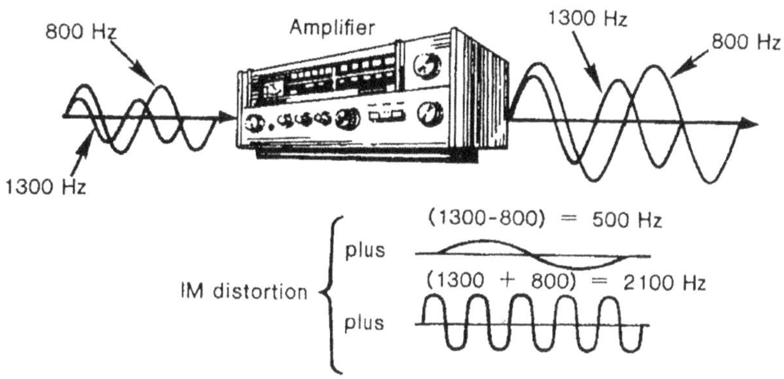

Figure 10.9 Amplifier producing IM distortion
From *Science of Hi-Fidelity* by Johnson, Walker, Cutnell. Reprinted by permission of the authors.

Continuous Average Power Output

The more powerful an amplifier is, the more expensive it is ordinarily. The continuous average power output of an amplifier is a measure of how much power the amplifier can deliver under a certain set of specifications. The power rating must specify the amount of power delivered on a **continuous basis per channel, both channels simultaneously driven.** If one channel was disconnected, the amplifier could deliver slightly more power to the speaker that remains connected. The power rating must also specify the **impedance of the speakers** that you plan to use. Smaller impedances allow more current to flow or more power to be delivered since $P = I^2R$. The rating must also include the **frequency range** along with the **THD** value at the rated power over this range. For example, the continuous average power output of an amplifier is given by

"100 watts (continuous) per channel into 8 ohm speakers, both channels simultaneously driven, from 20 Hz to 20 kHz at less than 0.5% THD at the rated power". [1]

Dynamic Headroom

The dynamic headroom of an amplifier specifies its ability to deliver more power than its rated continuous power for a short period of time on the order of 0.02 seconds. For example, if a 100 W amplifier can deliver 200 W for 0.02 seconds (200/100 would be the power ratio), the dynamic headroom measured in decibels would be 3 dB as shown in the table in Figure 10.13. A typical case would be when the amplifier has to handle the increase in loudness from a cymbal clash, where a short burst of power is needed. CD recordings require an amplifier with large dynamic headroom because of the high dynamic range of the recorded music on the CD.

Frequency Response

When we discussed speakers we saw that frequency response measures how different frequencies are amplified relative to each other over 20 - 20,000 Hz. Ideally, the frequency response of the amplifier should be flat or 20 - 20,000 Hz ± 0 dB. However, real amplifiers differ in their frequency responses with an overemphasis or underemphasis, measured in decibels, for a certain range of frequencies. For example,

Amplifier A: 20 - 20,000, + 03 dB, - 0.4 dB.

Amplifier B: 20 - 20,000 Hz, +0.1 dB, - 0.2dB.

If the frequency response does not include the ± decibel range, then what is given is the frequency range and is incomplete.

Signal to Noise Ratio (S/N)

The signal to noise measurement is the ratio of a test signal at 1 watt of power output, at 1000 Hz, to the A-weighted noise (which you will recall takes into account the response of the ear at low volume levels) and measured in decibels. We see that we can measure the sound intensity of the 1 kHz signal along with the noise produced by the amplifier (Fig.10.10).

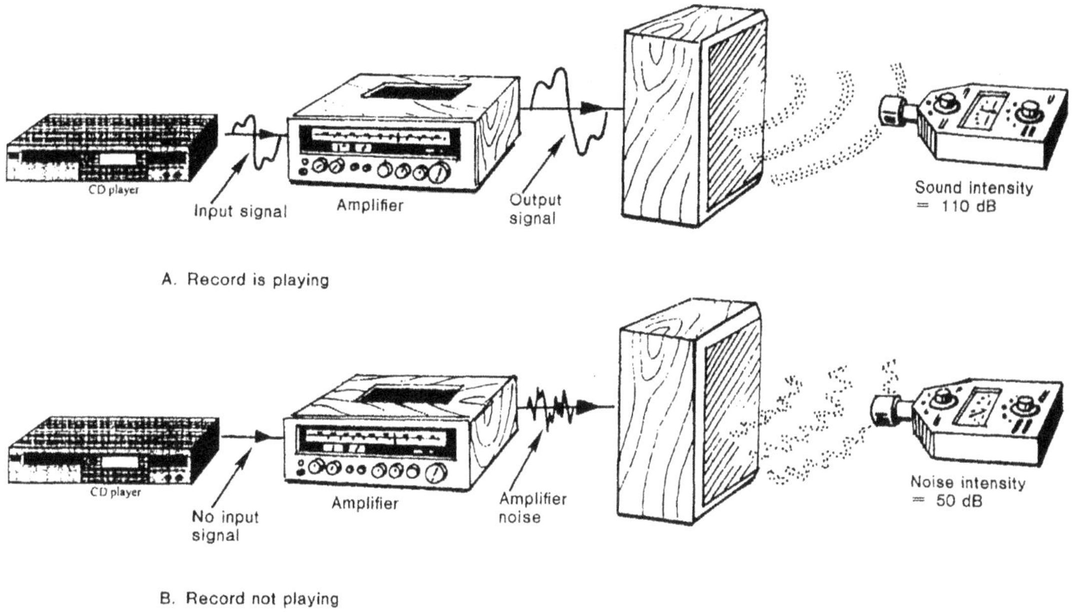

Figure 10.10 Measuring Signal to Noise Ratio
From *Science of Hi-Fidelity* by Johnson, Walker, Cutnell. Reprinted by permission of the authors.

If we now repeat the measurement at the output of the amplifier with no signal being sent through it, we get another measurement for the intensity using the A-weighted function on our sound level meter. We obtain the signal to noise ratio of the amplifier to be 110 dB – 50 dB = 60 dB (Fig. 10.13).

In other words, if one signal had an intensity of 110 dB, corresponding to an intensity level that is 10^{11} times the threshold of hearing, and the noise level was 50 dB, corresponding to a intensity level of 10^{5} times that of the threshold of hearing, the ratio of the signal intensity to the noise intensity would be 10^{6}or 60 dB (Fig. 10.13).

The signal to noise ratio is defined by the following formula. SNR = 10 $\log_{10}$ (I_s/I_N), where P_s is the intensity of the source and I_N is the intensity of the noise or 10 $\log_{10}(10^{11})/(10^{6})$ = 10 $\log_{10} 10^{6}$. Using a property of logs this can be written as SNR = 6(10) $\log_{10} 10$ = 60 dB, since, $\log_{10} 10 = 1$.

Input Sensitivity

Input sensitivity tells us the input voltage required to produce 1 watt output at 1 kHz with the volume control set at a maximum. If one amplifier had an input sensitivity of 0.5 mV and another of 0.05 mV, then the second amplifier could produce 1 W of output with a much smaller input than the first. Each amplifier input CD, Phono, Tuner, Tape, and Aux have different input sensitivities, with the Phono input being the most sensitive. The quality of the amplifier is not determined by its input sensitivity but it does provide us with a means of matching the source to the amplifier.

Maximum Input Signal

If the signal the amplifier receives is large to begin with, it tries to amplify it even more but because the carrying capacity of the amplifier has been exceeded, it clips (Fig. 10.11) the signal (shaves the top off). Clipping sometimes generates higher harmonics which will damage a tweeter. The largest voltage that the amplifier can accept as input without producing clipping with the volume set low is called the maximum input signal.

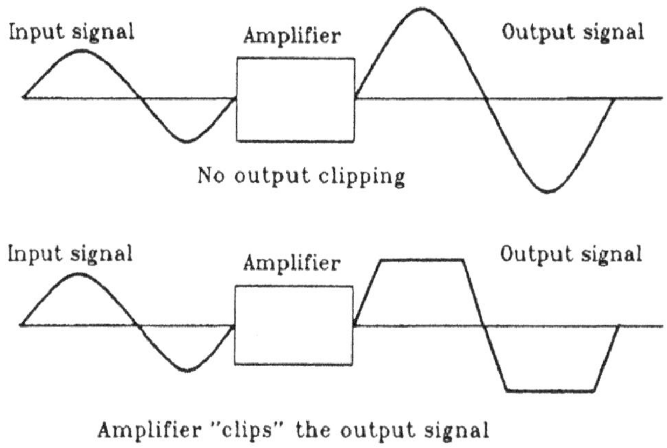

Figure 10.11 Clipping

From *Computer Music: Science and Technology of a New Art,* Stanionis, Berberich. Copyright © 1989 by Research Foundation of State University of New York. Reprinted by permission.

Stereo Separation

The ability of the amplifier to keep the two channels that it's amplifying separate is a measure of its stereo separation. There is always some leakage of the signal from one channel to the other because they are sharing the same power supply and the closeness of the wires carrying the signal from each channel. The stereo separation can be measured with a sound level meter by measuring the intensity of sound from each speaker with one channel from the source disconnected (Fig. 10.12). The stereo separation is the difference in decibels between the two readings.

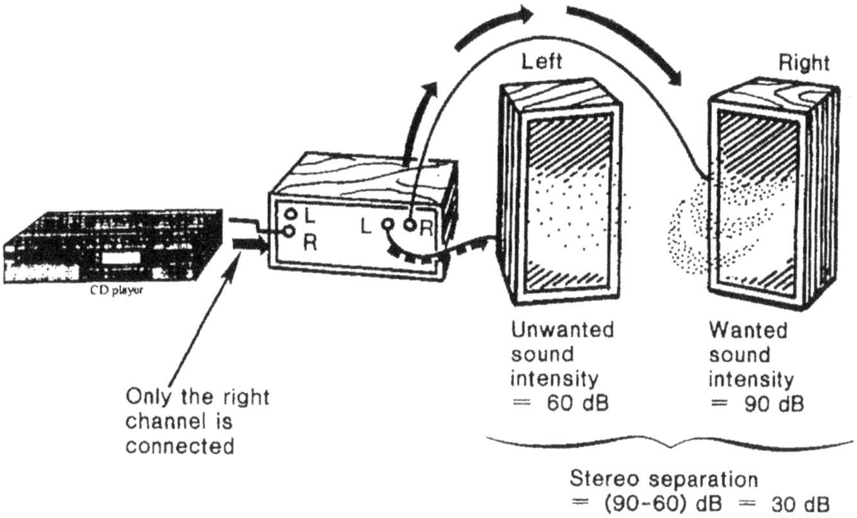

Figure 10.12 Measuring stereo separation

From *Science of Hi-Fidelity* by Johnson, Walker, Cutnell. Reprinted by permission of the authors.

Damping Factor

Damping factor is a specification that deals with the ability of the amplifier to electrically dampen the mechanical motion of vibration of the speaker after it receives a signal. The mechanical vibration of the speaker tends to blur the sound. Just like shock a absorber on a car dampens the motion of the car after it hits a bump, the amplifier's internal resistance plays a role in stopping the mechanical motion of the speaker. Damping factor may be computed by dividing the speaker's impedance by its internal resistance. Damping factors that fall between 20 and 100 are generally considered to be acceptable. The higher it is, the better.

Power Ratio or Intensity Ratio	Difference in Decibels
1.00	0 dB
1.26	1 dB
1.58	2 dB
2.00	3 dB
2.50	4 dB
3.16	5 dB
4.00	6 dB
5.00	7 dB
6.30	8 dB
7.90	9 dB
10.0	10 dB
20.0	13 dB
31.6	15 dB
50.0	17 dB
$100 = 10^2$	20 dB
$1{,}000 = 10^3$	30 dB
10^4	40 dB
10^5	50 dB
10^6 = 1 million	60 dB
10^7	70 dB
10^8	80 dB
10^9 = 1 billion	90 dB
10^{10}	100 dB

Figure 10.13 Power ratio or intensity ratio conversion to decibels and vice versa
From *Science of Hi-Fidelity* by Johnson, Walker, Cutnell. Reprinted by permission of the authors.

Review Questions

1. What are source pre-amps?
2. Why are source pre-amps needed?
3. What is the function of a system pre-amp?
4. What are the purposes of the tone controls?
5. How do the bass and treble controls operate?
6. How do the bass and treble controls differ from the low and high filters in their operation?
7. How does the loudness control work?
8. How does a power amplifier work?
9. What role does the transistor play in the operation of the power amplifier?
10. What is total harmonic distortion?
11. What is intermodulation distortion?
13. How is continuous average power output defined?
14. What is dynamic headroom?
15. What is meant by frequency response and how does it differ from frequency range?
16. Define signal-to-noise ratio.
17. What is input sensitivity?
18. Discuss maximum input signal.
19. Discuss stereo separation and describe how it is measured.
20. Define damping factor.

Multiple Choice Questions

1. The source pre-amp boosts the signal of the source to about _____ volts before passing it on to the system pre-amp.
a. 0.005
b. 0.05
c. 0.5
d. 5.0

2. The control center for the high fidelity system is the _____ .
a. source pre-amp
b. system pre-amp
c. power-amp
d. RIAA equalization-amp

3. To compensate for room décor when playing a hi-fidelity system one uses the ______.
a. source pre-amps
b. tone controls
c. cross over network
d. speaker dispersion

4. The high and low filters remove _____ music from the system than the bass and treble controls.
a. more
b. less
c. the same amount of
d. more mid-range

5. A ______ controls the amount of current that will be sent from the power supply by the power-amp to the speakers.
a. source voltage
b. variable thermistor
c. transistor
d. all the above

6. The device that regulates the amount of current the power amplifier will send to the speakers is a _____.
a. resistor
b. transistor
c. thermistor
d. capacitor

7. _____ is produced in an amplifier when multiples of the fundamental frequency (harmonics) are added to the fundamental.
a. THD
b. IM distortion
c. Dynamic headroom
d. Frequency response

8. The specifications for an amplifier states that the ______ is from 15 - 25,000 Hz, +2 dB, -1 dB.
a. frequency range
b. dynamic headroom
c. continuous power output
d. frequency response

9. Intermodulation distortion occurs when two waves passing through an amplifier produce not only enlarged versions of the original but also _______ of the originals.
a. multiples
b. sums and differences
c. harmonic
d. overtones

10. ______ is the specification that specifies the ability of an amplifier to deliver more power than its rated continuous power for a short period of time on the order of 0.02 seconds.
a. Overhead
b. Average power output
c. Continuous power output
d. Dynamic headroom

11. If the sound intensity coming from a speaker with music playing is 120 dB and is measured as 40 dB when no music is playing, we say the signal-to-noise ratio is_____.
a. 80 dB
b. 3 dB
c. 1/3dB
d. 160 dB

12. An amplifier control that compensates for the ears inability to hear low frequency and high frequency sounds at low volume levels is the _______ control.
a. loudness
b. volume
c. high filter
d. low filter

13. All amplifiers produce the sum and difference of the input frequencies. This is called _____.
a. total harmonic distortion
b. intermodulation distortion
c. frequency response
d. damping

14. If the dynamic headroom of an amplifier is 3 dB, how much extra power can it deliver if the output is rated at 200W? (See Figure 10.13.)
a. 66.6 W
b. 400 W
c. 600 W
d. 3 W

15. An amplifier has a stereo separation of 40 dB. If one channel is producing 80 dB, then the other channel is leaking ______ dB of sound.
a. 120 dB
b. 2 dB
c. 40 dB
d. 80 dB

16. The distortion level in the amplifier must be less than _____ to meet acceptable hi-fidelity standards.
a. 0.5%
b. 0.05%
c. 0.005%
d. 5%

17. It is important to minimize the distortion level of each electronic component in a hi-fidelity system because the distortion level is ______.
a. sensitive to the tone controls
b. cumulative
c. affects the THD
d. a strain on the speakers

18. If a 600 Hz signal is sent to an amplifier and the output consists of a 600, 1200, 1800, …, Hz signals, the amplifier is said to be producing _____.
a. IM distortion
b. THD
c. crosstalk
d. harmonic

19. If an amplifier amplifies all frequencies by the same amount, it is said to have a ______ response.
a. high
b. low
c flat
d. mellow

20. When we increase the output power of an amplifier beyond its rated value, we ______ distortion level.
a. decrease
b. modify
c. modulate
d. increase

21. The amplifier specification that relates the ability of the amplifier to control the vibrations of a speaker is called _____.
a. THD
b. IM distortion
c. input sensitivity
d. damping factor

22. Which of the following is the best frequency response for an amplifier?
a. 20 - 20,000, +0.1 dB, -0.2 dB
b. 15 - 20,000, +2 dB, -0.1 dB
c. 15 - 15,000, +0.1 dB, -0.2 dB
d. 10 - 20,000, ± 2.5 dB

23. The audio component that produces large currents is the _____.
a. preamplifier
b. volume control
c. power amplifier
d. loudness control

24. What is the stereo separation for an amplifier delivering 100 W to a speaker if the leakage from the other speaker is 0.1 W? (See Figure 10.13.)
a. 10 dB
b. 20 dB
c. 30 dB
d. 40 dB

25. Which of the following is a valid amplifier specification?
a. dispersion
b. frequency range
c. S/N value
d. amplification factor

26. If an amplifier increases the level of a sound by 20 dB, it appears to ____ times as loud.
a. ½
b. 2
c. 4
d. 20

27. If the input signal to amplifier produces an output signal of 110 dB with the source connected and 50 dB with the source disconnected, then the signal-to-noise ratio is _____ dB.
a. 160
b. 60
c. 3.3
d. 5/11

28. If an unwanted sound signal generates a sound intensity of 60 dB in the left speaker and 90 a dB wanted signal in the right one, the stereo separation is _____ dB.
a. 150
b. 30
c. 1.5
d. 2/3

29. A difference of 13 dB corresponds to an intensity ratio of _____.
a. 10
b. 13
c. 100
d. 20

30. If the impedance of a speaker connected to an amplifier is lowered, its continuous average power output _____.
a. increases
b. decreases
c. remains the same
d. doubles

Bibliography and Web Resources

[1]Johnson, Kenneth, and Willard Walker and John Cutnell. *The Science of Hi-Fidelity.* Kendall-Hunt Publishing Company, Dubuque, Iowa, 1995.

Symko, Orest. *Physics of Hi-Fi.* Kendall-Hunt Publishing Company, Dubuque, Iowa, 1995.

Stanionis, Victor and Hugh Berberich, *Computer Music: Science and Technology of a New Art.* Research Foundation of State University of New York, Stony Brook, NY 1989.

Chapter 11

Electromagnetism

"This velocity is so nearly that of light, that it seems we have strong reason to conclude that light itself (including radiant heat, and other radiations if any) is an electromagnetic disturbance in the form of waves propagated through the electromagnetic field according to electromagnetic laws."

James Clerk Maxwell, 1865
Scottish Physicist

Around 600 B.C. Thales of Miletus, often called the Father of Greek Science, discovered that lodestone (magnetite) attracts iron and other lodestones. He also found that iron itself became magnetized after touching magnetite, which was found in Magnesia along the Macedonian border. During the Han Dynasty (2 B.C. to 2 A. D.) in China, it was found that when a spoon made of lodestone was placed on a cast bronze plate (Fig. 11.1), its handle would point in a southern direction, a "south pointer". By the seventh-eighth century A.D. the Chinese had learned to magnetize iron needles by stroking them with magnetite and by the eleventh century A.D. had suspended them in water to form the first compasses that were used for navigation.

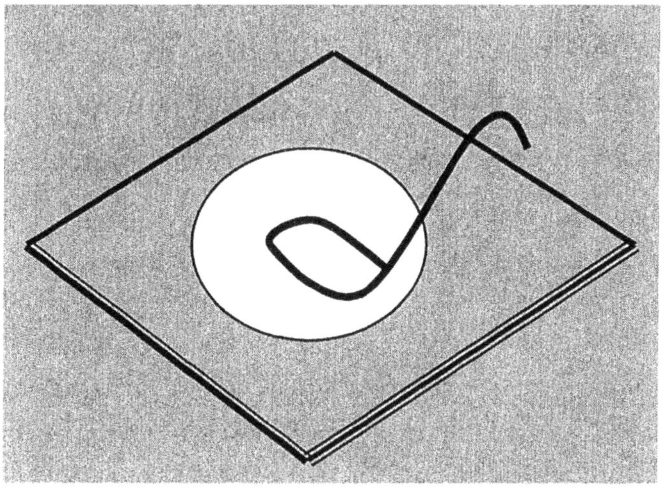

Figure 11.1 Lodestone spoon on cast bronze plate

Goals

After studying this chapter, you should be able to:

- Define a magnetic field.
- Know the relationship between the north and south poles on a bar magnet.
- Understand that electric currents produce magnetic fields.
- Understand how a bar magnet can have a permanent magnetic field.
- Understand why the earth has a magnetic field.
- Know some applications of electromagnetism in audio systems.
- Know what electromagnetic induction is.
- Know that changing magnetic fields generate currents.
- Understand the construction, operation, and purpose of a transformer.

Magnetic Fields

William Gilbert, a physician in the court of Queen Elizabeth I of England, presented the idea that the earth itself was a magnet in his published study of magnetism *De Magnete* in the year 1600. He was able to explain why the compass needle pointed in a north-south direction. He is said to have introduced the word "electricity", which is derived from the Greek word for amber ("elektron"). He is sometimes referred to as the Father of the Science of Electricity and Magnetism.

We all know that a small compass needle that is free to rotate around a vertical axis will line up in the north-south direction. In any region of space where the compass needle experiences a force causing it to face in a particular direction we know a magnetic field exists. We can use a compass to detect the presence of magnetic fields and explore them.

If we were to take a bar magnet (Fig. 11.2), and sprinkle some iron filings around it, we would see them form a pattern instead of being randomly distributed.

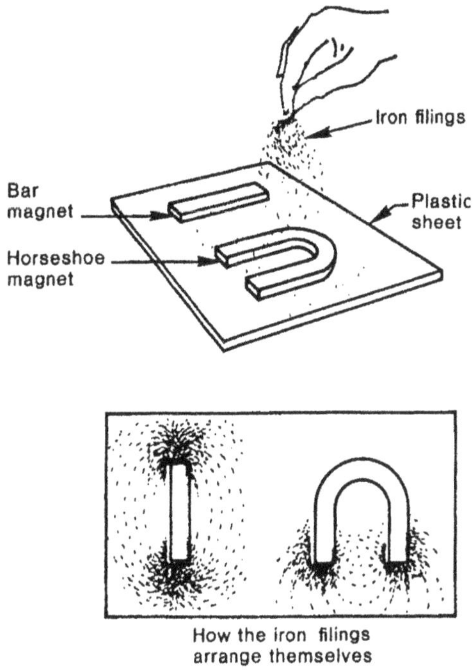

Figure 11.2 Iron filings around a bar magnet

From *Science of Hi-Fidelity* by Johnson, Walker, Cutnell. Reprinted by permission of the authors.

One end of the bar magnet is labeled N, for north pole and the other S, south pole. If it were made into a compass, the end pointing toward the geographical north pole of the earth would be labeled N. When we sketch the magnetic field of the bar magnet, we draw curved lines starting out on the north pole and following the path of the iron filings, or the exploring compass, to the south pole (Fig. 11.3).

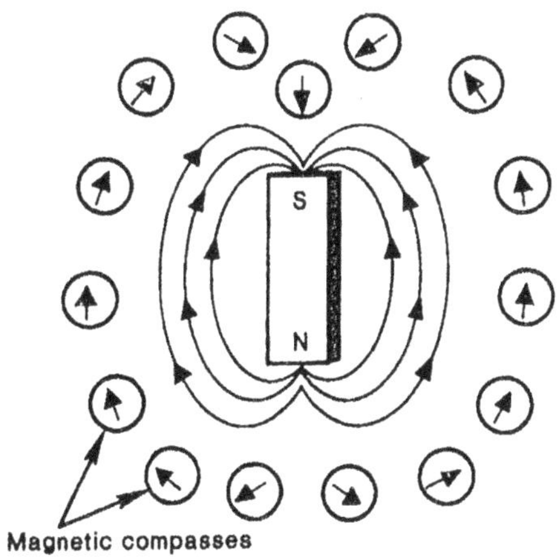

Figure 11.3 Bar magnet and compass headings

From *Science of Hi-Fidelity* by Johnson, Walker, Cutnell. Reprinted by permission of the authors.

Historically, some scientists believed that north and south poles were analogous to positive and negative charge and could be separated. However, experimentation revealed that when a bar magnet is cut into parts we always wind up with each piece having a north and south pole (Fig. 11.4).

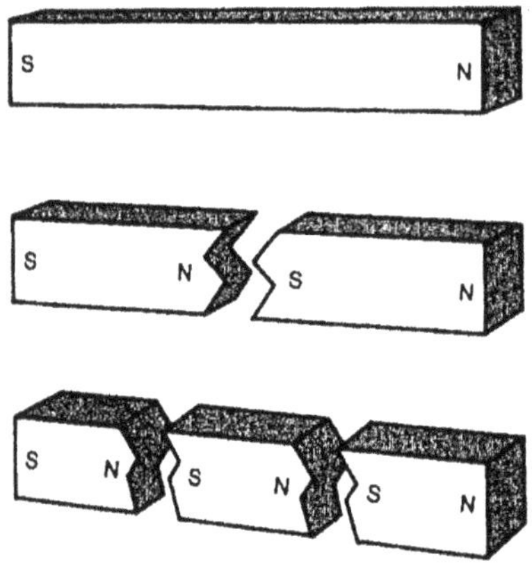

Figure 11.4 Bar magnet cut into several pieces
From *Science of Hi-Fidelity* by Johnson, Walker, Cutnell. Reprinted by permission of the authors.

Another property of bar magnets is that when two north poles or two south poles were in close proximity, they repelled one another. On the other hand, a north pole and south pole in close proximity seemed to attract one another. In other words, like poles repel and unlike poles attract (Fig. 11.5).

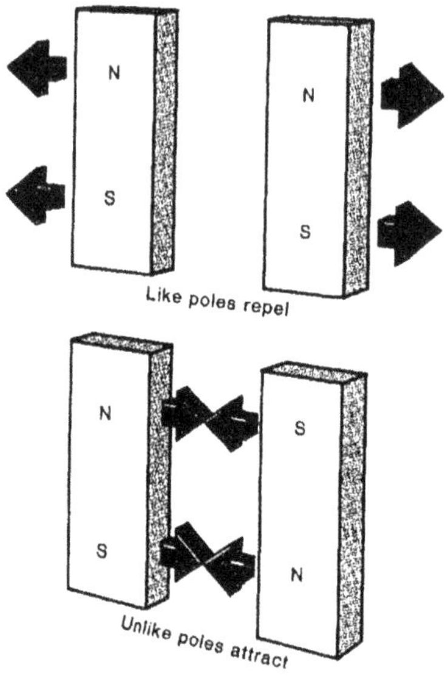

Figure 11.5 Repulsion and attraction of bar magnets
From *Science of Hi-Fidelity* by Johnson, Walker, Cutnell. Reprinted by permission of the authors.

In time it was discovered that magnetite and magnetized iron were not the only sources of magnetic field. It was found that wires carrying electric currents produce magnetic fields around them. Consider a simple experiment of running a stiff piece of wire with an electric current running through it through a piece of cardboard and sprinkling iron filings around it. We would find the iron filings lining up to form a pattern thus revealing the presence of a magnetic field Fig. 11.6).

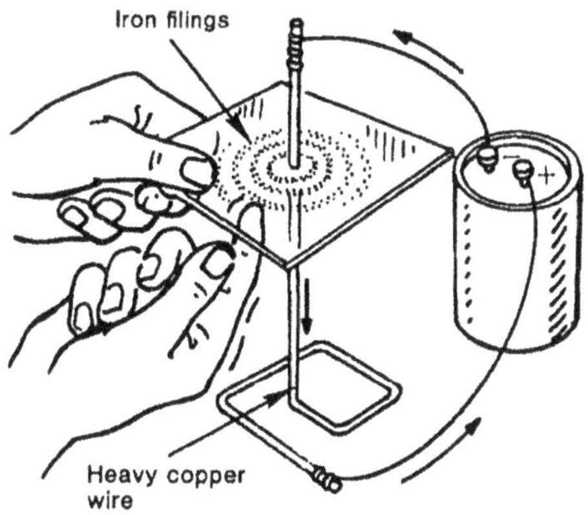

Figure 11.6 Electric current generates a magnetic field
From *Science of Hi-Fidelity* by Johnson, Walker, Cutnell. Reprinted by permission of the authors.

The direction of the magnetic field reverses when the current is reversed (Fig. 11.7). Note that the direction of the magnetic field around the wire may be obtained by pointing the thumb of your left hand in the direction of the current. Your fingers on your left hand when wrapped around the wire will define the direction of the magnetic field.

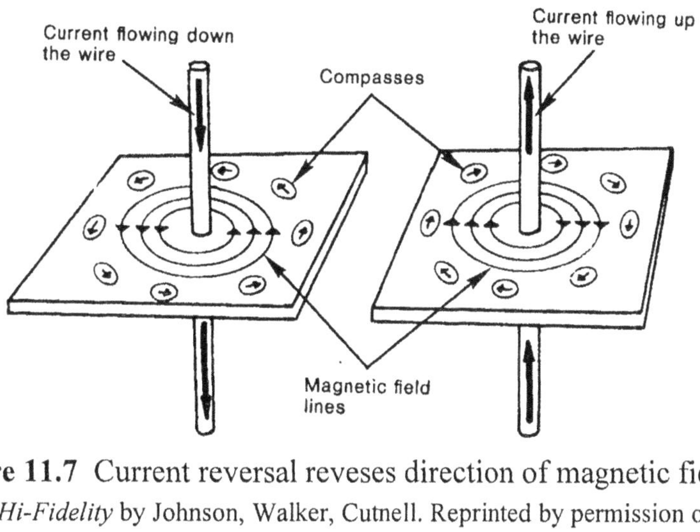

Figure 11.7 Current reversal reveses direction of magnetic field
From *Science of Hi-Fidelity* by Johnson, Walker, Cutnell. Reprinted by permission of the authors.

We can now explain why certain iron, cobalt, and nickel compounds have permanent magnetic fields. **The magnetic fields are produced by electric currents**. Where are the electric currents in a bar magnet? Remember the bar magnet is made up of atoms that have negatively charged electrons in orbit around a positive nucleus. The motions of the electrons constitute microscopic electric currents that give rise to microscopic magnetic fields. The superpositions of the microscopic magnetic fields in most substances cancel each other out. In iron, cobalt and nickel compounds, bunches of atoms are contained in separate domains, or regions. Within each domain, the atoms are aligned and their magnetic fields add up. In an unmagnetized piece of iron the domains are randomly distributed and there is no net magnetic field (Fig. 11.8).

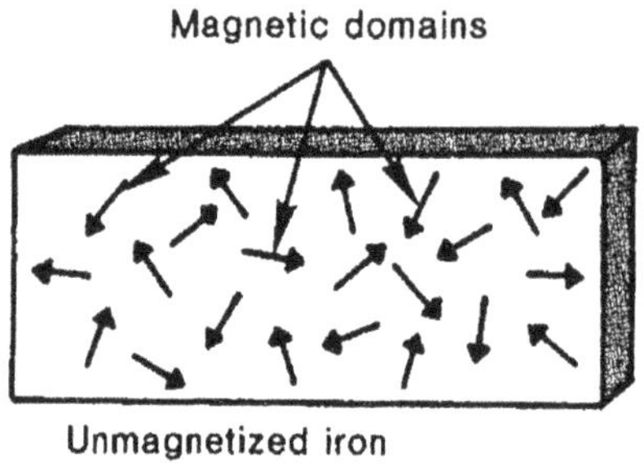

Figure 11.8 Magnetic domains

From *Science of Hi-Fidelity* by Johnson, Walker, Cutnell. Reprinted by permission of the authors.

However, if we apply an external magnetic field to the material, we can align the domains, giving rise to a strong external magnetic field (Fig. 11.9).

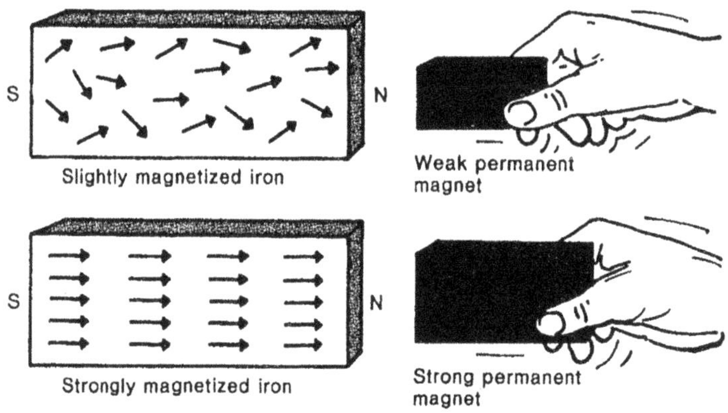

Figure 11.9 Magnetic domains align under action of external magnetic field

From *Science of Hi-Fidelity* by Johnson, Walker, Cutnell. Reprinted by permission of the authors.

Even when the external magnet is removed, the bar will retain its alignment because of friction between the walls of the domains. To demagnetize the bar we would have to overcome the friction by heating the bar or hammering it to cause the domains to orient themselves in a more random manner. It's the electronic structure of the iron, cobalt, and nickel atoms which allow them to group in domains and retain their magnetic fields while other atoms do not have this ability.

The Earth

We have known since Gilbert's time that the earth has a magnetic field. It behaves as if it contained a giant bar magnet in its interior with its north pole at the geographical south pole of the earth and its south pole at the geographical north pole (Fig. 11.10).

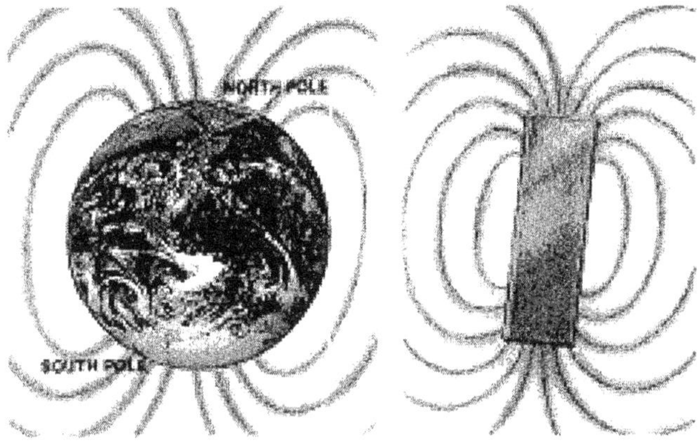

Figure 11.10 Magnetic field of the earth
http://science.nasa.gov/headlines/y2001/ast15feb_1.htm

The size of the Earth, which is about 12,750 kilometers (km), or 8,000 mi. in diameter, was known to the ancient Greeks, through Eratosthenes' (276-195 B.C.) work in which he measured the circumference of the earth using simple trigonometry. It was not until the turn of the twentieth century that scientists determined that our planet is made up of three main layers: crust, mantle, and core. They were able to determine the structure of the earth by observing earthquakes. Remember that longitudinal waves (P) can travel through a liquid but transverse (S) waves cannot. So by examining "shadow zones" emanating from the focus (origin) of an earthquake and found on the opposite side of the earth, it was concluded that there were certain areas where S waves could not travel and P waves were slowed down. The K waves shown in Figure 11.11 are S waves that have been converted to P waves upon entering the core.

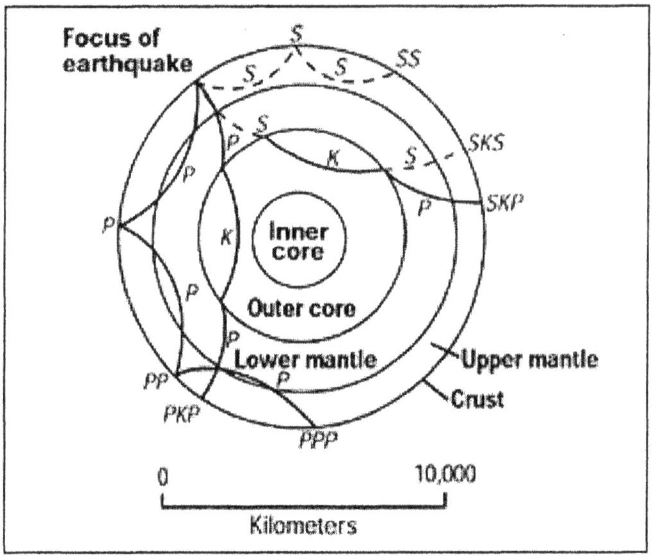

Figure 11.11 Earthquake waves
http://pubs.usgs.gov/gip/interior/

The crust of the earth is thought to have formed as the earth originally cooled from a molten state, sort of like the crust on the surface of a slowly cooling pudding. The mantle, below the crust is a dense layer of rock which becomes more plastic at the higher pressures and temperatures as we descend toward the core. The core lies at the center of the earth and is made up of two parts: the liquid outer core and solid inner core. The outer core is made up primarily of liquid iron and nickel while the inner core is largely solid iron and nickel (Fig. 11.12).

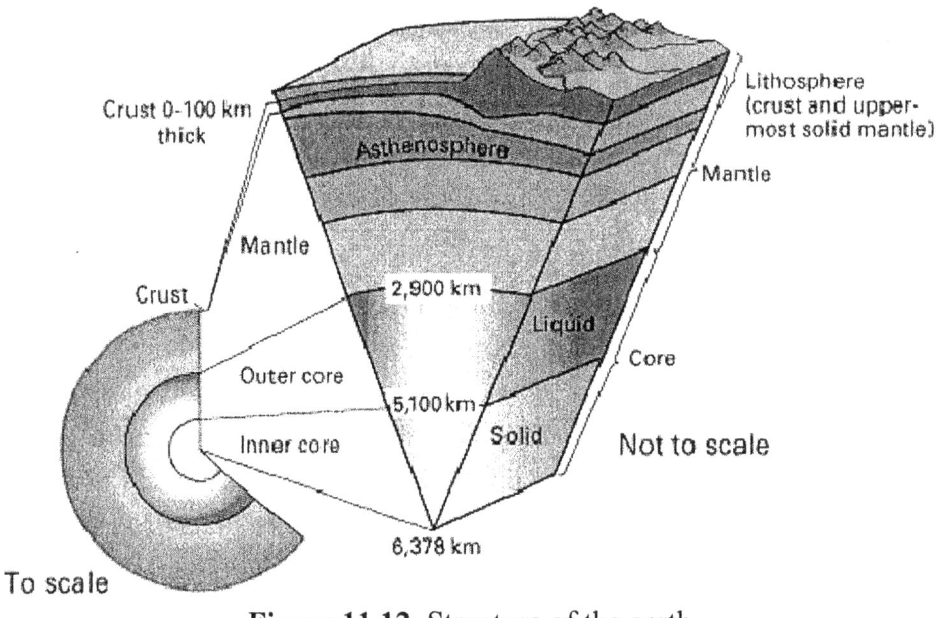

Figure 11.12 Structure of the earth
http://pubs.usgs.gov/publications/text/inside.html

As the earth spins on its axis the liquid core is also spinning and swirling at temperatures in the range of 5000-6000 °C. Because of the high temperatures there are many ions, charged atoms, moving in this swirling current. The motion of these charged particles is essentially an electric current creates the earth's magnetic field. Molten material that is emitted from deep inside the earth reaches the surface at the Mid-Atlantic Ridge and as the iron particles contained in these molten layers cool they align themselves with the earth's magnetic field (Fig. 11.13).

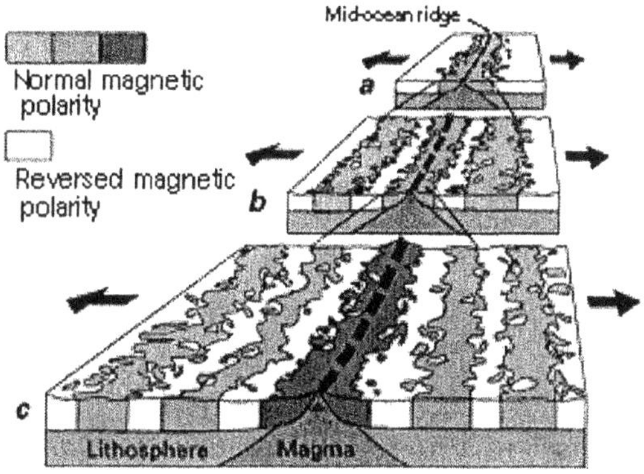

Figure 11.13 Mid-Atlantic ridge and magnetized strips
http://www.nasa.gov/vision/earth/lookingatearth/29dec_magneticfield.html

As these materials pile up at this crack in the earth's crust, its weight seems to generate a force that moves the earth's crust in opposite directions away from the crack at a slow rate. Suppose you stand with one foot in one rowboat in water and the other in another rowboat. What happens?

Observations on the material that has moved away from the crack over thousands of years reveals bands of magnetism in which the particles are aligned in a north-south direction and then later in a south-north direction. This seems to imply that the earth's magnetic field periodically reverses. It takes about 7000 years for the earth to complete a flip-flop and the flip-flops occur about every 10,000 years, with some intervals on the order of millions of years. The mechanism for reversal is not fully understood but might have to do with the wobble in the earth's rotation which seems to cause the direction of the current inside the earth to reverse periodically, allowing for the switch in directions of the earth's magnetic field (Fig. 11.14).

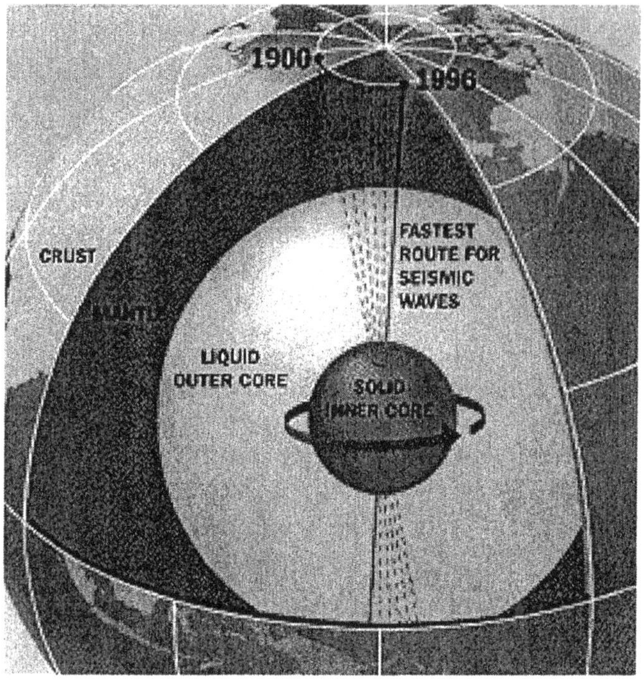

Figure 11.14 Wobble of earth
http://www.nasa.gov/vision/earth/lookingatearth/29dec_magneticfield.html

Contrary to popular opinion, during reversals the earth's magnetic field does not totally vanish but, rather, gets more complicated (Fig 11.15). The poles seem to wander.

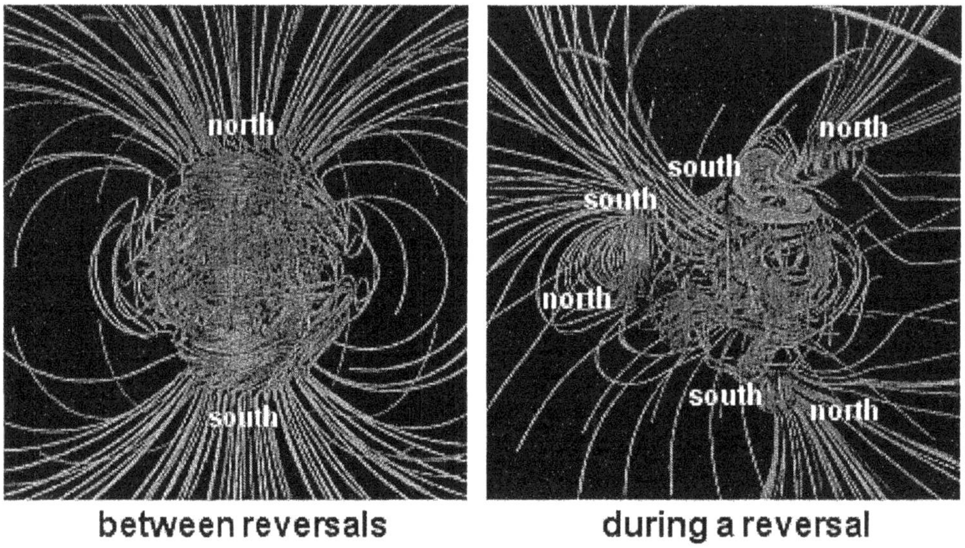

Figure 11.15 Reversal of earth's magnetic field
http://www.nasa.gov/vision/earth/lookingatearth/29dec_magneticfield.html

So what? Well the sun emits a wind of electrically charged particles which can cause damage to exposed human tissue. The magnetosphere, the outer magnetic field of the earth, acts as a force field to deflect these particles and other space radiation into outer space. Those particles that manage to get by the shield will usually enter the earth's atmosphere at high latitudes in the northern and southern hemispheres. This creates the aurora borealis (northern lights) in the northern hemisphere and the aurora australis (southern lights) in the southern hemisphere. The lights are produced by the ionization and excitation of nitrogen and oxygen air molecules by the electrically charged solar particles. As they lose energy and move to a more stable configuration, they generate light (Fig. 11.16).

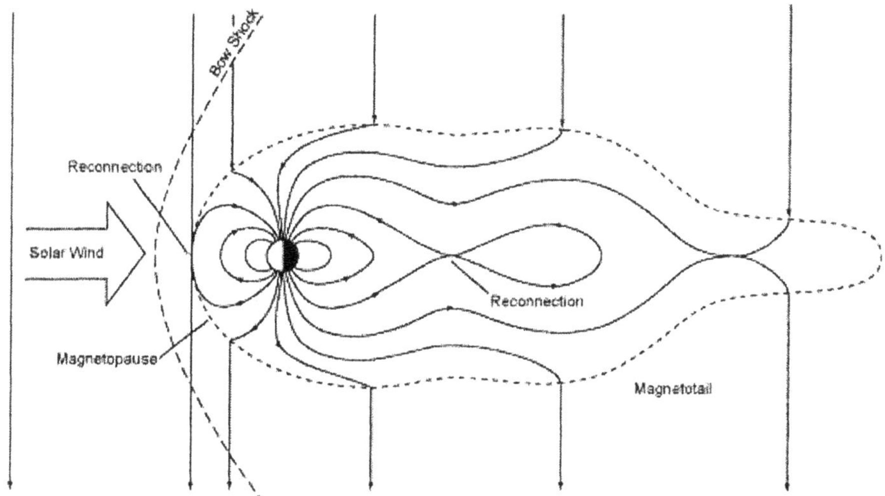

Figure 11.16 Earth's magnetic force shield and solar wind
http://geomag.usgs.gov/intro.html

Electromagnets

Since electric currents generate a magnetic field around a wire, we can duplicate the magnetic field of a bar magnet using a coil of wire, creating what is known as an electromagnet (Fig. 11.17).

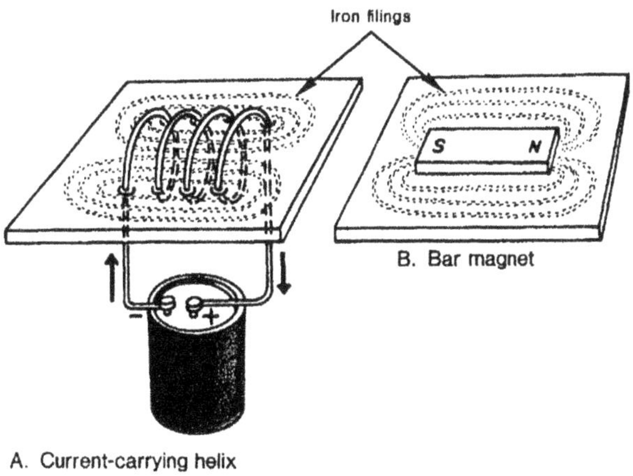

Figure 11.17 Magnetic field of a current coil and a bar magnet
From *Science of Hi-Fidelity* by Johnson, Walker, Cutnell. Reprinted by permission of the authors.

To determine which side of the coil is the north pole, the so-called "left hand rule" may be used. If you wrap the fingers of your left hand around the coil, with your fingers pointing in the direction the current is running through the coils, your thumb will point in the direction of the north pole of the coil (Fig. 11.18).

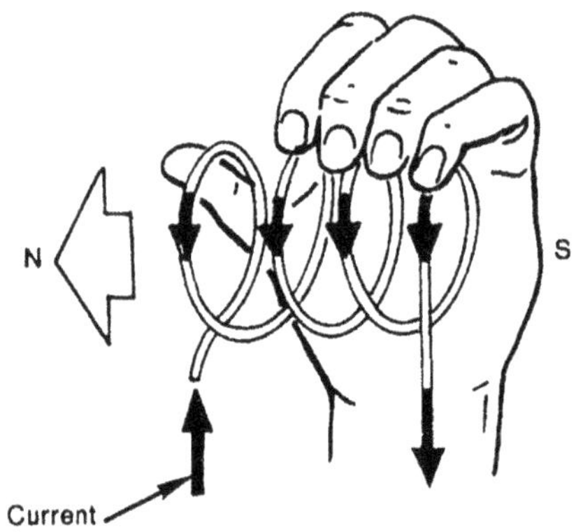

Figure 11.18 Direction of north pole for a current coil
From *Science of Hi-Fidelity* by Johnson, Walker, Cutnell. Reprinted by permission of the authors.

Experiments demonstrate that the strength of an electromagnet can be increased by increasing the amount of current flowing through the coil, increasing the number of turns of wire on the coil, and by adding a ferromagnetic core element. Placing a piece of iron or other ferromagnetic material inside the coil, aligns the domains in the iron causing its microscopic magnetic fields to add up and enhance the magnetic field produced by the current in the coil.

Two immediate applications of electromagnetism are evident. We saw that one of the main components of the cone speaker was the voice coil which would interact with the permanent magnet. Now it is easy to understand how the varying current through the voice coil generates a magnetic field around the coil proportional to the current passing through it. This varying magnetic field interacts with that of the permanent magnet generating attractive and repulsive interactions. The repulsive and attractive forces on the voice coil cause the diaphragm to move back and forth, generating the pressure wave we hear as sound (Fig. 11.19).

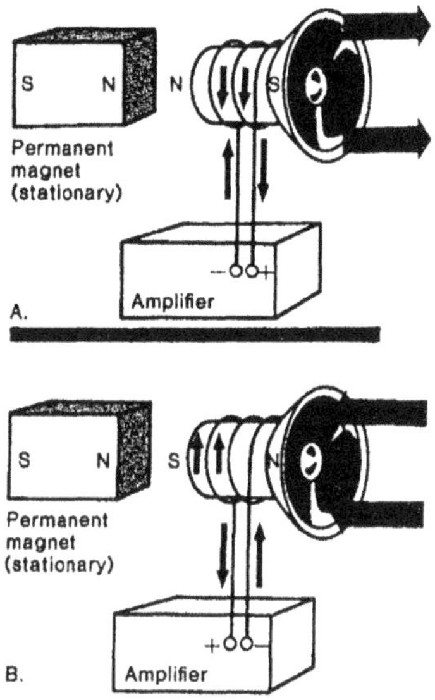

Fig**ure 11.19** Interaction of magnetic field of a speaker voice coil and permanents magnet
From *Science of Hi-Fidelity* by Johnson, Walker, Cutnell. Reprinted by permission of the authors.

Another application that comes to mind is the recording of information on magnetic tape by the record head on a magnetic audio tape machine. Magnetic tape is essentially is a sandwich of two pieces of plastic with iron particles in between. If we drag it past a circular electromagnet with a small gap, thus creating a north and south pole across the gap, a varying current in its coil creates a varying magnetic field at its poles. This causes the particles in the tape to align in a pattern according to the varying magnetic field. We essentially record information on the tape stored in the magnetic alignment of the particles (Fig. 11.20).

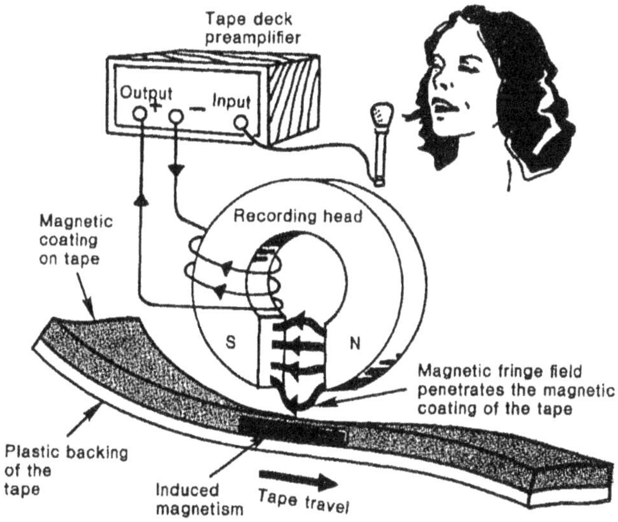

Figure 11.20 Recording on magnetic tape using a changing magnetic field
From *Science of Hi-Fidelity* by Johnson, Walker, Cutnell. Reprinted by permission of the authors.

Electromagnetic Induction

So far, we have considered two major ideas in electromagnetism:

1. Electric currents produce magnetic fields.
2. Magnetic fields can exert forces on moving electric charges and, therefore, on current carrying wires.

Now, the third major idea we want to consider is the generation of currents by changing magnetic fields. Michael Faraday is given credit for the discovery in 1831 of this phenomenon known as electromagnetic induction. Joseph Henry independently discovered this phenomenon in 1831, before Faraday. However, Faraday published his results before Henry.

Faraday was born to a poor family in South London. He had only a rudimentary education, learning the fundamentals of reading, writing and arithmetic. Working in a bookbinding shop as an apprentice, he read articles on electricity which fascinated him. He wanted to become a scientist, which seemed impossible until he sent a copy of his notes, recorded at a free chemical lecture series he had attended, to Sir Humphrey Davy at the Royal Institution of Great Britain in London asking for a job as lab assistant. At first there were no openings, but one of the assistants was dismissed and Faraday was hired. His lack of mathematical background and the fact he was self-trained led him to visualize scientific phenomena and to create the concept of field to describe electric and magnetic effects. He created diagrams representing electric and magnetic fields.

If we have a stationery bar magnet in front of a coil of wire with a voltmeter attached to the ends of the coil, the meter reads a zero voltage. As soon as we move the magnet towards the coil, the meter pointers swings in one direction and as we move it away from the coil it swings in the opposite direction. Faraday found, as illustrated in this simple experiment, that whenever there was relative motion between the coil and magnet, a voltage was induced which caused a current to flow (Fig. 11.21).

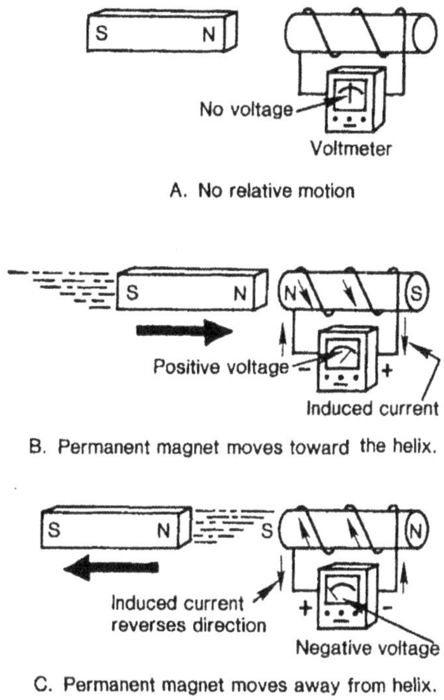

Figure 11.21 Moving magnet induces a current in a coil

From *Science of Hi-Fidelity* by Johnson, Walker, Cutnell. Reprinted by permission of the authors.

Experimentation revealed that the current became larger if a stronger magnet was used or if the magnet was moved slower or faster. The number of turns on the coil also played a role in how large a current would be induced. Emil Lenz was able to give a definite direction to the induced current when the field was changing. Lenz's Law states that the induced current will be in a direction to oppose the change in the applied magnetic field.

So when we move the magnet towards the center of the loop, the induced current sets up its magnetic field opposing the increase of magnetic flux lines through the center of the loop. It tries to keep them the same as they were when the magnet was stationery (Fig.11.22). Remember, your left thumb will point in the direction of the induced current when the fingers on your left hand wrap around the wire in opposition to the increasing magnetic flux lines.

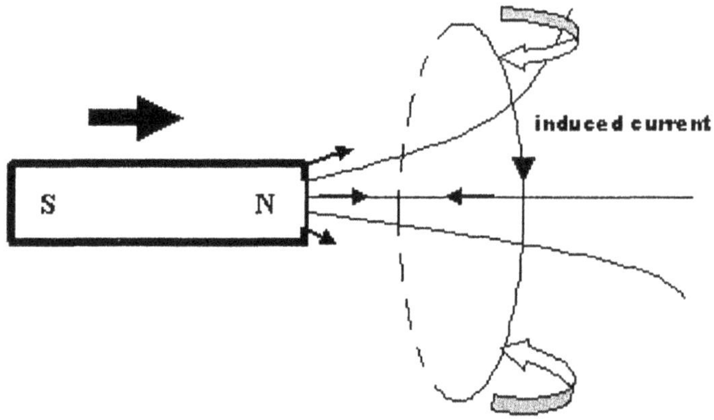

Figure 11.22 Approaching magnet and magnetic field causes induced current to add its field

When we move the magnet away from the center of the loop, the induced current now flows in such a direction so as to keep the density of magnetic flux lines and the magnitude of the field through the loop the same. It adds its flux lines to the receding flux lines of the bar magnet (Fig. 11.23). Your left thumb again indicates the direction of the induced current.

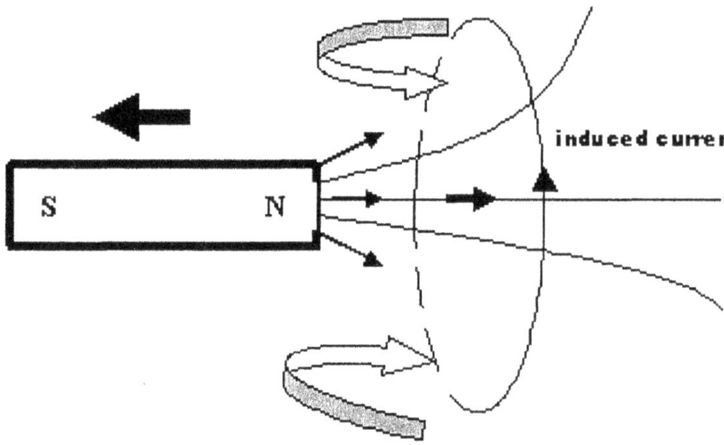

Figure 11.23 Retreating magnet and magnetic field causes induced current to add its field

We can now have some understanding of how the magnetic phono cartridge works. In the moving magnet phono cartridge the stylus is physically connected to a magnet. As the stylus moves in the grooves of a vinyl record, the motion moves the magnet in a coil of wire inducing an AC current in the coil. This varying current is sent on to the preamplifier (Fig. 11.24).

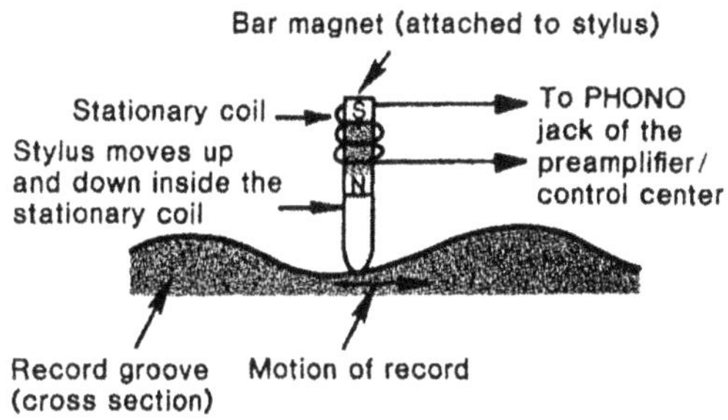

Figure 11.24 Magnetic cartridge moving in groove of record induces a current in a coil
From *Science of Hi-Fidelity* by Johnson, Walker, Cutnell. Reprinted by permission of the authors.

In a moving magnet microphone the process is basically the same. As a sound wave strikes the diaphragm of the microphone it moves it back and forth. Attached to the back of the diaphragm is a magnet which is forced to move back and forth through a coil of wire, inducing an AC current that is sent on to the preamplifier. This is called a moving magnet (dynamic) microphone (Fig. 11.25).

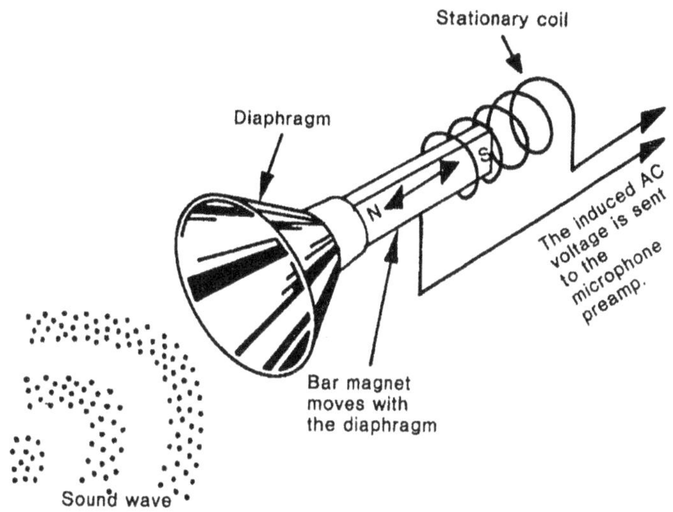

Figure 11.25 Moving magnet in a microphone induces a current in a coil
From *Science of Hi-Fidelity* by Johnson, Walker, Cutnell. Reprinted by permission of the authors.

We could also have attached the coil to the diaphragm and have it move back and forth on a stationery magnet. Then we would have a moving coil dynamic microphone.

Suppose we were to drag a piece of magnetized tape by the slit in the cylindrical head of a tape recorder. As the magnetic field embedded in the tape passes the slit in the iron core of the coil, the changing magnetic field will induce a current to flow in the coil. This head is called the playback head (Fig. 11.26).

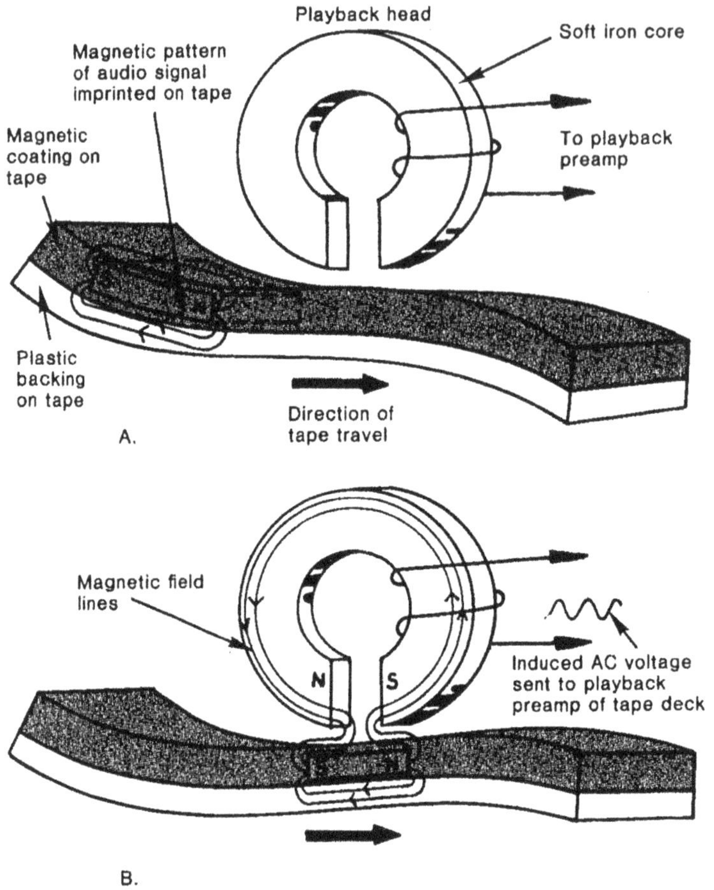

Figure 11.26 Magnetic field on magnetic tape induces a current in a coil as it moves past gap
From *Science of Hi-Fidelity* by Johnson, Walker, Cutnell. Reprinted by permission of the authors.

During World War II, there was great concern that the enemy would attempt to destroy shipping in New York harbor using submarines. So a large loop of wire was laid across the entrance to the harbor with electric current running through it. The wire set up a magnetic field which was monitored by the armed services and when a submarine would try to sneak into the harbor it would cause a change in the steady-state magnetic field, signaling the Coastguard to send out launches with depth charges to destroy the submarine. Over time the submarines tried to sneak into the harbor by hiding under large freighters entering the harbor. However. the Coastguard knew the change in magnetic field that should be produced by each ship and could detect the extra magnetic disturbance caused by a submarine.

Airport security metal detectors are now multi-zone, which means they have multiple transmit and receive coils. It is similar to having several metal detectors at different heights. The airport detectors use a pulse induction technique in which several hundred pulses per second are sent. Detectors measure the time between pulses and if it changes, sets off an alarm. The presence of metal objects disturbs the magnetic field enough to cause the timing change.

Each pulse generates a magnetic field which collapses in a few microseconds causing a back current, or self-induced current, to flow in the coil. The back current or reflected pulse only last for about 30 microseconds before the process repeats. The number of pulses per second can range from a couple dozen to over a thousand per second depending on the manufacturer. If the person is carrying a metal object, the initial pulse induces a magnetic field in the opposite direction. As the initial pulse's magnetic field collapses giving rise to the reflected pulse, the magnetic field of the metal object causes the reflected pulse to last longer. A sampling circuit in the detector compares the length of time the reflected pulse lasts to its expected value and set off an audible alarm if it is longer than normal (Fig. 11.27).

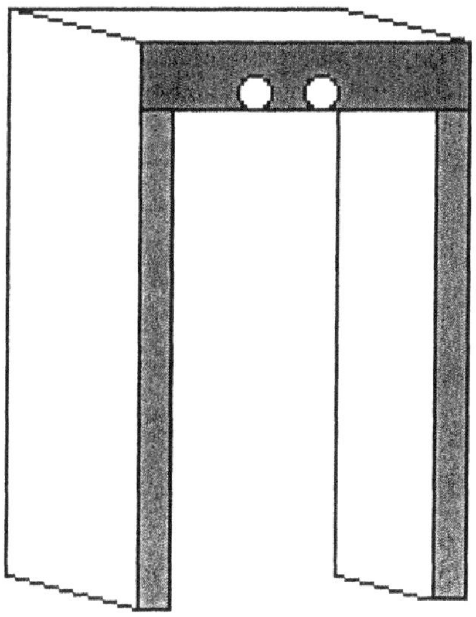

Figure 11.27 Airport metal detector

Transformer

Finally, the most important application of electromagnetic induction lies in the operation of the transformer. This device makes possible the transmission of electric energy over long distances and allows us to change voltages in AC circuits rather easily. A transformer is made up of three main parts: an iron core, a set of primary turns of wire or windings, and a set of secondary turns of wire or a secondary (Fig. 11.28).

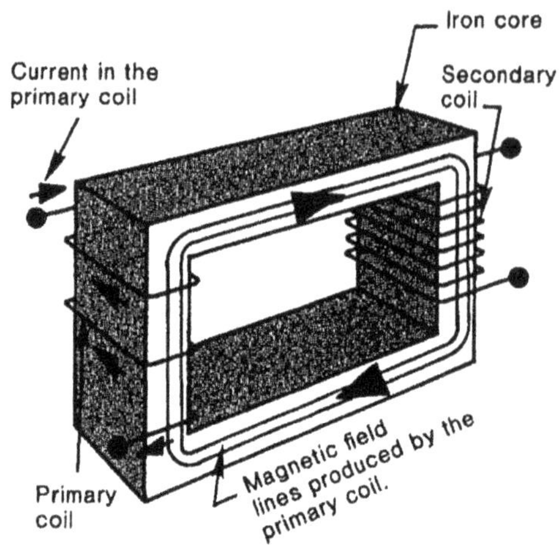

Figure 11.28 Transformer
From *Science of Hi-Fidelity* by Johnson, Walker, Cutnell. Reprinted by permission of the authors.

As current flows through the primary coil it sets up its magnetic field whose lines flow around the magnetic core through the center of the secondary windings. Faraday's law tells us that the changing magnetic field through the center of this coil will induce a voltage in the coil.

According to Faraday's Law of electromagnetic induction the voltage in the primary is proportional to the rate of change of magnetic flux, ϕ, flowing through the coil and also depends on the number of turns, N.

$$V_P = -N_P \Delta\phi/\Delta t.$$

And for the secondary coil:

$$V_S = -N_S \Delta\phi/\Delta t.$$

Dividing the second into the first and remembering that $\Delta\phi/\Delta t$ is the same in both formulas, we get:

$$V_P/V_S = N_P/N_S.$$

So, if the number of turns in the primary is 100 and the number of turns in the secondary is 400, then the voltage in the secondary will be:

$$V_S = (400/100)\ V_P$$

or

$$V_S = 4\ V_P.$$

The voltage in the secondary will be four times the voltage in the primary. This is called a step-up transformer.

Note that the power produced in the primary is given by

$$P = V_P I_P$$

and the power in the secondary (assuming no power losses in the transformer due to the resistance of the windings, hysteresis losses due to friction in the iron walls of the core that generates heat as the direction of the magnetic field changes, or eddy current losses due to the interaction of the applied and induced magnetic fields), is given by

$$P = V_S I_S.$$

If we now set them equal to each other, noting the power in both is the same:

$$V_P I_P = V_S I_S$$

or $$V_P / V_S = I_S / I_P.$$

So, $$V_P / V_S = N_P / N_S = I_S / I_P.$$

Considering a step-up transformer that boosted the voltage by a factor of 4, $V_P / V_S = 4$.

$$I_S / I_P = 1/4$$

and $$I_S = 1/4\ I_P.$$

So boosting the voltage reduces the current in a corresponding way.

Review Questions

1. How do we define a magnetic field?
2. Discuss the interaction of two bar magnets.
3. How are all magnetic fields generated?
4. What are magnetic domains?
5. Discuss the magnetic field of the earth. Does it change? Explain.
6. What is an electromagnet?
7. Discuss two applications of electromagnetism in high fidelity.
8. Discuss electromagnetic induction.
9. What does Lenz's law tell us?
10. Discuss two applications of Faraday's law in hi-fidelity.
11. Describe pulse induction and describe one application.
12. What is a transformer and what is it used for?

Multiple Choice Questions

1. Thales of Miletus discovered that ______ attracts iron.
a. positive charges
b. amber
c. lodestone
d. bronze

2. The "south pointer" was discovered by _____.
a. Thales
b. the Persians
c. the Chinese
d. the Assyrians

3. Gilbert came up with the idea that the earth itself was _____.
a. a sphere
b. a magnet
c. a lodestone
d. rotating

4. On a bar magnet one end is labeled north and the other south. Experiment reveals that two north poles in proximity _____.
a. attract
b. repel
c. unite
d. cancel in each other's magnetic field

5. We may say ______ poles attract and ______ poles repel.
a. north, south
b. unlike, like
c. south, north
d. like, north

6. To sketch a magnetic field we use ______.
a. straight lines
b. curves that go from the north pole to the south pole
c. curves that go from the south pole to the north pole
d. circles

7. If a current is flowing upward in a wire in the plane of the paper, the magnetic field flows _____as viewed from above.
a. downward
b. in a circle but clockwise around the wire
c. in a circle but counter clockwise around the wire
d. from the north pole to the south pole

8. All magnetic fields are produced by _____.
a. electric currents
b. lodestones
c. magnetite
d. bar magnets

9. Certain materials can be magnetized permanently. They are usually _____, cobalt, or nickel compounds.

a. copper
b. silver
c. iron
d. aluminum

10. The reason certain materials can acquire a permanent magnetic field is because of the existence of magnetic ______.
a. currents
b. domains
c. habitats
d. regions

11. The earth's magnetic field is thought to be caused by _____ in its interior.
a. lodestone
b. electric currents
c. iron
d. a bar magnet

12. The magnetic field of electromagnet can be increased by ______.
a. more current
b. less coils of wire
c. smaller coils
d. thicker wire

13. If an iron core is inserted into an electromagnet, the magnetic field_____.
a. decreases
b. increases
c. is unaffected
d. becomes thinner

14. A speaker is able to move air with its diaphragm because of the interaction of the permanent magnet with the _______.
a. magnetic field of the voice coil
b. basket
c. iron core of the voice coil
d. flexible edge

15. A changing magnetic field in a wire loop induces a(n) ______.
a. attractive force
b. current
c. resistance
d. repulsive force

16. Faraday discovered the phenomenon of _____.
a. magnetism
b. current electricity
c. electromagnetic induction
d. voltage

17. In order to induce a current in a coil of wire, the magnetic field has to be _____.
a. stationery
b. changing
c. constant
d. large

18. A moving magnet magnetic cartridge induces a current in a _____ wire.
a. long
b. short
c. coil of
d. magnetized

19. A moving magnet dynamic microphone has a fixed _____.
a. coil
b. core
c. current
d. grill

20. An airport metal detector operates in such a fashion that a _______ magnetic field induces a current.
a. steady
b. changing
c. oscillating
d. reflected

21. An electrical transformer has three major parts: the primary coil, the secondary coil and the _____.
a. permanent magnet
b. iron core
c. digital processor

d. generator

22. Step up transformers have _____ turns in the primary than the secondary.
a. less
b. more
c. the same number of
d. stepped

23. Step-down transformers ______ the voltage in the secondary.
a. lower
b. raise
c. transform
d. rectify

24. In a D.C. circuit the transformer _____ in the secondary.
a. raises the voltage
b. lowers the current
c. lowers the voltage
d. doesn't work

25. The power delivered to the primary is 300 W with 100 turns in the primary and 400 in the secondary. Assuming no losses the power from the secondary will be _____.
a. 0 W
b. 300 W
c. 1200 W
d. 75 W

26. In a step-up transformer the current in the secondary will be _____ primary.
a. lower than
b. higher than
c. the same as
d. a multiple

27. A step-up transformer has 20 turns in its primary and a primary voltage of 200 V. The secondary has 40 turns and a voltage of _____ volts.
a. 100
b. 200
c. 400
d. 80

28. If the current in the secondary is 20A and its voltage is 40 V, then the voltage in the primary is _____ volts when its current is 2 A.
a. 4
b. 40
c. 400
d. 80

29. 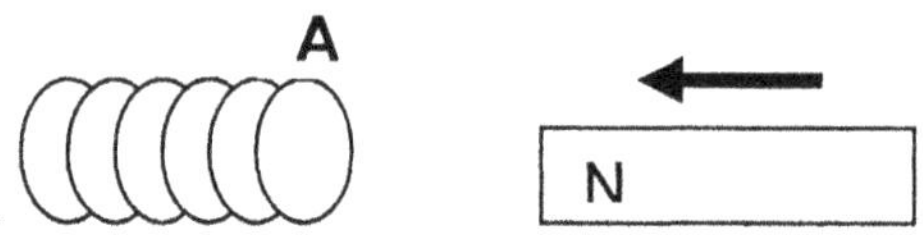

Moving the magnet in the direction shown induces a _____ pole at A in the coil.
a. north
b. south
c. neither

30. Doubling the number of turns of wire in an electromagnet _____ the strength of the magnetic field.
a. doubles
b. quadruples
c. halves
d. does not effect

Bibliography and Web Resources

Johnson, Kenneth, and Willard Walker and John Cutnell. *The Science of Hi-Fidelity.* Kendall-Hunt Publishing Company, Dubuque, Iowa, 1995.

Symko, Orest. *Physics of Hi-Fi.* Kendall-Hunt Publishing Company, Dubuque, Iowa, 1995.

Stanionis, Victor and Hugh Berberich. *Computer Music: Science and Technology of a New Art.* Research Foundation of State University of New York, Stony Brook, NY 1989.

Thales of Miletus =
http://www.mlahanas.de/Greeks/ElectroMagnet.htm

Chinese compass –
http://www.smith.edu/hsc/museum/ancient_inventions/compass2.html

Earth as a Magnet –
http://www.phy6.org/earthmag/demagint.htm

Pulse metal detection-
http://engineeringworks.tamu.edu/episodes/2004/metaldetector/

Pulse Induction –
http://www.thomasathomas.com/How_Metal_Detectors_Work.htm

Airport screening door –
www.controlscreening.com/brochures/9000.pdf

CHAPTER 12

MIDI Coding

"MIDI, the Musical Instrument Digital Interface, has revolutionized the world of professional music."

Craig Anderson, 1987
Editor, *Electronic Musician*

Computers and synthesizers both use microchips for their operation. Information inside the computer is stored as bits or binary digits. Bits and combinations of bits, called bytes, are the building blocks of the MIDI code which allows many electronic instruments to work together under the control of the personal computer to produce music. Machine language that computers use for their operation is coded in binary. Binary code uses the two symbols "0" and "1" to code meanings. Our decimal number system has ten symbols that we can use to code meanings.

Goals

After studying this chapter, you should be able to:

- Understand binary coding.
- Know how information and meaning can be stored in a binary form.
- Understand MIDI coding.
- Know the difference between status and data bytes.
- Distinguish between system messages and channel messages.
- Know what the MIDI channel reception modes are.
- Know how a device receives messages and responds.

Binary Coding

If the length of a word was restricted to one character in length, we could code only two meanings using the zero and one of binary code.

0 1

If the words in our language were allowed to be two characters in length, then the number of meanings we can code would be:

00 01 10 11.

Four meanings can now be coded.

Three character length words allow us to code eight meanings.

000	011
001	101
010	110
100	111

The basics of a pattern begin to emerge.

For a one character word, we associate 2 meanings, or 2^1.
For a two character word, we associate four meanings, $2 \times 2 = 2^2$.
For a three character word, we associate eight meanings, or $2 \times 2 \times 2$ or 2^3 .

A little thought reveals that a four character word would allow $2 \times 2 \times 2 \times 2 = 2^4$ or 16 meanings. Exponential notation allows us to express the results compactly.

How many characters long would our word have to be if we wanted to code each of the letters of the alphabet in terms of a one and a zero? Now, it will be necessary to write a one and zero code for twenty-six meanings. A four character long word allows us to code 2^4 or sixteen meanings. For a five character long word, the number of meanings would be 2^5 or 32 meanings, more than enough to code the twenty-six letters of the alphabet.

The first microcomputers used words which were eight binary digits (eight bits) long. Eight bits is called a byte. Now 2^8 or 256 meanings can be coded which means all the letters of the alphabet, the ten decimal digits, the dollar sign, the plus and minus sign, etc. can be coded in terms of ones and zeros. The 8-bit code has been replaced by a larger set of characters so as to provide additional codes for all of the world's characters. It is called Unicode and resembles a 16 bit code in which 2^{16} or 65,536 meanings can be encoded.

Binary Numbering System

The binary numbering system works just like our everyday decimal system. The word decimal itself takes its root from the Latin number for ten, decem. It implies that there are ten different symbols that may be used in expressing numbers in our system. Their value is intimately connected with their position in the number. For example, 287 with a 7 in the units position, an 8 in the tens position and a 2 in the hundreds position tells us we have two hundreds, eight tens and seven units for a total of two hundred and eighty-seven. Using powers of ten, we can write this as:

$$287 = 2 \text{ x } 10^2 + 8 \text{ x } 10^1 + 7 \text{ x } 10^0\text{, where } 10^0 = 1.$$

Any number in our base 10 or decimal system can be decomposed in this way.

In the binary system, the base is two instead of 10. We have only two symbols to work with, 0 and 1. Under these circumstances a number like 101_2, when decomposed, becomes:

$$1 \text{ x } 2^2 + 0 \text{ x } 2^1 + 1 \text{ x } 2^0 = 5_{10} \text{ in the decimal system.}$$

An algorithm or recipe for converting any decimal number to binary involves dividing it by two and keeping track of the remainder after each division.

Convert 14 to binary. The recipe is:

14 divided by 2 = 7 with remainder 0.

7 divided by 2 = 3 with remainder 1.

3 divided by 2 = 1 with remainder 1.

1 divided by 2 = 0 with remainder 1.

So, $14_{10} = 1110_2$ reading the remainders from the bottom up.

Notice to convert 1110_2 to base 10 we would write:

$$1 \text{ x } 2^3 + 1 \text{ x } 2^2 + 1 \text{ x } 2^1 + 0 \text{ x } 2^0$$

or $8 + 4 + 2 + 0 = 14_{10}$.

Convert 12_{10} to binary.

12 divided by 2 = 6 with remainder 0.

6 divided by 2 = 3 with remainder 0.

3 divided by 2 = 1 with remainder 1.

1 divided by 2 = 0 with remainder 1.

Therefore, $12_{10} = 1100_2$.

Checking $1 \times 2^3 + 1 \times 2^2 + 0 \times 2^1 + 0 \times 2^0 = 8 + 4 + 0 + 0 = 12_{10}$.

Consider the trip odometer, the mileage indicator, on a car. Suppose it has room for four digits. When the car is delivered, the odometer reads: 0000.

After one mile it reads:	0001.
After two miles,	0002.
After it travels nine miles,	0009.

After the tenth mile, the last digit becomes a zero and the second digit in the tens position reads one or

0010.

If we had only two symbols to work with as in the binary system, after the first mile it would read 0001.

And after the second mile the last digit would become a zero and the next to the last, a one. So it becomes,

0010.

After three miles,	0011,
and after four,	0100 and so on (Fig. 12.1).

Decimal Numbers	Binary Numbers
0	0000
1	0001
2	0010
3	0011
4	0100
5	0101
6	0110
7	0111
8	1000
9	1001
10	1010
11	1011
12	1100
13	1101
14	1110
15	1111

Figure 12.1 Binary number table

From *Computer Music: Science and Technology of a New Art,* Stanionis, Berberich. Copyright © 1989 by Research Foundation of State University of New York. Reprinted by permission.

DNA

In the 1950s, one of the most important patterns of information in determining the biochemical reactions in the cell was analyzed. The problem that faced researchers at the time was the question of how heredity information carried by the genes was implemented in the formation of the organism. Researchers had already discovered that the cell contained many molecules in the form of fats, sugar, proteins, and genes made-up of deoxyribonucleic acid (DNA). They had already determined that the genes carried the information needed to fabricate the proteins.

In 1954, Watson and Crick discovered the structure of DNA. They found that DNA is made up of four chemicals called bases which are adenine, thymine, cytosine, guanine. These bases were arranged in a double helix. The next question to be answered was how the DNA specified a strand of protein. It was known that all protein molecules could be fabricated from twenty different amino acids. Using a coding system that has four symbols, what is the minimum length a word has to be to code twenty different meanings?

If the length of a word was one character, we could code four different meanings:

A, G, T, C.

If the length was two characters, the number of meanings would be 4 x 4 = 16. Still, it is not enough to code the twenty amino acids needed to make-up the genetic code.

If we use words three characters long, we have 4 x 4 x 4 = 64 meanings, more than enough to code the twenty needed. The sequence of three letter words made-up of A, C, T, G along a strand of DNA carry the information in a cell. Amino acids are called by names or words such as ACG or ATT. Whether it is a biological system or a computer system, information, which is a form of knowledge in transit, is transferred in a code.

MIDI Code

The most important function of the MIDI system is the transmission of information from instrument to instrument and to the computer. Instructions are coded, telling each musical instrument when to play, what note to play, how to play it, etc. The MIDI code is a binary code, like the computer code, that carries data in terms of ones and zeros.

The MIDI interface transmits data in a serial fashion (Fig. 12.2) which allows transmission over long distances in an accurate fashion in which ones and zeros are represented by means of voltages. The transmission rate is 31,250 bits per second.

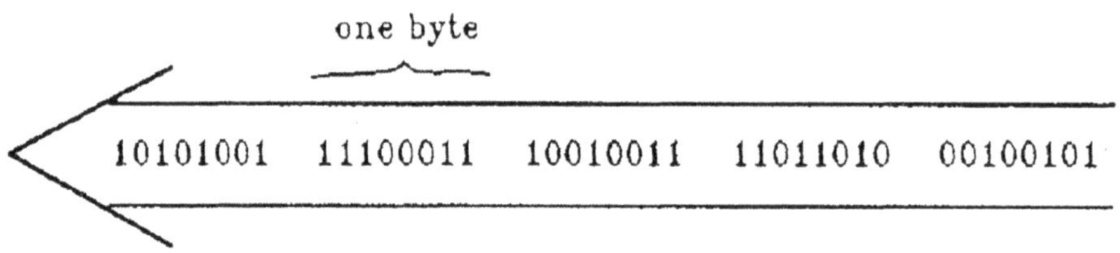

Figure 12.2 MIDI data coded in binary

From *Computer Music: Science and Technology of a New Art,* Stanionis, Berberich. Copyright © 1989 by Research Foundation of State University of New York. Reprinted by permission.

Each group of eight bits (a byte) carries a part of the MIDI message. MIDI also uses a start bit and a stop bit to alert the computer of the beginning and end of a byte. Ten bits are used in a MIDI byte and MIDI transmits 3125 MIDI bytes per second. A MIDI command is usually made-up of one, two, or three bytes sent one after the other.

The MIDI Specification 1.0 (Document Version 4.0) distinguishes between two different types of bytes: status bytes and data bytes. Status bytes identify the type of information being transmitted. They tell what kind of message it is (e.g. note on, note off). Following the status byte are data bytes, which carry the information describing what note is to be played and at what velocity. A velocity sensitive keyboard can transmit to the microprocessor in the keyboard how fast a key has been struck and thus govern how loud the note should be played by the synthesizer. MIDI messages can have a status byte followed by two data bytes, one data byte or even none (Fig. 12.3). Once a status byte has been received by an instrument it can continue running under that command until a different status byte is received.

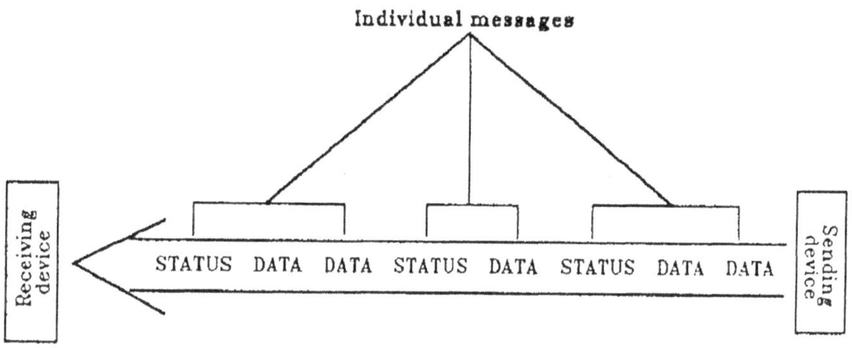

Figure 12.3 MIDI messages
From *Computer Music: Science and Technology of a New Art,* Stanionis, Berberich. Copyright © 1989 by Research Foundation of State University of New York. Reprinted by permission.

Groups of notes can be played one after the other by first sending a status byte with note-on message followed by data bytes containing note number and velocity for each actual note (Fig. 12.4). A new status byte would have to be transmitted if a new command such as pitch bend is to be executed. Remember, the pitch bend wheel on many synthesizers allows one to change the pitch of a MIDI device by half-steps, up and down the keyboard for up to a whole octave.

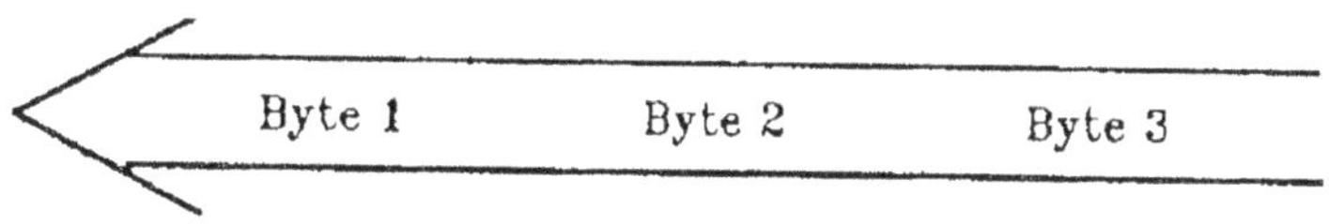

Byte 1 = Note-on status

Byte 2 = Note number

Byte 3 = Velocity

Figure 12.4 MIDI data
From *Computer Music: Science and Technology of a New Art,* Stanionis, Berberich. Copyright © 1989 by Research Foundation of State University of New York. Reprinted by permission.

In the 8 bit computer code it is possible to code 2^8 or 256 meanings. Since the MIDI code uses words which are 8 bits long or one byte long, excluding the start and stop bits, we have 256 bytes we can assign. The MIDI specification assigns bytes from 0 to 127 as data bytes and bytes from 128 to 255 as status bytes. Once a status byte is received by a MIDI instrument it knows how to interpret the bytes that follow. If a note-on status byte is received, it interprets the following data byte as pitch and the second data byte as velocity. A protocol that is used in distinguishing status bytes from data bytes is to let the leftmost or most significant bit be equal to 1 in a status byte and 0 in a data byte (Fig. 12.5).

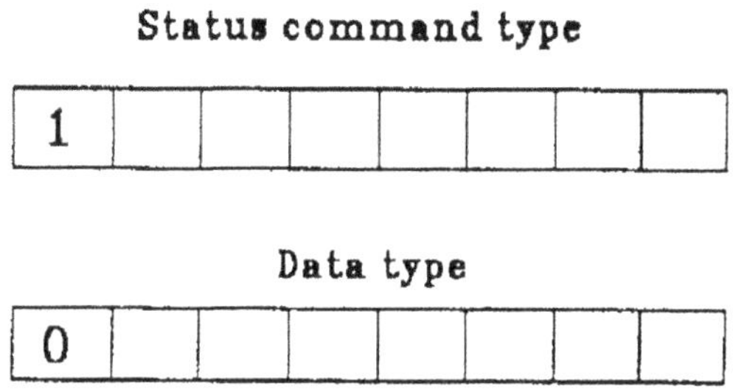

Figure 12.5 Status byte vs data byte

From *Computer Music: Science and Technology of a New Art,* Stanionis, Berberich. Copyright © 1989 by Research Foundation of State University of New York. Reprinted by permission.

Running Status

Such a protocol allows a musical instrument to continue running under a status command until a different status command is received. In this way, once the note-on command has been received the instrument can interpret the pairs of data bytes that follow as note number and velocity. A lot of notes can then be played without having to include a status byte for each (Fig. 12.6).

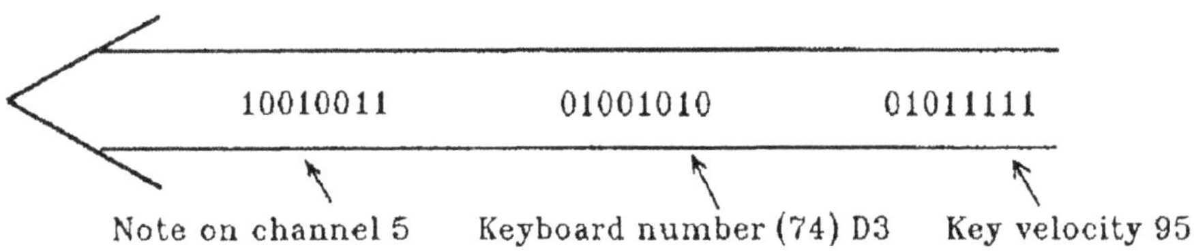

Figure 12.6 MIDI code

From *Computer Music: Science and Technology of a New Art,* Stanionis, Berberich. Copyright © 1989 by Research Foundation of State University of New York. Reprinted by permission.

Networks

In MIDI, information can be sent out over 16 different channels which allow several MIDI equipped instruments playing different parts to play together in time. There are two main ways in which several MIDI instruments may be connected to one another: daisy-chain and star-network (Fig. 12.7). In a daisy-chain, a master keyboard will usually control a relatively small number of synthesizers.

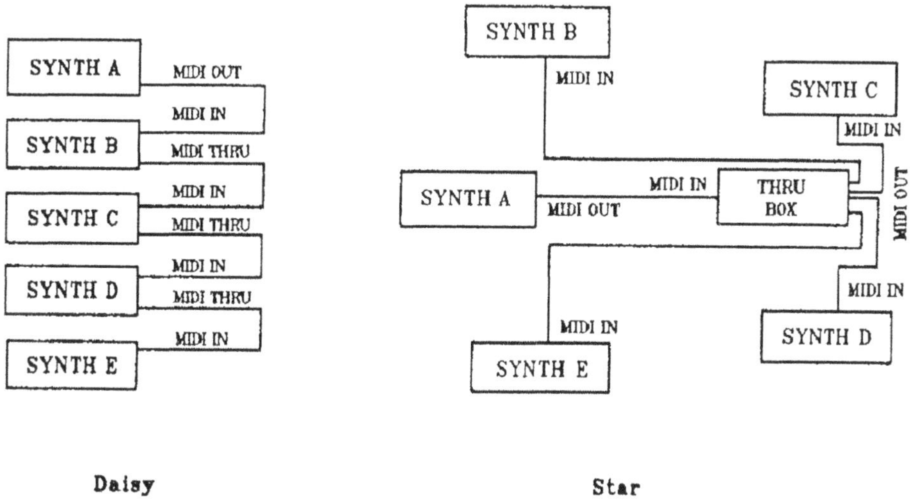

Figure 12.7 Networks
From *Computer Music: Science and Technology of a New Art,* Stanionis, Berberich. Copyright © 1989 by Research Foundation of State University of New York. Reprinted by permission.

The main problem with this way of networking the instruments is the time it takes for the data to travel to the last instrument in the chain. The MIDI signal is not only delayed but degraded leading to data distortion. In the star-network, a MIDI THRU box expands a single MIDI Out into several buffered Outs. No distortion takes place and the time problem is eliminated.

Messages

All MIDI messages can be subdivided into two types: system messages and channel messages. System messages are received by all devices while channel messages go to only certain ones. When a MIDI device receives the status byte of a channel message, it identifies it as a channel message and it checks to see if it is allowed to read that message. If so, it will respond accordingly. Otherwise, it ignores it.

System Messages

There are three types of system messages: system common messages, system real time messages and system exclusive messages.

System Common Messages

System common messages include: song select, song position pointer, tune request, and end of exclusive. System common messages are sent without channel numbers to all units on all channels. A song select message loads a specific song from 0 to 127 from the sequencer's memory. The song position pointer allows the song to be played starting at an arbitrary point. The tune request message asks all MIDI devices to tune themselves. The end of exclusive (EOX) message informs the MIDI device of the end of a system exclusive message.

System Real-Time Messages

There are several types of system real-time messages: timing clock, start, stop, continue, active sensing, and system reset. System real-time messages are controlled by a timing clock that synchronizes the timing among different MIDI instruments during performance. They are one byte long and are dropped into the data stream at regular intervals. When it receives such a byte, a MIDI device acts on it first and then resumes its activity. The time messages interrupt the data stream when necessary. The timing is based on a 24 pulse-per quarter note clock.

Once the tempo has been set and a sequencer has to play a series of notes such as three quarter notes and two eighth notes, it would count 24 pulses for the first quarter note, the same for the second and third. For the first eighth note, it would count 12 pulses before proceeding to the second eighth note.

The start message causes the sequencer to start playing from the beginning. The stop message tells it to stop playing. The continue starts it playing from whatever point it last stopped.

Active sensing helps to determine whether a MIDI device is properly connected and is on-line. Once the instrument is connected into the MIDI system, it is sent an active sensing message. It then receives an active sensing message every 300 milliseconds or about three messages per second. If it does not, it turns off all the notes it has been playing.

System reset messages will usually cause the device to reset all parameters to what they were when first turned on.

System Exclusive Messages

System exclusive messages allowed manufacturers to develop custom messages they could distribute among instruments without interrupting the normal flow of MIDI messages. The status byte denotes it as a system exclusive message and the first following data byte identifies the manufacturer of the device. The MIDI device first reads the manufacturer's code to see if it should read the rest of the message or not. If it identifies itself with the manufacturer, it reads the rest of the data bytes, for which each manufacturer is allowed different meanings.

Channel Messages

MIDI channel messages can be subdivided into two categories: voice messages and mode messages. Since the most significant bit is set aside to signal a status byte or data byte, seven bits remain and we can code 2^7, or 128, meanings. The leftmost bit will be a 1 indicating it's a status byte. The next three will define which channel command (Fig. 12.8) is being transmitted and the last four will specify one of the sixteen channels.

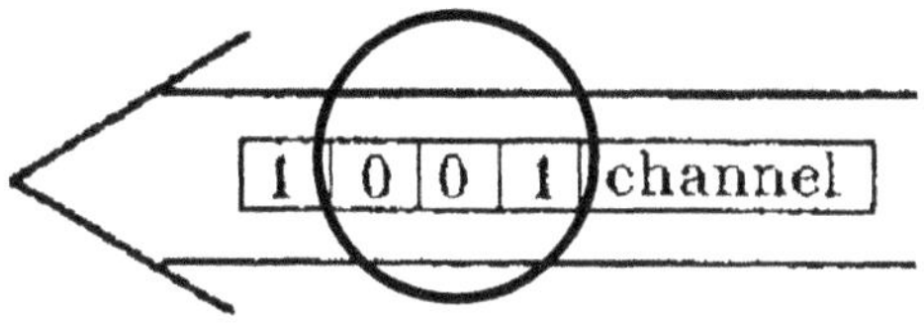

Figure 12.8 Channel messages

From *Computer Music: Science and Technology of a New Art,* Stanionis, Berberich. Copyright © 1989 by Research Foundation of State University of New York. Reprinted by permission.

How many bits are required to convert 15 to binary notation? In binary, $15_{10} = 1111_2$. So, we need only four bits to code the sixteen channel numbers. Only instruments for the channel specified will respond to the message (Fig. 12.8). Since most MIDI commands are channel messages which require four bits to define one of the sixteen channels, only three bits remain to send as channel commands (Fig. 12.9).

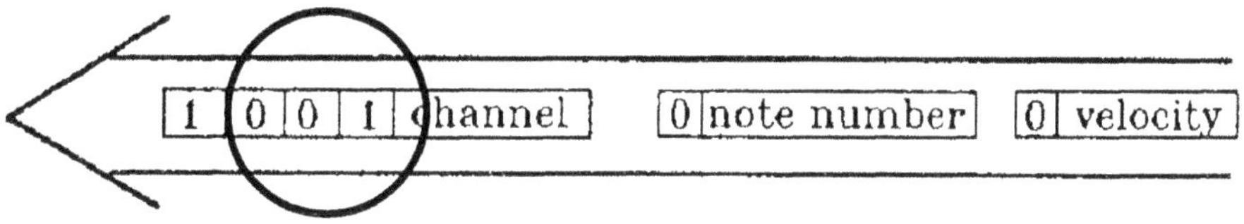

Figure 12.9 Three bits free for channel commands

From *Computer Music: Science and Technology of a New Art,* Stanionis, Berberich. Copyright © 1989 by Research Foundation of State University of New York. Reprinted by permission.

Three bits allow 2^3 or eight different meanings to be coded. They are: note-off, note-on, polyphonic key after-touch, controller, program change, channel-after touch, pitch bend, and system. Remember, system commands have no channel number attached. There are actually sixteen different non-channel commands available.

Channel Voice Messages

The note-off command has a status byte indicating the channel to be chosen followed by two data bytes. The first data byte defines a note number from 0 to 127 and the second denotes the release velocity of the note (Fig. 12.10).

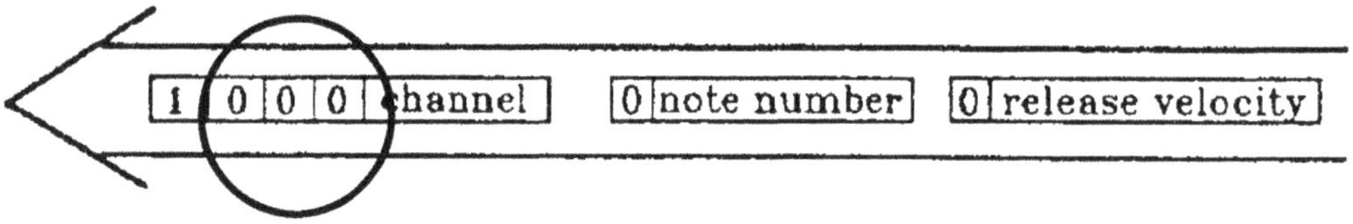

Figure 12.10 Note-off command

From *Computer Music: Science and Technology of a New Art,* Stanionis, Berberich. Copyright © 1989 by Research Foundation of State University of New York. Reprinted by permission.

The note-on command also has three bytes: status byte, note number, and velocity. If the velocity is zero, this is interpreted as a note-off command. Polyphonic after-touch transmits electrical signals denoting the pressure being applied to each note of a chord being played at the keyboard. Pressure transducers are needed for each key. The MIDI message requires three bytes: channel number, key number, and pressure value.

The controller command also has two data bytes associated with it. The first defines which controller is being used (e.g. foot controller, volume, etc.) and the second gives a value for the position of the controller.

Two bytes are used to accomplish a program change: status byte with channel number and program number byte. The program number byte changes the program number of a peripheral MIDI device. Channel after-touch is made up of two bytes: channel byte and pressure byte. The pressure byte contains information about pressure applied to the keyboard which can be translated by the MIDI device to volume level, brightness, etc.

Pitch bend is a separate controller command that has two data bytes associated with it (Fig. 12.11). Basically the pitch bend wheel raises or lowers the pitch of notes as it is turned.

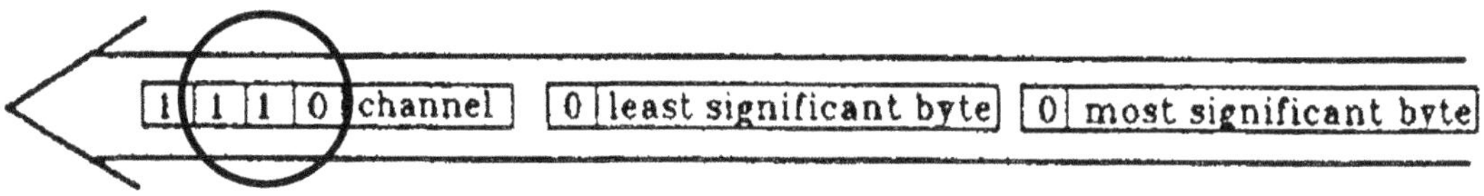

Figure 12.11 Pitch-bend

From *Computer Music: Science and Technology of a New Art,* Stanionis, Berberich. Copyright © 1989 by Research Foundation of State University of New York. Reprinted by permission.

Channel Mode Messages

MIDI allows many different messages to travel, one after the other, down the same wire without interfering with one another. In the case of television broadcasting many different signals are broadcast simultaneously over the air or through a cable. The viewer selects the channel he wishes to view by tuning it in. In MIDI there are sixteen channels that are carrying information from the computer along a cable to various instruments. If the instrument is set for a particular channel, it reads the information coming over the cable and plays the notes indicated.

How do synthesizers assign note information received on a particular channel to various voices? Polyphonic synthesizers contain multiple voices to allow them to play multiple notes simultaneously. The four different modes of assigning data to the synthesizer voices are made up of combinations of the three MIDI mode messages: omni, poly and mono. Mode messages are different for receivers and transmitters.

MIDI mode messages are: omni on, omni off, poly, and mono.

The receiving instrument in ***omni on*** responds to messages on all sixteen channels, as if they were one channel. It doesn't matter on which channel it is sent, the MIDI instrument will respond. It automatically turns off all notes being played by the receiving instrument.

In ***omni off***, the receiving instrument responds to messages on certain MIDI channels. It will respond to that channel alone and ignore messages on other channels. It also turns off all notes being played by the receiving instrument.

In ***mono,*** the instrument responds to messages coming in on one channel. Note messages are assigned monophonically or only one note per channel. Inherently monophonic instruments such as most brass and woodwinds have this property. It turns off poly mode and stops all notes being played by the receiving instrument.

In ***poly,*** the musical parts are played polyphonically (more than one note per channel). It can play chords on that channel. It turns off mono mode and stops all notes that the receiving instrument is playing.

MIDI Modes

The first part of the mode indicates how the MIDI device is monitoring incoming messages, while the second half of the mode designates how the MIDI device will play incoming notes.

Mode 1: Omni on/Poly

By definition, the instrument receives notes on all sixteen channels and responds polyphonically. If the synthesizer has multiple voices, each voice being a sound generated by one of the synthesizer's tone generators, it can play multiple notes. If a synthesizer has eight voices, it can play eight notes simultaneously (Fig. 12.12). Voice messages are assigned to voices in the synthesizer until they are all used up. Any additional voice messages will be assigned to the first voice played. In this Mode 1 every part of a musical piece sounds like it is being played on a single instrument. This mode is sometimes referred to as "omni".

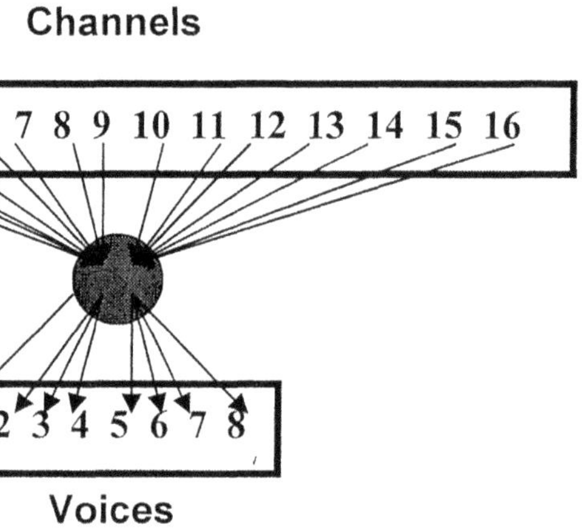

FIGURE 12.12 Omni on/poly

Mode 2: Omni on/Mono

Information coming in on any channel is assigned to one voice. The instrument will play only one note no matter how many messages are being received to play more than one note at a time (Fig. 12.13). This mode is usually referred to as "mono".

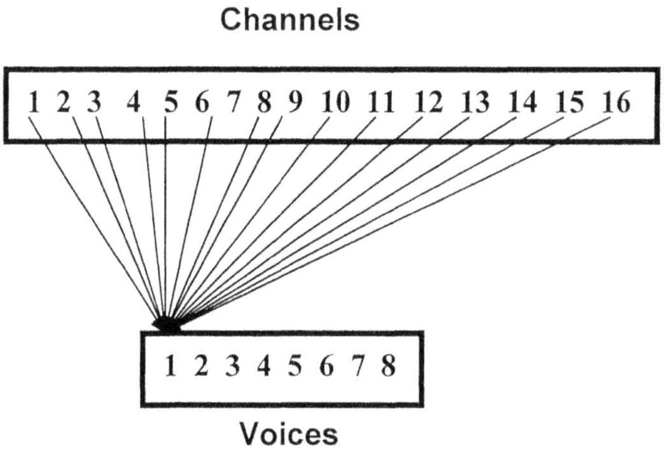

Figure 12.13 Omni on/mono

Mode 3: Omni off/Poly

The instrument only reads information coming in over the channel it is assigned to. The instrument assigns voices polyphonically but all voices play the same preset sound like a musician playing one particular instrument from a score (Fig. 12.14). This mode is called "poly".

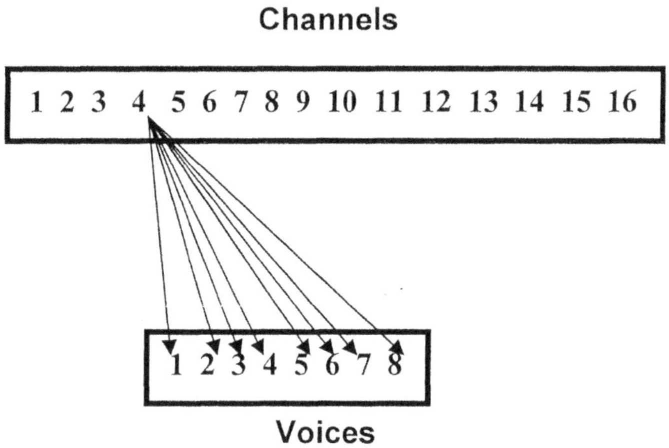

Figure 12.14 Omni off/poly

Mode 4: Omni off/Mono

Now each synthesizer voice is assigned to a particular channel number. Each channel activates its own voice. The instrument can now respond to commands from the sequencer producing several different independent melody lines coming in over different channels. Data can now be sent from track one to the synthesizer's first voice which could be set for a trumpet, sent from track two to the multitimbral synthesizer's second voice which would be programmed for a bass sound and so on. Each voice plays monophonically (Fig. 12.15). However, chords can be created by using more than one sequencer track and more than one instrument voice. This mode is also referred to as the "other mono".

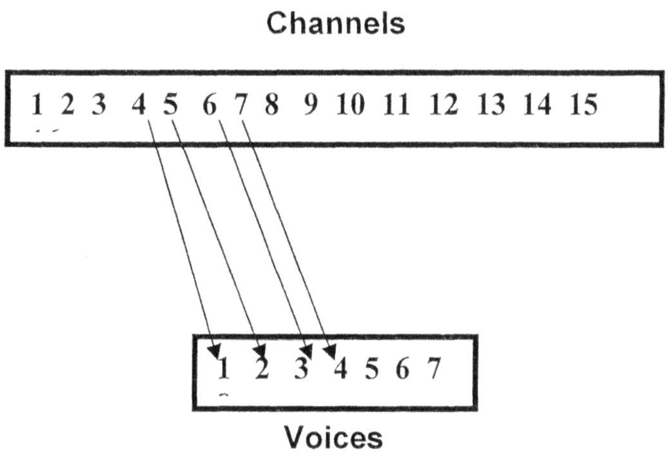

Figure 12.15 Omni off/mono

Most instruments will default to omni/poly when turned on. In transmission mode, omni on/poly transmits all voice assignments on the basic channel. On omni on/mono, voice data for a single voice are transmitted on the basic channel. For omni on/poly voice messages, all voices are transmitted on the basic channel while omni off/mono voice messages for different voices are transmitted one voice per channel. Information is transmitted over the same channels assigned to receive data.

Review Questions

1. Discuss how many meanings can be coded in binary if a word is restricted to 4 bits.
2. How does one convert a decimal number to binary and vice versa?
3. How does MIDI coding differ from ordinary binary coding?
4. What are the roles played by status bytes and data bytes in MIDI messages?
5. Discuss the use of networks in a MIDI system.
6. What is the difference between system messages and channel messages?
7. Discuss system common messages, system real time messages, and system exclusive messages.
8. Discuss the difference between voice messages and mode messages.

MultipleChoice Questions

1. In the decimal number system there are _____ symbols used for numbers.
a. 2
b. 5
c. 9
d. 10

2. In the binary numbering system, there are _______ symbols used for coding purposes.
a. 2
b. 5
c. 9
d. 10

3. If words are restricted to four characters in length, the number of meanings that can be coded is ______.
a. 2
b. 8
c. 16
d. 32

4. If there are three ways to get to the bottom of a mountain and then two ways from there to get to the top, the total number of ways to get to the top from far away is _____.
a. 5
b.6
c. 7
d. 8

5. The number 1001_2 in binary is the same number as ______ in decimal.
a. 7
b. 8
c. 9
d. 10

6. What is the number equivalent of 1111_2 in decimal?
a. 4
b. 10
c. 13
d. 15

7. The number 12 in decimal would equal _____ in binary notation.
a. 1111
b. 1001
c. 1110
d. 1100

8. Writing the decimal number 7 in binary we obtain _____.
a. 1111
b. 0111
c. 1011
d. 1101

9. Write the number 54 in binary notation.
a. 110110
b. 111010
c. 100111
d. 101101

10. The binary number 101010. In decimal notation this is the same as _____
a. 32
b. 42
c. 23
d. 24

11. You have four symbols, A, C, T, and G. What is the minimum length of a word in order to code 22 meanings?
a. 3
b. 4
c. 5
d. 6

12. Given five symbols. What is the minimum length of a word needed to code 200 meanings?
a. 2
b. 3
c. 4
d. 5

13. In binary coding each word consists of 8 bits or a _____.
a. decem
b. byte
c. nibble
d. chunk

14. In MIDI coding the word length is _____ bits or a MIDI byte.
a. 6
b. 8
c. 10
d. 12

15. In MIDI 31,250 bits per second are transferred or _____ bytes per second.
a. 312
b. 3125
c. 391
d. 31.2

16. The two different kinds of bytes used in MIDI coding are _____.
a. empirical and decimal
b. 8-digit and 10 digit
c. numerical and alphabetical
d. status and data

17. A MIDI byte is similar to an 8 bit computer byte but has an extra _____ bit at the beginning and an extra _____ bit at the end.
a. two, two
b. start, stop
c. begin, end
d. 4, 4

18. In MIDI coding a status byte is always followed by at least one _____ byte.
a. numeric
b. start
c. data
d. signal

19. One example of a status byte could be _____.
a. velocity
b. key number
c. note on
d. command

20. Excluding the start and stop bits, MIDI uses 8 bytes to convey information and there fore _______ meanings may be coded in a MIDI byte.
a. 64
b. 128
c. 256
d. 512

21. Two types of networks that may be used in connecting synthesizers are _____.
a. star and token ring
b. bus and Ethernet
c. star and daisy-chain
d. token ring and Ethernet

22. The two main kinds of MIDI messages are ______.
a. system and channel
b. system common and system exclusive
c. omni on and poly
d. voice and system

23. Data bytes always begin with a _____.
a. 1
b. 0
c. 10
d. 01

24. In a MIDI status byte only three bits are left for a channel command like note on. This means that ______ meanings may be coded.
a. 3
b. 9
c. 8
d. 16

25. In the MIDI mode omni on/poly, the instrument receives notes on all sixteen channels and responds______.
a. monophonically
b. polyphonically
c. multitimbrally
d. omni off

26. The first part of the mode message indicates how the MIDI device is monitoring incoming messages, while the second half of the mode designates how the MIDI device will play incoming notes. An example of a mode message would be_____.
a. omni/on
b. poly/on
c. omni off/poly
d. mono/poly

27. To code the channel number 16 requires _____ bits.
a. two
b. three
c. four
d. five

28. A MIDI byte which begins with a 1 is called a ____ byte.
a. MIDI
b. data
c. channel
d. status

29. Most MIDI instruments default to ____ when turned on.
a. omni on/mono
b. omni on/poly
c. omni off/mono
d. omni off/poly

30. Active sensing in a MIDI program determines whether a MIDI device is _____.
a. on-line
b. in channel mode
c. omni on/poly
d. reset

Bibliography and Web Resources

Anderson, Craig. *MIDI for Musicians.* Amsco Publications, New York, NY, 1986.

Baird Joel, editor, "*Understanding MIDI*", *Musician*, special edition, Amordian Press, Inc, Gloucester, MA 1986.

Bassett, Rick and Jonathan Matykin. "*A MIDI Musical Offering*", *PC Magazine*, V7, No.20, pp. 229-272, November 1988.

Boom, Michael. *Music Through MIDI.* Microsoft Press, Redmond, WA, 1987.

De Furia, Steve and Joe Scacciaferro. *The MIDI Resources Book*, Third Earth Publishing Inc., Pompton Lakes, NJ 1987.

De Furia Steve and Joe Scacciaferro. *The MIDI Book.* Third Earth Publishing Inc., Pompton Lakes, NJ 1988.

Milano, Dominico editor. *Mind over MIDI*, Hal Leonard Books, Milwaukee, WI, 1987.

Stanionis, Victor and Hugh Berberich. *Computer Music: Science and Technology of a New Art.* Research Foundation of State University of New York, Stony Brook, NY 1989.

History of MIDI - **http://www.physics.smu.edu/~olness/www/02fall1320/TrevorChapman/MIDI%20Web%20Page2/midi/history.htm**

How does MIDI work? - **http://www.physics.smu.edu/~olness/www/02fall1320/TrevorChapman/MIDI%20Web%20Page2/midi/how.htm**

MIDI note number for different octaves **http://www.harmony-central.com/MIDI/Doc/table2.html**

Short MIDI Overview - **http://www.rbjones.com/rbjpub/music/mus007.htm#what**

APPENDIX A
Laboratory Activities

All illustrations in the laboratory activity worksheets marked by a single asterisk (*) are taken from *Record Producer Deluxe* and are used with permission from Voyetra Technologies, Inc. Illustrations from *Computer Music: Science and Technology of a New Art*, Stanionis, Berberich. Copyright © 1989 by Research Foundation of State University of New York are marked with a double asterisk (**). Reprinted by permission.

ACTIVITY 1: LABORATORY ORIENTATION

Laboratory Activities

The laboratory activities are designed to acquaint the student with how science works. They are meant to reinforce the concepts developed in lecture and also to provide the students with hands-on skills in exploring and exercising those concepts.

Rules of Order

1. Safety is the keyword in describing the rules and procedures for laboratory work. Rules of order exist to insure the safety of all, to protect the equipment from damage, and to provide an effective learning environment. The primary concern is personal safety. Attached at the end of this activity is a safety checklist sheet which the instructor will review with you and ask you to sign acknowledging that you are aware of the General Rules of Order and Safety for this laboratory.

Laboratory Procedures

1. Each laboratory session will take approximately two class periods and you should be prepared to spend the full time in the laboratory. You may leave the laboratory with the permission of the instructor if you have finished the work and checked your results with the instructor. Leaving early and without permission will be recorded as an absence.

2. Your seats will be assigned by the professor at the beginning of the semester and you will work with the same lab partner during the semester. It is important to be on time for each session. You will miss essential instructions if you are late and may not be allowed to participate in the activity. Lateness and absences are recorded for each session and missing 25% (usually three) of the scheduled laboratory sessions will result in a failing grade for excessive absences.

3. Do not bring unnecessary articles to the lab. Place your coats on the coat racks provided in the lab and bags where instructed in the lab. You are responsible for the lab equipment at your lab station and should report missing and broken equipment immediately. Do not operate or handle any of the equipment at your station until the lab begins and you have been given proper instructions for its operation.

4. Laboratory reports are normally to be handed in at the beginning of the next scheduled lab session and placed at the front of the lab on the instructor's desk. It is important that you bring your own copy of the laboratory work sheets appropriate for the week's laboratory activity.

Laboratory Reports

Each student is expected to report on the laboratory activity in their own words in a printed report done on a computer using a word-processing package.

Format

1. The font type should be Times Roman, 12 pt font, using 1.25" margins with the text fully justified. The text should be in English paragraph form using correct English and punctuation and should be double spaced. Outlines and bullet formats are to be avoided.

2. Each page should have a header with the student's name appearing at the left and the date of the laboratory on the right of each page. Pages should be numbered at the bottom and centered. The laboratory worksheets are a guide and are not to be copied and cut-and-pasted into the report.

3. Make sure you fully describe the activities performed in your report. Use English paragraphs, writing in the past tense and impersonal (i.e. do not use the personal pronouns I, you, he, she, it, we, you or they).

4. Include appropriate diagrams which may be drawn using computer tools or hand drawn in an appropriate space and placed in the text. Start with a title page with Activity Number centered and centered below it the Activity name. In the right lower corner, type your name and that of your partner in parentheses. Type the date the laboratory activity was performed below.

5. At the top of each page use a header with your name on the left and the date of the lab on the right. Type the **Title**, **Purpose,** and **Summary** as given on the worksheets on separate lines. Next type the **Procedure** and insert all necessary diagrams drawn by hand or using computer tools. As stated before the text should be fully justified, that is, an even margin on the right and left.

6. Many of the worksheets have discussion questions at the end which should be answered by the student in the laboratory report.

LABORATORY SAFETY RULES

1. Familiarize yourself with the location of the safety items in the laboratory: eye-wash fountain, first aid kit, and fire extinguisher. Make sure your pathway to them is never blocked.

2. Your work area (including the floor) should be absolutely clear. Remove all book-bags, books (except the lab worksheets) and briefcases from the lab bench area to the sides or back of the room. Use the coat racks and stands provided.

3. If you should get a chemical in your eyes, notify your instructor, and wash them in the eye-wash fountain for 15 to 20 minutes.

4. Pay attention to the written instructions, and the instructor's oral instructions, and follow them carefully. If you do not know what to do, ask! Even "safe" experiments can be dangerous if instructions are not followed. DO NOT CONNECT TO ANY ELECTRICAL POWER SOURCE UNTIL YOUR SET-UP HAS BEEN CHECKED BY THE INSTRUCTOR.

5 No food or drink is allowed in the laboratory. They should be consumed outside the lab. Never taste any chemicals in the laboratory. (Even common compounds like salt and sugar in the lab are not fit to eat.) Never use laboratory glassware to drink from.

6. Smoking is prohibited in the building, including the classrooms, labs and stairways. Violators may be reported to security and barred from attending.

7. Report all accidents and glass breakages to your instructor. Ask to be taken to the Infirmary for treatment of cuts, burns, or inhalation of vapors. Do not attempt to clean up broken glass or spilled blood. (Because of HIV risks, blood must be removed by approved methods.)

8. Avoid breathing vapors of any kind.

9. Protect yourself! Wear loose, comfortable clothing to lab.

10. Never block aisles with chairs or stools. Class chairs should be moved to the side of the lab. Never push or shove others in the lab. Violators may be asked to leave or be debarred

I have read and understand the above rules and agree to obey them. I understand I may be removed from the lab room for failing to follow the rules.

Signed:______________________ CourseNo:_______ Instructor____________________

RECORD PRODUCER DELUXE

LOADING THE SOFTWARE:

Click on the **Record Producer Deluxe** icon to boot the Track/View window. Make sure it is maximized on the desktop. Note the Menu Bar, Transport Bar, Status Bar and Quick View Buttons at the bottom of the screen.

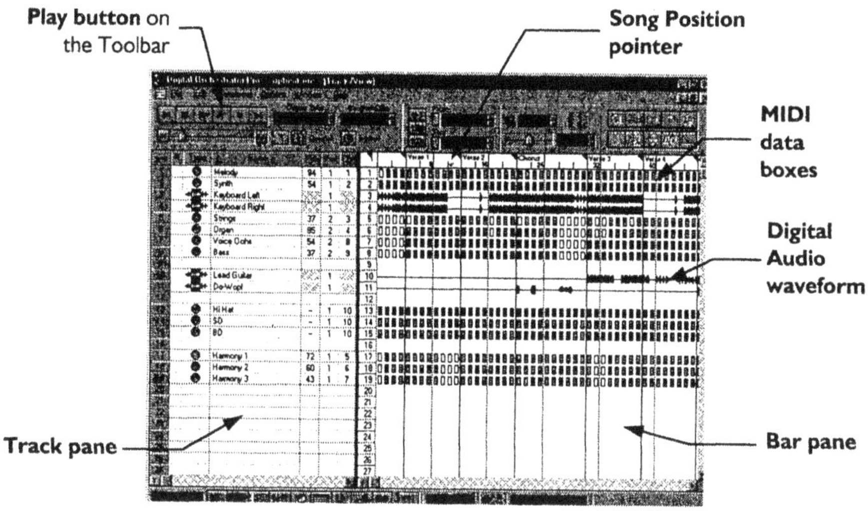

*

LOADING A SONG:

Go to the Menu Bar and click on File. From the drop-down menu click on Open. Click on Voyetra and then double click on the file named by the instructor to load it. The left side of the Track/View window is the Track Pane, the right side of the window is called the Bar Pane, and the small rectangles (MIDI data boxes) represent bars of recorded MIDI music for the tracks in this piece.

To **PLAY** the file click on the **PLAY** button in the Transport Bar or tap the spacebar on the keyboard. The red triangle or Play Position Pointer moves from bar to bar as the piece is playing indicating which bar is playing. To stop play and reset to the beginning of the song, press the spacebar or click on the **STOP** button in the Transport Bar. To start playing from a different point, move the mouse to the appropriate bar and click with the **RIGHT** button. Tap the spacebar and it will start playing from that point.

TRANSPORT CONTROLS

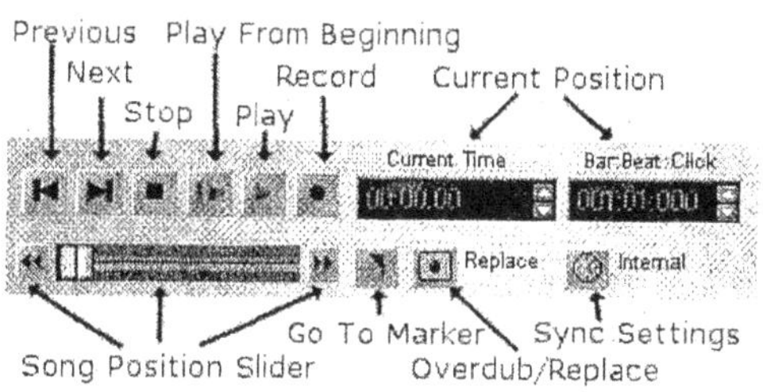

*

TO CHANGE INSTRUMENTS

Put the cursor into the Pch column in a track and change the number to play a different instrument by clicking on the up or down arrows on the spin wheel. You can also type in a number and hit the enter key. To **NAME** a track move the cursor to the Name column and click. Type in the name.

COPY and PASTE

Select a track by starting at bar 1 and holding the left button on the mouse as you drag it to the end of the data, or, click on the track number next to the Bar Pane. Go to Edit in the Menu Bar and select Copy. Move to an empty track and put the mouse at bar 1. From the Menu Bar choose Edit, Paste. Change the Chn number and Pch number to add a new instrument to the piece.

RANGE and LOOP

Use the Play From to set the starting point of a range within a song or digital audio file. Play From uses the format: Bar: Beat: Click. Play Until is then set for the ending point of a selected range. Once these parameters are set press the Play Range button to play a selected range or the Loop button to play a selected range repeatedly.

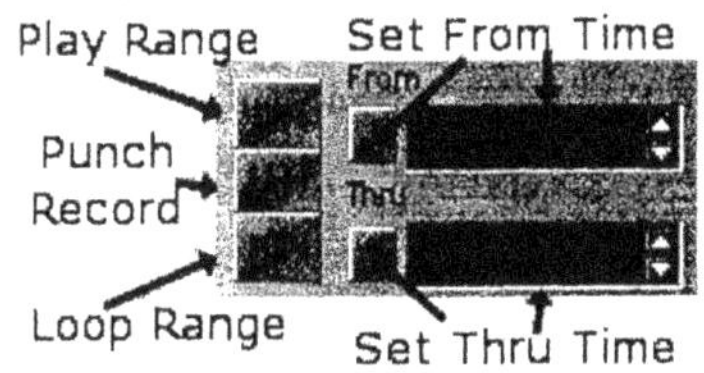

MUTE and SOLO

Go to the S and M columns. Move the mouse to a particular track and click on S. You should only hear that track playing. Click on it again to turn the S off. Try clicking on M. It will silence that track.

CHANGING TEMPO

The Toolbar Bar has the Current Tempo control display with spin buttons and also a slide bar control. Clicking on the buttons or typing a value in the window changes the tempo of a song from any value between 16 and 500 beat per minute. Tempo Offset control displays the amount by which a MIDI song's tempo can be increased or decreased in beats per minute.

RECORDING A NEW TRACK

Click the mouse on the Record column of an empty track. Click the mouse on the Type column until the MIDI plug symbol appears. Now you are ready to record on this track. Click on the channel column to select a new channel number and change the instrument sound as desired. Click on the Record button in the Transport Bar, and play something on the piano keyboard. To end the recording, press the spacebar on the keyboard. Click the Play button to hear it.

PIANO ROLL WINDOW TO RECORD OR EDIT

The Piano Roll Window lets you work with MIDI data note-by-note. You can add or delete notes or change their duration and pitch. The thin rectangles represent the notes. Their length is the duration and the vertical position is the pitch. To insert a note the **Pencil** tool in the Piano Roll toolbar allows you to insert a note at the desired location by clicking it at the desired location. It will insert the note with the current Insert Note parameters. You can click and drag the pencil to draw any length of note. **Eraser** tool deletes notes. **Arrow** tool sets the cursor to the normal arrow shape. Later we shall learn to use the step-method for recording.

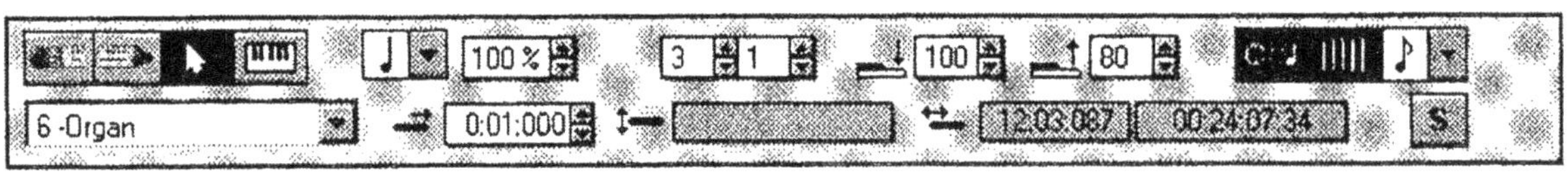

ACTIVITY 2: THE MIDI SYSTEM

PURPOSE

This activity seeks to familiarize the student with the hardware and software of the MIDI music system.

SUMMARY

This activity requires the students to review the hardware used in the operation of the computer and the Windows system software. It also introduces the students to the sequencer software. Wiring-up the hardware, loading and using the software will also be learned.

PROCEDURE

Part I:

The hardware components of the computer and its operating system will be reviewed. The role and use of the following commands will be also be reviewed.

a. Start
b. Loading
c. Minimize, Maximize, Restore, Close
d. Formatting a disk
e. Switching Windows
f. Manipulating Windows
g. Organizing Windows

Part II:

The hardware will be wired-up according to the following schematic using the **gray** MIDI interface cables:

One MIDI cable runs from the OUT on the MIDI interface at the rear of the computer to the IN on the synthesizer

The second MIDI cable goes from the OUT on the synthesizer to the IN on the MIDI interface of the computer.

NOTE: Using the **black** MIDI interface cables it is IN to IN and OUT to OUT.

The Y cable has its two ¼" plugs leaving the rear of the SYNTHESIZER (MT-32) and the other single ¼" plug connected into the single port on the preamplifier.

The voltage converter for the synthesizer plugs into its back panel and the other end is plugged into the electrical power outlet on the table. The power converter for the preamplifier is also plugged into the power outlet.

BEFORE PLUGGING ANY DEVICES INTO THE ELECTRICAL MAINS, YOUR CIRCUIT MUST BE CHECKED BY THE INSTRUCTIOR.

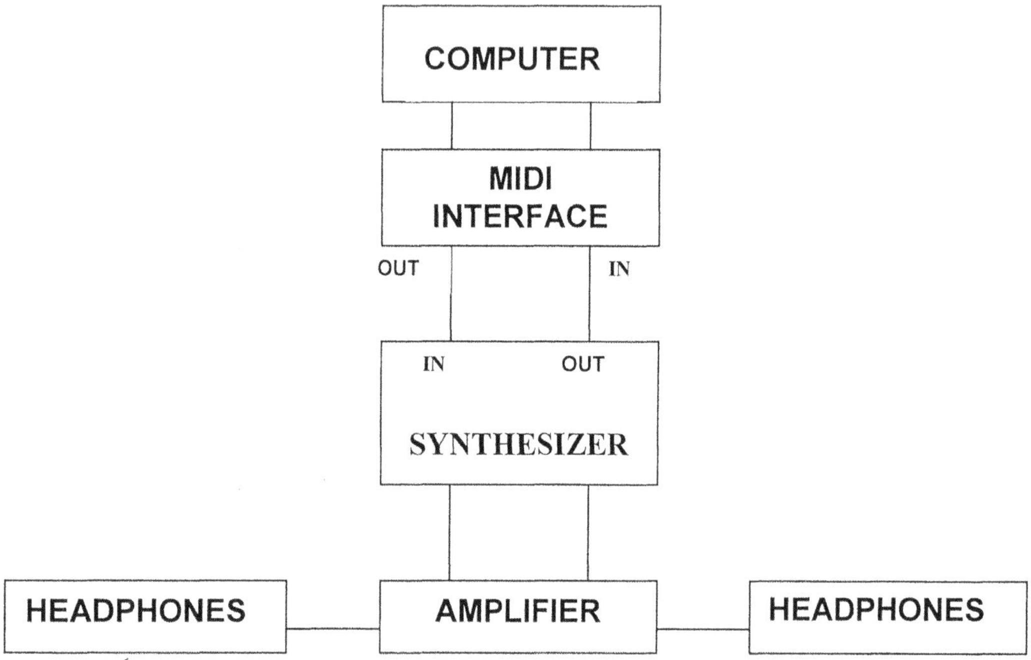

Part III:

Use the attached ***Record Producer Deluxe*** summary instruction sheet to:

1. Load the software.
2. Load a file.
3. Play a song.
4. Use the Piano Roll Screen.
5. Use the Mixer Screen.
6. Change a patch number.
7. SOLO an instrument.
8. MUTE an instrument.
9. LOOP a line.
10. Copy a track.
11. Set the tempo.
12. Explore the program.

ACTIVITY 3: WAVE MOTION AND SOUND

Purpose

In this activity we shall learn something about the characteristics of transverse and longitudinal waves.

Summary

This activity has three parts. In the first part we study the relationship between a vibrating string that is tuned to the sound of a tuning fork. In the second activity, we study resonance in a closed tube, and in the third, standing waves.

Part I:

The first part is designed to study the relationship between the frequency of a tuning fork and a wire under tension.

Procedure

The monochord consists of a tightly stretched wire extended over three bridges. The center bridge is moveable and may be used to vary the effective length of the wire. The wire is always plucked halfway between the fixed and moveable bridges so as to have the wire vibrate in one segment corresponding to its fundamental mode of vibration. Since the wire may break, please take care not to bend over it while plucking.

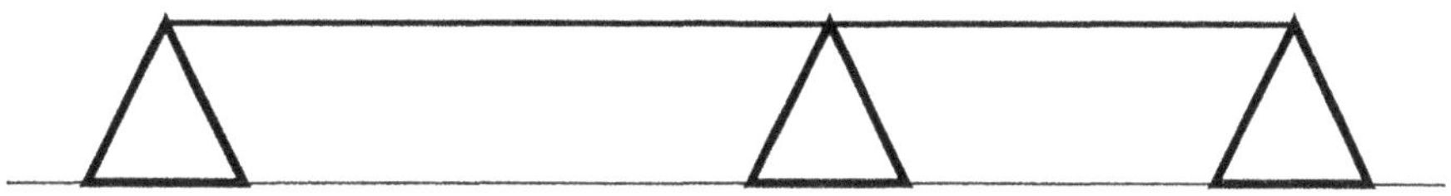

Choose one of the four tuning forks and strike it with the rubber hammer. Place its end on the monochord's wooden platform so as to force it to vibrate at the same frequency as the fork making the sound louder. Now pluck the wire to have it vibrate and sound the same as the fork. You will have to adjust the moveable bridge until the wire sounds the same frequency.

When the wire vibrates at the same frequency as the fork, its center should be an antinode. To check your measurement: place a small piece of paper on the wire at its center between the moveable and fixed bridges. Strike the fork and hold it against the board. If the paper jumps off, the fork and wire are vibrating at the same frequency. Measure the length of the wire from the fixed bridge to the moveable one and record in the table along with the frequency of the fork. Repeat for the other three forks.

Fork frequency f Hz				
Length L cm				
f x L cm/s				

Average (f x L) = ___________

From the average value of f x L we may determine the frequency of an unknown fork, f_x, by striking it with the rubber hammer and determining the length of wire L_x on the monochord that vibrates with the same frequency.

$$f_x = \text{Average } (f \times L)/L_x$$

$$f_x = \_\_\_\_\_\_\_\_ \text{ Hz}$$

Part II:

The second part of this activity is to study the relationship between the frequency of a tuning fork and the length of a resonant air column.

Procedure

Strike the tuning fork and hold it over the resonance tube. Move the inner tube from the lowest position upwards until the sound is loudest and then clamp it in place. Measure the length of the tube (using the meter stick provided) from the top of the water to the top of the tube. Record the frequency and length L of the tube in the table.

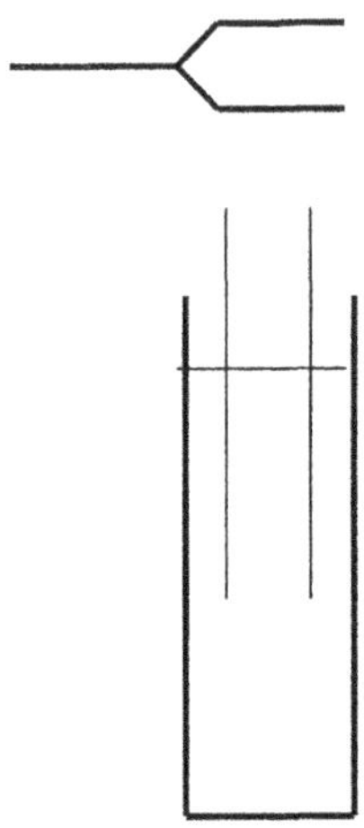

Diameter of tube used,..........cm. ΔL,...........cm.

Readings

Freq. F	L cm	$L_o = L + \Delta L$	$F \times L_o$
		Average $F \times L_o$	

NOTE: A correction factor must be added to the length of the tube L since the reflection at the interface takes place slightly above the tube. The correction factor, ΔL, is equal to 1/6 the diameter of the tube.

In a closed tube at resonance, $\lambda/4 = L$. Therefore, $\lambda = 4L$. Now the velocity of a wave motion such as sound is given $v = f\lambda$. So in this case, $v = f(4L)$ or $v = 4$ times the average value of $(f \times L_o)$.

Velocity of sound at room temperature will be

$$v = 4(f \times Lo)_{avg}$$

v =__________ cm/s

Part III:

The third activity involves a study of the modes of vibration of a fixed length of string at constant frequency as the tension is varied. (Meldes' Experiment)

Procedure

A long string has one end fastened to an electromechanical vibrator and the other end hangs over a pulley with a small box attached. The vibrator oscillates with a fixed frequency.

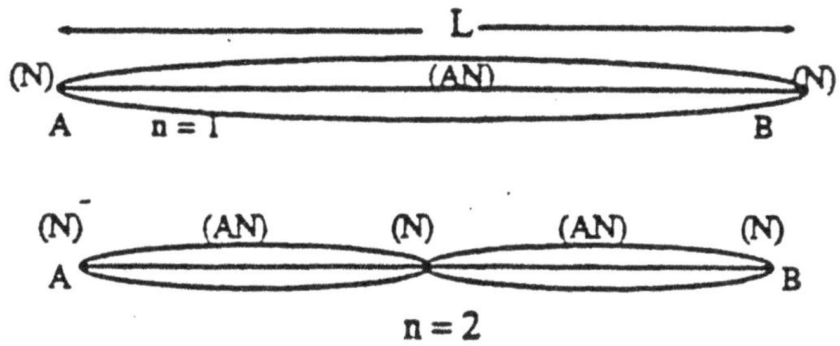

**

We would like to verify that

$$f = nv/21 \text{ but } v = \sqrt{(T/\rho)}$$

$$f = (n/21)\sqrt{(T/\rho)}$$

squaring both sides $\quad f^2 = (n^2/41^2)\, T/\rho$

multiplying both sides by $41^2\rho$ $\quad n^2T = 41^2f^2\rho = K$

or $\quad n^2T = K$

By turning the electromechanical vibrator on, we cause the string to vibrate in many segments. By adding weights to the attached box we increase the tension, causing the string to vibrate first in one segment (its fundamental). As we decrease the number of weights in the box, we can get it to vibrate in two segments, then three, four and so on. We record the value of the tension for each number of segments.

1. Determine the mass needed to cause the string to vibrate in one segment.

2. Use the value obtained in the first measurement to predict the value needed to cause the string to vibrate in two segments and then determine it exactly by adjusting the masses put into the box.

3. Continue the above procedure determining the tension to cause the string to vibrate in three, four, five, etc. segments. For each case, record the values in the table.

RESULTS:

n	T gm	n^2 T
1		
2		
3		
4		
5		

FOR CONSIDERATION AND DISCUSSION

1. What is a forced vibration?

2. What is resonance?

3. What are harmonics and overtones?

4. What are standing waves?

5. How do open tubes differ from closed tubes in terms of resonance?

6. How does a flute work? How is it similar to a resonance tube?

7. How does a guitar work? How is it similar to a monochord?

ACTIVITY 4: ORCHESTRATION

PURPOSE

The purpose of this activity is to familiarize the student with the sequencer software so that it may be used to orchestrate a given musical composition.

SUMMARY

The student will be asked to re-orchestrate the musical composition provided using the sequencer software.

Play button on the Toolbar

Song Position pointer

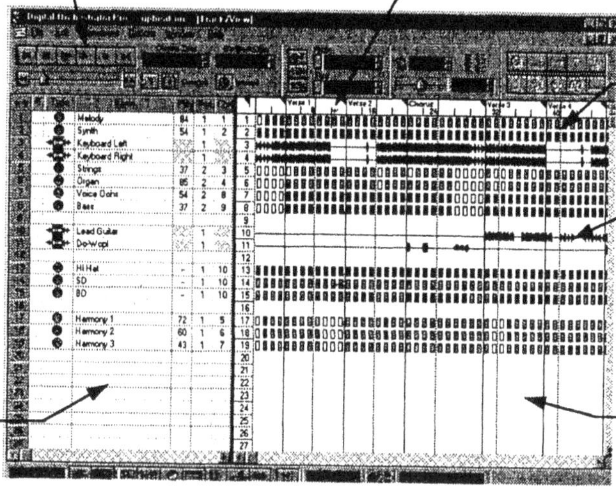

MIDI data boxes

Digital Audio waveform

Track pane

Bar pane

PROCEDURE

1. Wire up the parts of the MIDI system as in Activity 2. Be sure to have the instructor check your wiring before connecting to any power source.

2. Load the sequencer software and the musical piece suggested by the instructor. Play the piece to see that everything is operating, as it should. Use the **one page summary sheet** on how to use ***Record Producer Deluxe*** to perform the requested operations.

TRANSPORT CONTROLS

Previous Play From Beginning
Next Record Current Position
Stop Play

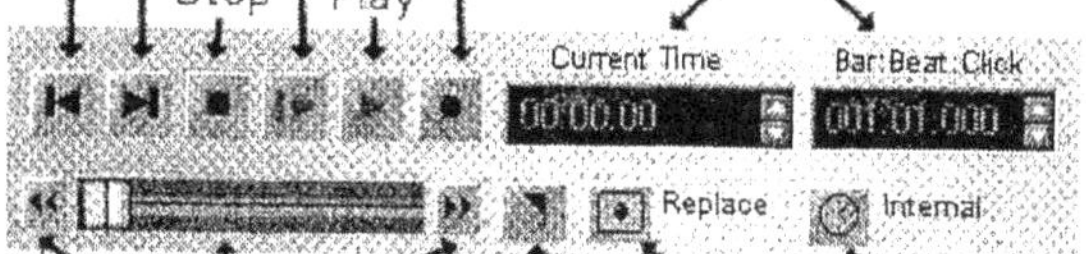

Go To Marker Sync Settings
Song Position Slider Overdub/Replace

3. Re-orchestrate the selection using the COPY command from EDIT in the formatting tool bar so as to have at least six instruments playing.

4. Make a permanent tempo change.

5. Use the MUTE or SOLO option to audition or selectively listen to certain tracts.

6. Use the TRACK OFFSET VELOCITY function to create the illusion of several of the same instruments playing simultaneously. The + and – keys are used to change the values.

7. Make a transposition of an interval using the TRANS column of an octave (12 half steps) on one track, which was copied from another an play just the original track with the transposed track against it.

8. Save your piece on your floppy disk and also on the Class Flash Drive. Obtain a printout of the Track/View screen recording all the changes you have made. Be sure to include this printout with your report.

9. Record your name and your partner's name on the notepad, which you will find at the bottom of the Track/View screen among the Quick View Buttons, with a one or two sentence summary of the activity.

QUICK VIEW BUTTONS

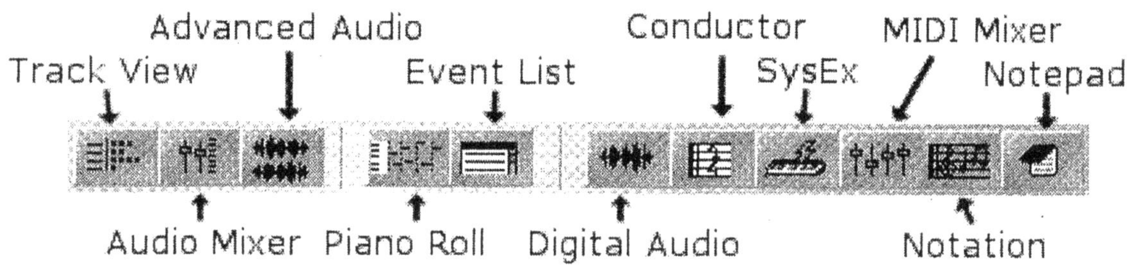

10. Be sure to exit the program properly. Unwire the system after turning all power switches off and return each component to its proper place. Remember, you are responsible for your set-up.

FOR CONSIDERATION AND DISCUSSION

1. Define tempo, offset, transposition, and half step.

2. What effects did all the changes lead to in terms of the re-orchestration?

ACTIVITY 5: PROPERTIES OF WAVES AND SOUND DEMONSTRATIONS

PURPOSE

The purpose of this activity is to examine and illustrate some of the properties of waves and sound.

SUMMARY

Some of the properties of waves are reflection, refraction, interference, diffraction, beats and polarization. In this demonstration activity an attempt will be made to illustrate these properties in concrete form using mechanical and electrical demonstration apparatus. A mixer allows us to switch our sound sources from the microphone to the sound generator and finally to the Fourier synthesizer allowing us to hear the sounds and view the waveforms on an oscilloscope.

Sound waves may be converted to electrical signals by means of a microphone, the amplitude of the wave at any instant being proportional to the voltage developed by the microphone. These electrical signals may be fed through an amplifier (to make the voltage bigger) to an oscilloscope which then displays the waveform as a vertical deflection against a horizontal time axis. If the signal is exactly repetitive and time base is synchronized with the signal, a steady wave form is obtained on the oscilloscope.

The second part of this activity is a **Nova** videotape entitled "What Is Music?" There are a number of questions you will have to answer based on your observations.

Part I:

PROCEDURE

A. Voice patterns and other demonstrations on the oscilloscope

1. Voice patterns such as vowel sounds, a tuning fork signal, a whistle, a horn, etc., are shown when sounded in front of a microphone. Note the shape of the complete cycle of the waveform when a tuning fork is struck and held up to the microphone. Pure notes have a sine wave, and a smooth wave shape. The shriller the note is, the greater the content of octaves of the fundamental (called harmonics) and the more distorted the wave shape.
2. Note what happens when two identical tuning forks are struck when one has a rubber band tied on one prong. Listen to the beats that are generated between the two forks and note the pulsating wave form on the oscilloscope.
3. Note the regularity in the wave pattern when the musical horn is blown as opposed to the type of pattern obtained when one generates a noisy clapping sound.
4. A definite pattern may be seen as each vowel sound is spoken into the microphone. What does this imply about voice recognition systems?

B. Sound frequencies and human frequency range

A sound generator is used to slowly sweep through frequencies ranging from 10 Hz to 30,000 Hz with the corresponding wave pattern shown on the oscilloscope and the sound played through a receiver and a set of four external speakers. The ear's sensitivity is noted by keeping the volume control on the receiver fixed and noting the increase in loudness for frequencies around 1,000 to 2,000 Hz. Try to determine the lowest frequency you can hear and the highest. With the sound generator set at about 200 Hz, you can observe the standing waves set up in the room by the four speakers by moving your head to and away from the speakers over about a two foot range and note the change in volume due to the presence of the standing waves.

C. Fourier Synthesizer

Note the pure waveform of the fundamental. What happens to the waveform as each harmonic is added? What happens to the sound you hear?

D. Multimedia – Tacoma Narrows Bridge

After watching the film clip on the Tacoma Narrows Bridge collapse, why do you think the bridge failed? What role did resonance play?

FOR CONSIDERATION AND DISCUSSION – Part I

1. Define each of the properties of wave motion. What is sound?
2. What is resonance? What is reverberation?
3. Why do instruments sound different when playing the same note?
4. What would happen if all the harmonics except for the fundamental were removed from a note played on a musical instrument?
5. What is a mixer?

Part II:

A **Nova** videotape on music and sound: **What is Music?**[1]

1. What is Music? Compare some of the different definitions of music mentioned by the participants.
2. What was the significance of Jean-Claude Risset's discovery at Bell Labs? What were Xavier Rodet's contributions, which enabled computers to imitate instrumental and vocal sounds more realistically? What is involved in realistic synthesis?
3. Compare and contrast how the organ, the Moog synthesizer, and the computer make music. Why is the organ referred to in the video as a thirteenth-century synthesizer? Why is the Moog synthesizer less adept at synthesizing sound than a computer?
4. Why is it so difficult to analyze how music and the mind interact? Cite an example from the videotape of how the mind creates acoustic illusions.
5. Why is Joseph Nagyvary, a chemistry professor at Texas A & M, considered a controversial figure in the world of music? What discoveries does he claim to have made about the nature of seventeenth and eighteenth-century violins? What different scientific tools has he used to make these discoveries?
6. Discuss why the neuroscientist Manfred Clynes believes that gestures may represent the emotional message of music. Describe Clynes's experiments. Why does Clynes conclude that music is a universal language of emotions? Do you agree or disagree with Clynes' conclusions?[1]

[1]**What Is Music**? 58 min. Coronet Film & Video, 108 Wilmot Road, Deerfield IL 60015.

ACTIVITY 6: USING THE COMPUTER MOUSE TO ENTER A MELODY

PURPOSE

This activity allows the student to enter a melody and orchestrate it using the computer mouse.

SUMMARY

In this activity we shall learn how to enter a melody and chords into the computer and then orchestrate it for six instruments.

PROCEDURE

1. Wire-up the MIDI system. Load ***Record Producer Deluxe***. In the Track/View screen click on the Piano Roll Window button. The Track View window is your overview of the song. In the left hand pane of the window you set the parameters associated with each track while the right hand pane has all of your song data.

QUICK VIEW BUTTONS

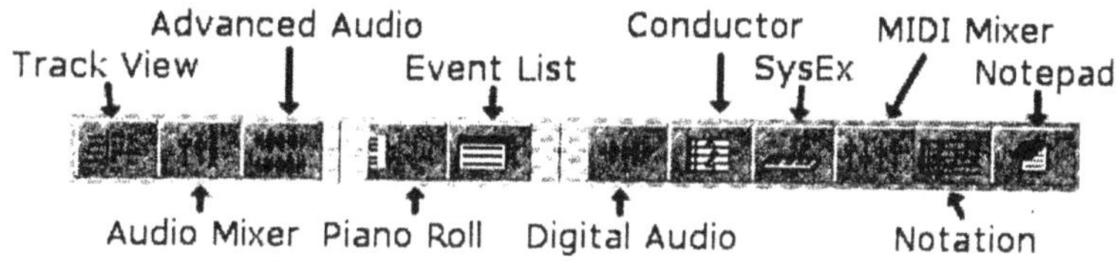

2. Click the **Snap-to-Grid** button to turn it on and set its value to the appropriate note by clicking on the control until the note appears. Check to see if the **Articulation** is set to 100%.

PIANO ROLL VIEW

The piano roll view gets its name from the old player piano paper rolls with spaces cut out to play notes. In this view you can edit any MIDI track in a graphical manner.

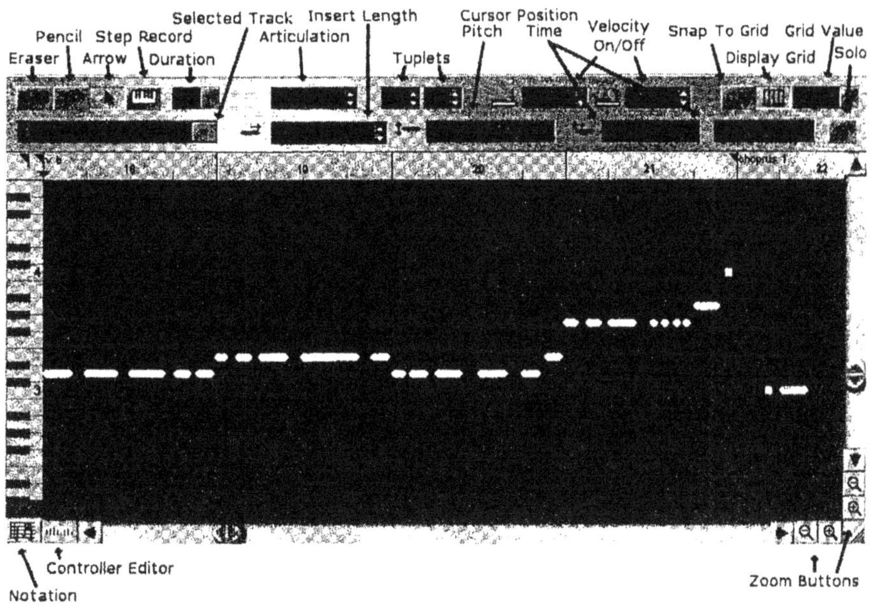

3. Using the music provided, use the **Pencil Tool** and the mouse to enter each note with the appropriate pitch and the correct duration as shown on the music sheet. Notes can be inserted with preset lengths or duration by first clicking on the **Note Button** in the Piano Roll Tool bar.

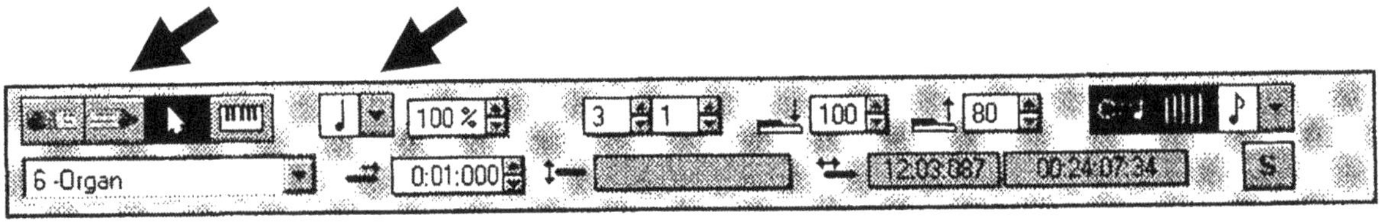

The pitch, length and start-time of each note may be edited using the **Arrow Tool**. To change the **pitch**, place the cursor on the middle of the note until the up and down arrow appears and using the mouse the note can be dragged to the appropriate pitch line. To change the **start time** of the note, place the cursor at the beginning of the note on the screen until the horizontal double arrow appears and drag the note to the appropriate start position. To change the **length or duration** of the note, place the cursor on right end of the note until the single-headed horizontal arrow appears and extend it by dragging it to the right. The **Eraser Tool** deletes a note.

Pencil Tool

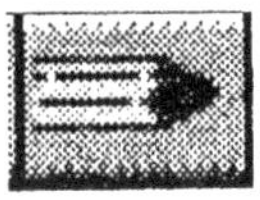

Eraser Tool

Arrow Tool

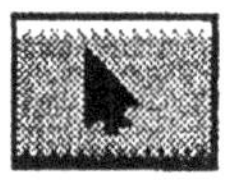

The **Notation button** turns the Notation pane on and off. It darkens when it is selected. The current track then appears in conventional music notation at the bottom of the Piano Roll Window.

Notation Pane

4. Enter the chords below the melody line but on a new track. The notes for the chords are given on the chord sheet.

5. Orchestrate the song for six instruments and save it on your floppy disk.

6. Obtain a printout of the Track/View screen to be turned in with your report. Your name and your partner's name should be printed on it.

7. Document your work on the built in notepad. Save your work on your own disk.

8. Save your work on the Class Flash Drive provided by the instructor.

ACTIVITY 7: USING STEP RECORDING AND A PIANO KEYBOARD TO ENTER A MELODY

PURPOSE

In this exercise the efficiency of entering a melody using a piano keyboard and step recording is revealed.

PROCEDURE

1. Wire up the system with the piano keyboard and the synthesizer according to the diagram below.

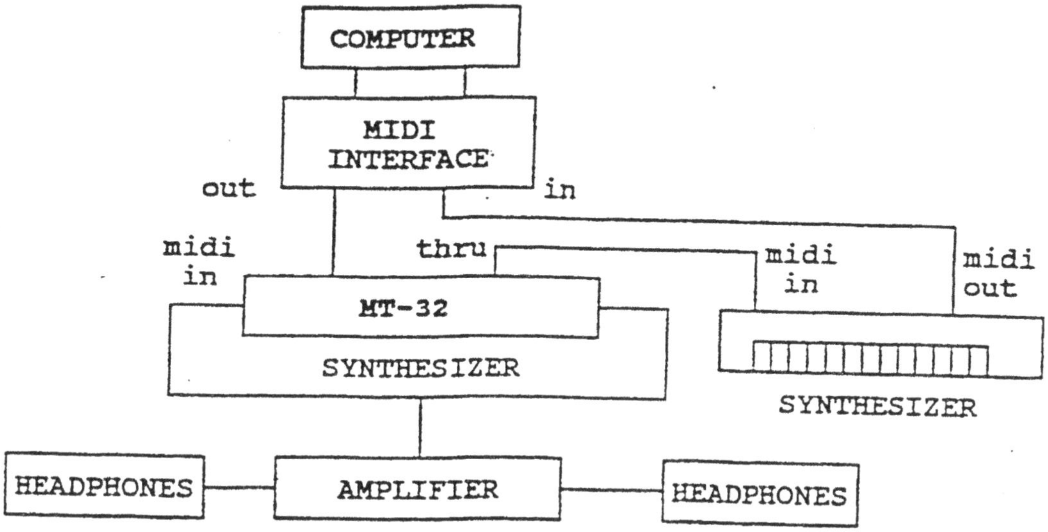

2. Connect one of the gray MIDI cables from the OUT of the MIDI interface on the computer to the IN of the synthesizer and from the Thru on the synthesizer to the IN of the keyboard. The other MIDI cable is connected from the OUT port of the keyboard to the IN of the computer. (Remember for the black MIDI interface cables it is IN to IN and OUT to OUT.)

3. Connect the split wires to the back of the synthesizer. Make sure that the "dead" plug without any wire lead is plugged into the headphone jack of the keyboard to silence its speakers.

4. Boot-up ***Record Producer Deluxe*** and switch to the Track/View window. Step recording is the process of entering notes one step at a time (in non-real time).

5. Select a channel number other than one. Select a Patch (instrument) number for the instrument you will be using to record.

6. Open the Piano Roll Window and Click the Step Record (keyboard icon) button to enable the Step Record mode.

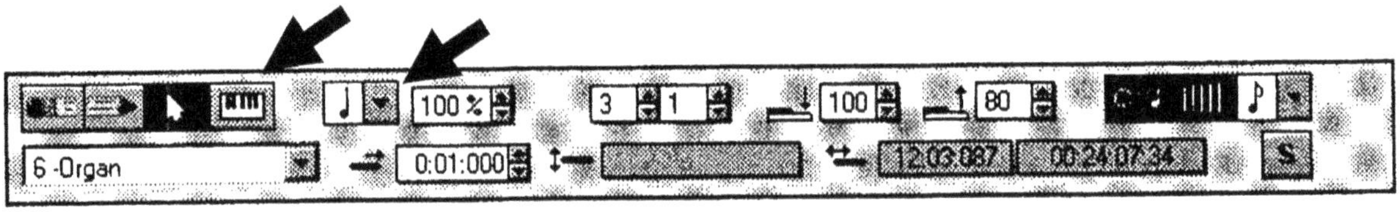

7. Click the location on the Piano Roll screen where the first note should be input and set its duration in the Note Duration controls.

8. Play the note on the keyboard and continue inserting notes in a similar fashion by changing the durations and pressing the correct key on the piano keyboard.

9. Input the notes from the song provided using the piano keyboard.

10. Enter the chords on a new track below Middle C by holding down the first note in a chord while entering all of the other notes in succession.

11. Orchestrate the musical piece for six instruments as was done in the previous laboratory activity.

12. Use the Notepad to document your work and save it on your disk. Also record your work on the Class Flash Drive provided by the instructor and print a copy of the Track/View window,

FOR CONSIDERATION AND DISCUSSION

1. How does this method of recording compare to using the computer mouse to enter a melody?

ACTIVITY 8: ENTERING A SONG WITH A DRUM RHYTHM TRACK

PURPOSE

In this activity an attempt will be made to develop a rhythm track that can be used with a newly entered melody using the Casio keyboard.

SUMMARY

The Casio keyboard has 61 touch response keys. It has five built in tones (piano, harpsichord, vibraphone, electric piano and pipe organ) provided by an advanced PCM sound source using 12-bit sampling technology. All tones are clear and true to life since they are based on multiple samples of instrument sounds at various key attack levels. It is a 16 note polyphonic keyboard and when the power is turned on the MIDI basic channel is automatically set to "1".

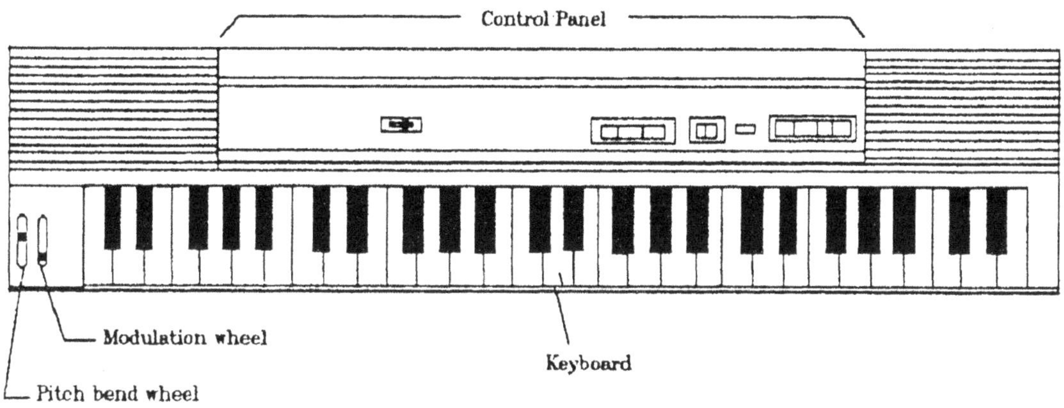

PROCEDURE

1. Wire up the system with the MT-32 and the Casio CPS-300 as done previously. Connect the headphone amplifier to the MT-32 output jacks and plug in the headphones into the preamplifier. Insert the ¼" jack provided into the Casio headphone outlet so as to silence the built-in speakers on the Casio keyboard.

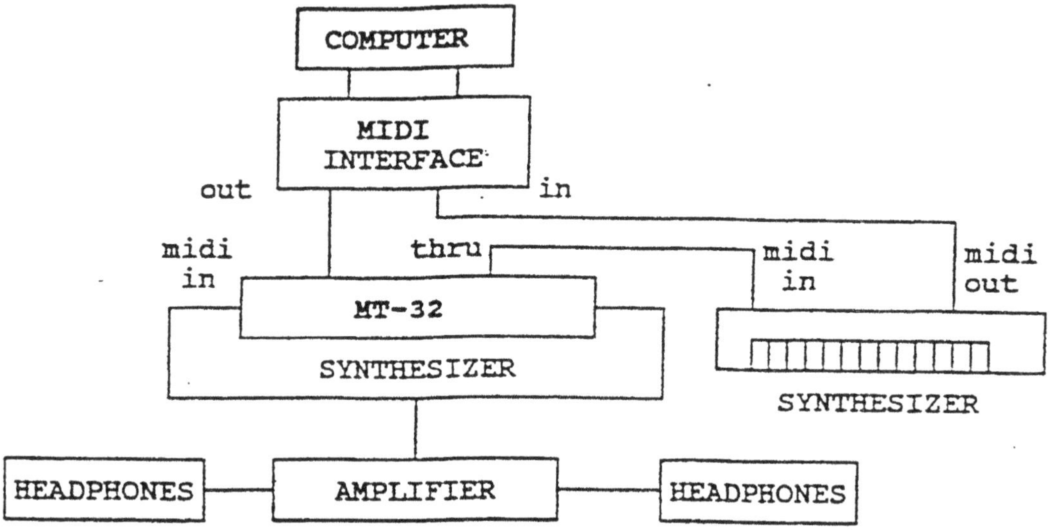

2. Enter the song provided by the instructor using the piano keyboard and orchestrate it for six instruments.

3. Switch the CPS-300 power switch on while holding down the PIANO tone selector and pressing the key on the keyboard corresponding to channel 10 (i.e. the tenth key counting both black and white keys from the left). Channel numbers are assigned to the keyboard keys as shown below. Drum sounds should be heard as the keys on the keyboard are pressed.

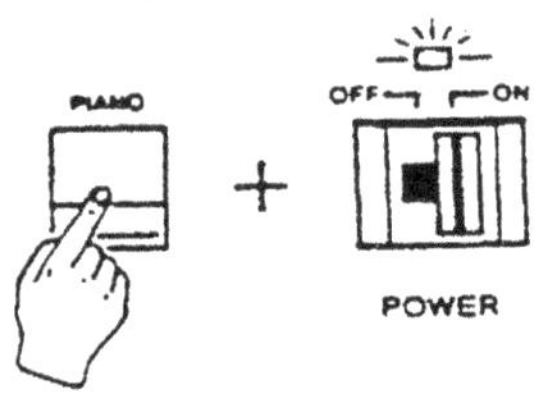

CH	1	2	3	4	5	6	7	8
Keyboard	C2	C♯2	D2	E♭2	E2	F2	F♯2	G2

CH	9	10	11	12	13	14	15	16
Keyboard	A♭2	A2	B♭2	B2	C3	C♯3	D3	E♭3

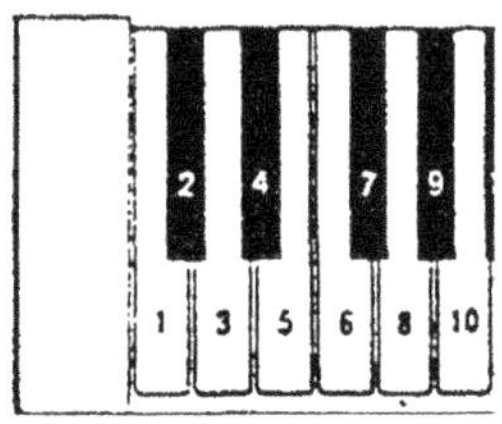

4. To record a drumbeat, set the channel number on a blank track to 10. This allows one to access the rhythm section of the MT-32 synthesizer, which has 30 sounds in the rhythm part. Remember, the MT-32 is 32 note polyphonic meaning it can play 32 notes simultaneously. It is multitimbral, meaning it can play 8 sound parts of the 128 voices it contains in its sound library (and one rhythm part simultaneously).

5. Below is a list of rhythm instrument parts contained in the MT-32 with the note number assigned to each voice. The appropriate keys on the Casio keyboard have been labeled with the drum sounds available.

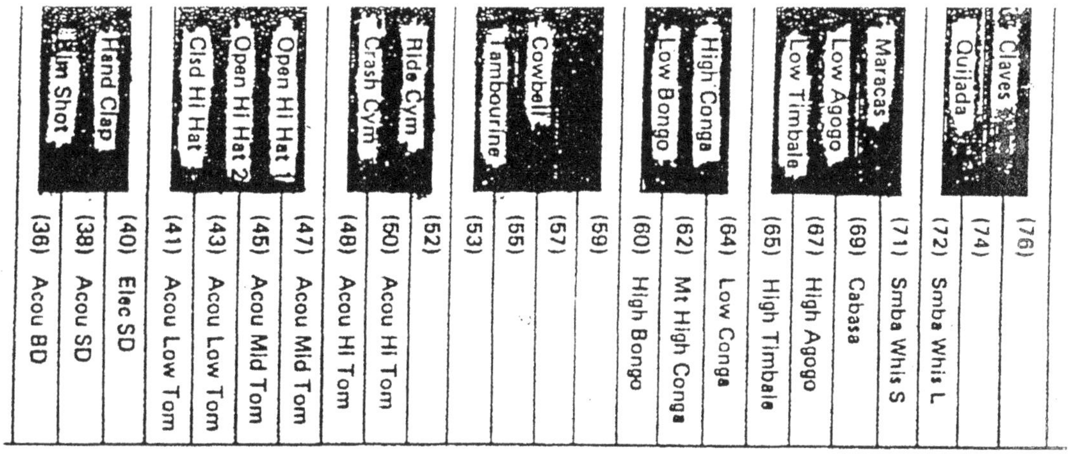

6. With the Channel number on the track set to 10, try generating a repetitive beat or drum sound appropriate to the melody it's to accompany by playing it first alone. Press the spacebar to play your song and accompany it on the keyboard with the beat you like.

7. Once you have found the correct sound record it on the empty track by setting the Type column to the MIDI plug symbol and clicking the R column to enable the track for recording. Check to see that the Channel number is set to 10 on the track, Next press the circular red Record button on the Transport Bar to start recording. The computer will give you a four beat count in after which you may accompany your song by playing and recording it in real time.

8. Using the built-in notepad document your orchestration, save it and place a copy on the Class Flash Drive circulated by the instructor. Be sure to obtain a printout of the Track/View screen for your lab report.

ACTIVITY 9: DIGITAL AUDIO RECORDING

PURPOSE

In this activity a digitized vocal track will be added to a song.

SUMMARY

After recording a song with melody, chords and rhythm section and orchestrating it for six instruments, a vocal track will be added using the stick microphone. In addition, the editing features of ***Record Producer Deluxe*** will be used to cut and replace pieces of the song if necessary.

PROCEDURE

1. Wire up the system with the MT-32 and the Casio CPS-300 as done previously. Connect the headphone amplifier to the MT-32 output jacks and plug in the headphones into the preamplifier. Insert the ¼" jack provided into the Casio headphone outlet so as to silence the built-in speakers on the Casio keyboard.

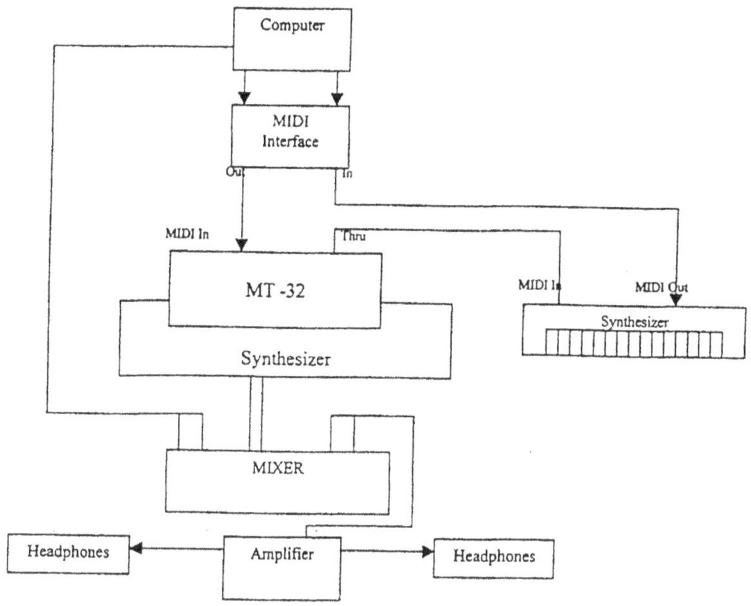

2. Enter the melody, chords, and rhythm sections for the song provided and orchestrate it for six instruments.

A. Digital Audio Recording

Select the next available free track and click the mouse in the Type column until the icon changes from the MIDI plug symbol to the Digital Audio waveform. Click the Enable Record column until the letter R appears.

Digital waveform

MIDI plug symbol

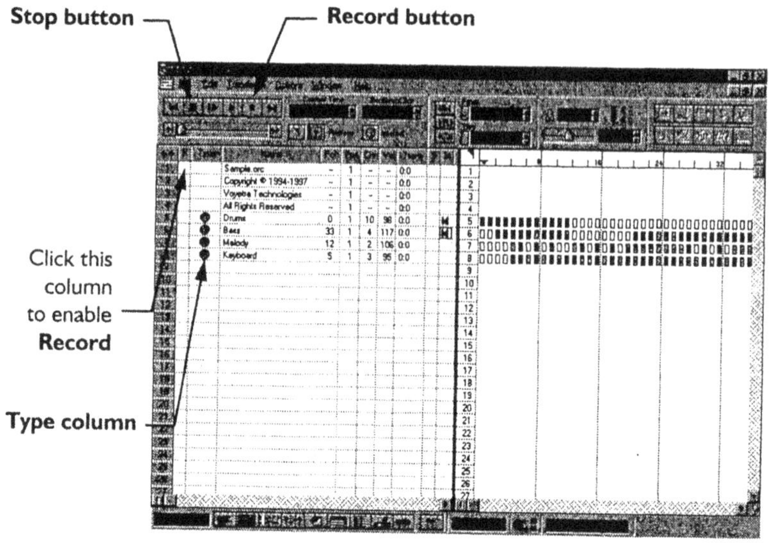

1. Clear your throat. Click the Record button on the Transport Bar to start at the beginning and after the four beat count-in, start singing into the microphone. If necessary, make-up the lyrics as you go. Don't use up precious disk space by recording from the beginning of the song. Hit the record button when the vocal begins.

TRANSPORT CONTROLS

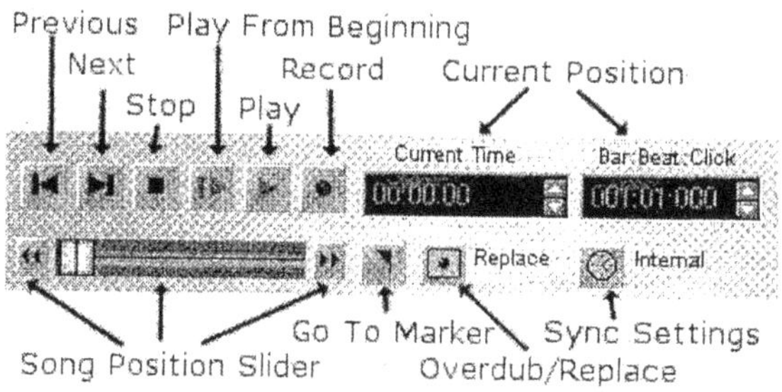

2. To end the recording, click the STOP button or press the Spacebar. ***Record Producer Deluxe*** draws a waveform of the audio in the Track Pane.

B. Punch In Recording

1. If you want to rerecord a small section, use "Punch In" recording. Suppose you want to redo bars 8 to 40. Click and drag in the Track Pane to select the range. Place the Position Pointer (upside down red triangle) over the first selected bar and click *From* on the Range Control. Then put the position pointer over the last selected bar and click on *Thru* on the Range Control. Click on Play Range to hear part of song selected or click on Record to record over selected bars.

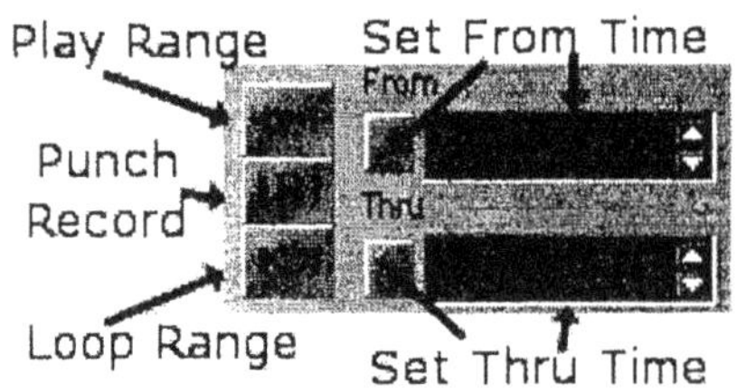

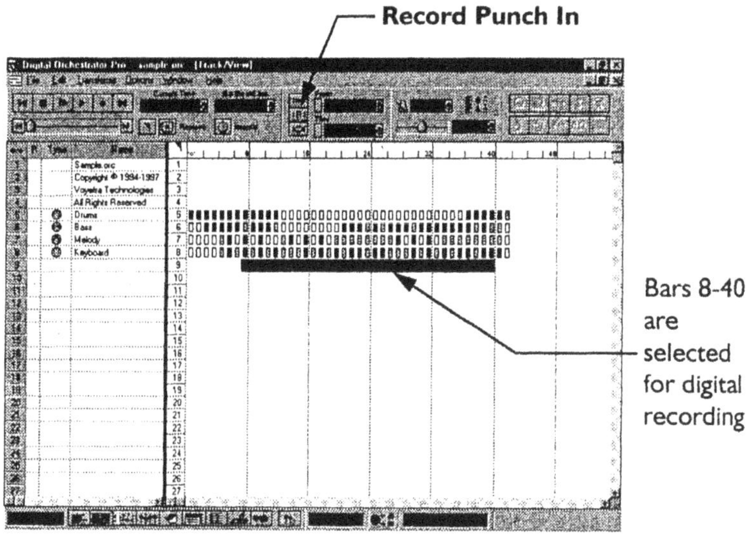

C. Adjusting Recording Volume

The input volume level is set in the PC Audio Mixer application for the sound Use the mixer software to adjust.

D. Editing

1. In the Track/View window, double-click on the vocal track, just to the left of the 4th bar as an example. This opens the Digital Audio editing window with bar 4 at the left of the screen. You can use the Zoom tool for an even closer look.

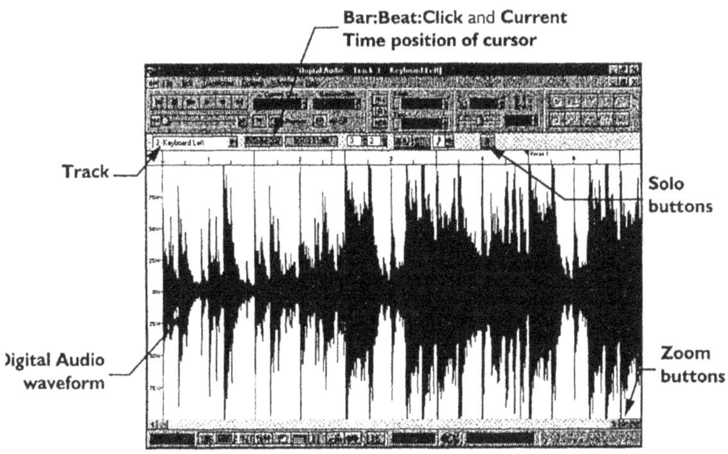

Digital Audio Window

2. Edit out silences to save disk space. For example select a small amount of data in the area of bar 8 and zoom, choosing Range. Select silent data (flat black line) and use the Cut command from the Edit menu. Unrecorded areas of silence are shown as a red line and do not take up disk space.

3. After the data is cut, go back to the Zoom menu and select Default to return to the original view. Select data around another bar and cut any area of silence. Hit the Play button to hear the leaner digital track.

Record a copy of your work on the Class Flash Drive. Be sure to document your work in the Notepad and obtain a printout of the Track/View screen showing the digital audio in recorded form. Be sure to include a copy of the printout in your lab report.

ACTIVITY 10: ELECTRIC VOLTAGE, CURRENT, RESISTANCE

PURPOSE

This activity familiarizes the student with using the Digital Multimeter in measuring potential difference, current, and resistance in a simple circuit illustrating the use of Ohm's law.

SUMMARY

Electric current is a measure of the flow of charge in an electric circuit. Electric current tells us how much charge flows by any point in circuit in one second. Electric potential or voltage is a measure of the energy differences that electric charges have between any two points in a circuit. Resistance measures the ability of the conducting material to inhibit the flow of charge when a current is passing through it.

PROCEDURE

A. Measuring Voltages

1. Set the dial on the Radio Shack Digital Multimeter to the DC volts setting with the black banana patch cord attached to the common (COM) or center socket on the meter and the red patch cord in the DC 1000 V right hand socket.

Observe that the power supply has several numbered and colored sockets. Plug the black patch cord from the multimeter into the common socket on the power supply and then attach the red patch cord successively to the +5 V, -5 V, 12 V and 2-20 V sockets recording your reading of the voltages below.

+5 V = __________; -5 V = __________; +12 V = __________;

2-20 V (min) = __________; 2-20 V (max) = __________

B. Measuring More Voltages

1. Wire up the circuit as shown below. Use the resistor marked A , the miniature light bulb and connect them to the power supply in a series circuit.

2. Have your circuit checked by the instructor. Turn the variable voltage knob on the power supply to minimum (fully counterclockwise). Turn the power on and set the power supply to 5 volts.

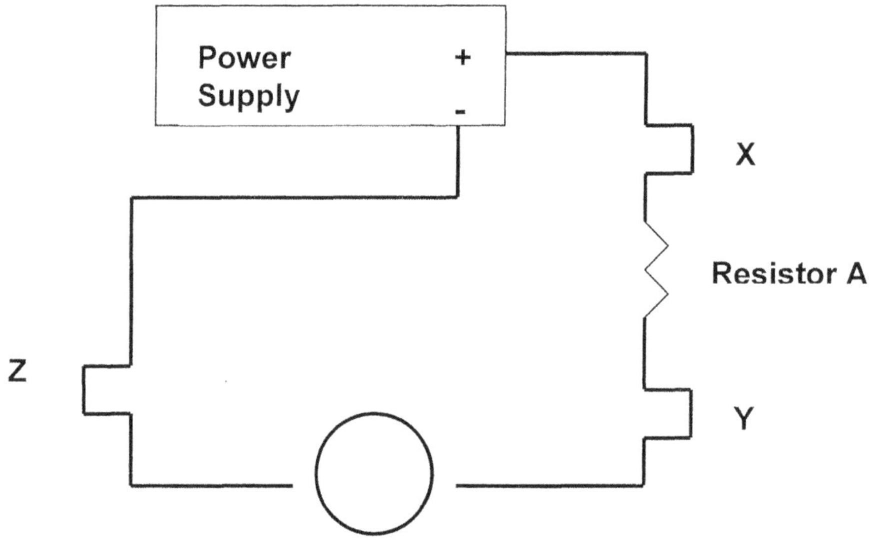

3. Using the multimeter as before with the black probe plugged into the common socket and the red probe in the DC 1000 Vvolts socket set the dial to DC V. Measure the voltage on either side of the resistor A (between **X-Y**) by pressing the probes to the terminals where it is connected to the circuit and record in the table. Repeat by measuring the voltage across the miniature light bulb (between **Y-Z**). Finally measure the voltage across the power supply (between **Z-X**) and record.

4. Remove resistor A and replace it with resistor B. Once again with the light bulb lit measure the voltage across resistor B, the light bulb, and the power supply and record the values in the table.

Table 1

Voltage across resistor A	Voltage across light bulb	Voltage across power supply

Voltage across resistor B	Voltage across light bulb	Voltage across power supply

When you add the voltages in the first two columns they are approximately equal to the voltages in the third column. Why?

C. Measuring Currents

1. Set the probes in the multimeter with the black one in common and the red in the current DC mA/KΩ port. Turn the dial on the meter to DC mA. Connect the meter into the circuit at point **X** in series by removing the jumper wire at that point and replacing it with the probes from the multimeter. The light bulb should be on. With the meter dial set to DC mA read and record the current in the table.

2. Remove the meter probes and replace the jumper at **X**. The light bulb should be on. Now remove the jumper at **Y** and replace it with the meter probes. Read and record the current. Replace the jumper and repeat the operation for **Z** recording all currents in the table with resistor A in the circuit.

3. Replace resistor A with resistor B and repeat the three current readings as before recording all the current readings at **X, Y, Z** in the table for resistor B.

Table 2

Current at **X** with resistor **A**	Current at **Y**	Current at **Z**

Current at **X** with resistor **B**	Current at **Y**	Current at **Z**

What do you observe about the currents at **X, Y, Z** for each resistor? What may you conclude about the current in a series circuit at any point?

D. Measuring Resistances

1. Turn the power supply off and unplug it. Remove the bulb and resistors A or B from the circuit board. Set the red probe in the multimeter to DC Ω and set the dial to k Ω. The black probe should be in the common port.

2. Place the probes across the ends of resistor A as it sits on the table and read the meter and record below. Next do the same for resistor B.

Measured resistances

Resistor A = __________ Ω Resistor B = __________ Ω

Now, calculate the resistance of A and B using Ohm's Law in the following way. Since

$$I = V/R$$

$$R = V/I.$$

Use the average value of the current measured at **X,Y,Z** with resistor A. Use the voltage V across the resistor A (between **X-Y**) and calculate R. Repeat the calculation for B.

Calculated resistances from data:

Resistor A = __________ Ω Resistor B = __________ Ω

What do you notice?

E. Measuring More Resistances

1. Using the multimeter, measure the following resistances by placing the probes across each end of the wires A, B, C, D and record in the table.

Table 3

Resistance in Ω	Resistance Resistor in Ω
Long 100 cm wire	Short 50 cm wire
Thick 50 cm wire	Thin 50 cm wire

2. Measure the resistance of the cold 75 W light bulb in the same way and record below.

R = ______Ω (cold)

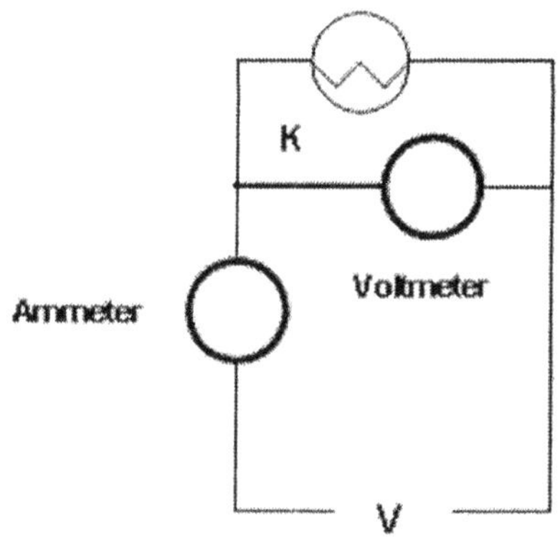

3. Screw the cold 75 W bulb into the circuit with the AC ammeter and voltmeter. Plug it into the electrical socket being careful not to touch any exposed terminals. **DANGER.** Next read the value of the current and record. Read the voltage and record. Using Ohm's law, compute the value of the resistance of the hot bulb in the following way: I = V/R so R = V/I.

I = ______ A V = _______ V R = V/I = _________ Ω

Calculate the Power: P = IV = ___________A x___________V = _________ W.

FOR CONSIDERATION AND DISCUSSION

1. What general statements can you make about current at various points in a series circuit and voltage across different elements in the circuit?
2. Why does the intensity of the light bulb vary when resistor A or B is in the circuit?
3. Notice the difference in resistances between the short and long wires. Which has the greater resistance? Why?
4. Notice the difference in resistances between the thin and thick wires? Which has the greater resistance? Why?
5. Compare the resistances of the cold bulb and the hot bulb. Which is greater? Why?

ACTIVITY 11: PRINTING AND MIXING

PURPOSE

In this activity an attempt will be made to develop a familiarity with various options, such as printing music in sheet music form, and using the mixer window.

PROCEDURE

Wire up the system with the MT-32 and the Casio CPS-300 as done previously. Connect the headphone amplifier to the MT-32 output jacks and plug in the headphones into the preamplifier. Insert the ¼" jack provided into the Casio headphone outlet so as to silence the built-in speakers on the Casio keyboard.

1. Enter the melody and chords of the song provided by the instructor. Enter a drum rhythm. Orchestrate the song for six instruments.

A. To Print the Music

1. Open the Notation window by clicking its Quick View button (second button from the right on the Status Bar). Maximize the window.

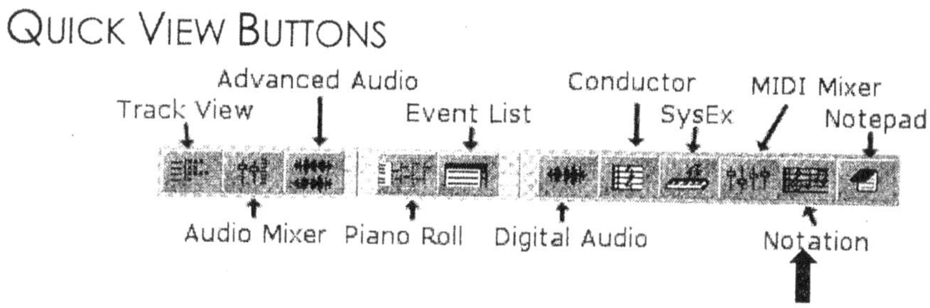

2. Click the Transcribe button on the Notation window control bar.

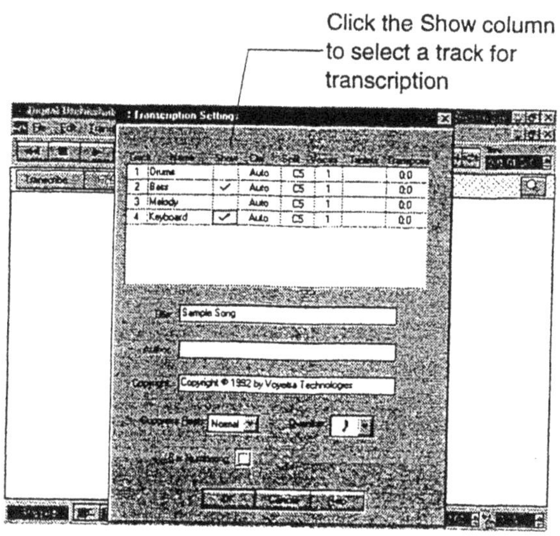

3. Click the Show column for one melody track and one chord track to select them. Click OK.

4. After a few seconds, ***Record Producer*** displays them in standard musical notation. Each part is identified at the top of the first bar.

5. Click the Print button on the Notation window control bar and use the standard settings of 8 1/2 x 11 and portrait orientation. Print only one page to hand in with your report. Click OK.
6. Document your work in the Notepad, and print a copy of the Track/View screen. Be sure to save your work on your disk and the Class Disk.

B. Mixer Window

1. Click the Mixer Window Quick View button at the bottom of the screen. The mixer window gives you a quick and easy way to balance volumes and try out patch and controller setting.

2. You can create an optimum mix by fine-tuning the relative volume of instruments. You can audition different sounds (patches) for a particular part. You can adjust controller settings such as reverb and chorus.

a. Channel Info

By clicking on the channel module's Info button, you can access the Channel Info dialog box.

Tracks on Channel lists the names of all the tracks assigned to the MIDI channel for the selected channel module. You can then select tracks for naming or transposing.

Selected Tracks allows you to edit the track name in this box.

Patch allows you to choose one of the 128 instruments available.

Transpose allows you to change keys for a selected track by semitones or octaves.

b. MIDI Density Meters

The LED meters indicate the relative density of the notes and other MIDI activity on each channel. As the song plays, the MIDI density meters blink to show which channels are active. The number of LEDs lit gives a relative indication of how much MIDI data is being played on a track.

c. Controller A/Controller B Settings

Above the row of Solo buttons are two rows of knobs labeled chorus and reverb and, farther down, another row labeled Pan. These knobs operate the MIDI controllers. By selecting Controllers from the Options menu, you can reassign the two red controller knobs so that they will operate the MIDI controllers used most frequently.

d. Solo

The Solo buttons allow you to solo a particular MIDI channel, which silences all the other channels. When a channel is soloed, its LED will be lit. If the Ctrl key is held while clicking on the Solo buttons for each of the desired channels, a group of channels may be selected.

e. Mute

The Mute buttons allow you to silence a channel on playback. Just click the Mute button on the channel to be silenced.

f. Pan

Pan is used for stereo playback. It sets the channel's left/right positioning. Turn the knob to the left or the right with the mouse to set that channel in the stereo field.

g. Patch Selector

This control allows you to select an instrument sound for each channel. It is one of the most valuable aspects of MIDI synthesis. To change the patch, click the patch selector. The number will increase if you click the top half of the window and decrease if you click the bottom half.

h. Volume Sliders

The volume sliders let you adjust the volume of each channel so you can create a mix for the song. Drag them up and down with the mouse to raise or lower the playback volume of each channel independently.

i. Port

The Port control allows you to select the output port that is to be controlled with the Mixer window. In our case we have only one port: Port 1.

j. Track Name

The display shows the track name for each channel. By clicking on the module's Info button you can select a different track name to be displayed here.

Use the Mixer window to modify and enhance the song previously recorded. Be sure to try each control to see how it operates.

Save a copy on the Class Flash Drive and print a copy of the Track/View window..

FOR CONSIDERATION AND DISCUSSION:

1. How does the Mixer window help to customize your song?

ACTIVITY 12: ELECTROMAGNETIC FIELDS, ELECTROMAGNETIC INDUCTION DEMONSTRATIONS, AND SYNTHESIZERS

PURPOSE

In this activity an attempt will be made to develop a familiarity and understanding of the effects of electromagnetic fields and electromagnetic induction effects through a series of observation. In addition a videotape on synthesizers will be shown.

SUMMARY

A number of demonstrations will be performed emphasizing Faraday's law of electromagnetic induction. A taped lecture on synthesizers will be shown and a number of questions about the tape will be posed.

PROCEDURE

Part 1: Demonstrations and Explanations

1. Electromagnetic induction of current by a wire crossing or linking with a magnetic field.

2. AC-DC generator principle

3. Induction coils: coil and galvanometer, Rumkorff coil (small and large versions).

4. Transformer principles and its parts.

Part II:

1. Videotape on synthesizers. **Science and Music : Scales, Synthesizers and Samplers.**[1]

FOR DISCUSSION:

1. What is the difference between a computer playing "Twinkle, twinkle little star, …" and a pianist? What goes on when you are playing an instrument?
2. Define and discuss the three types of synthesizers.
3. What is Theramin?
4. Why do electronic instruments sound electronic?
5. How do we make the electronic sound more realistic?
6. What is vibrato?
7. What is frequency modulation?
8. What is oral illusion?
9. What is a sampler?
10. Why bother with synthesizers at all?

[1] **Science and Music : Scales Synthesizers and Samplers.** 60 min. Films for the Humanities, Inc., Box 2053, Princeton, NJ 08540.

ACTIVITY 13: ENTERING A SONG IN REAL TIME USING A PIANO KEYBOARD

PURPOSE

In this exercise the efficiency and difficulty of entering a melody using a piano keyboard is revealed.

SUMMARY

In this activity we shall use a third way of entering music into the computer in real time using a piano keyboard.

PROCEDURE

1. Wire up the system with the piano keyboard and the MT-32 synthesizer according to the diagram below.

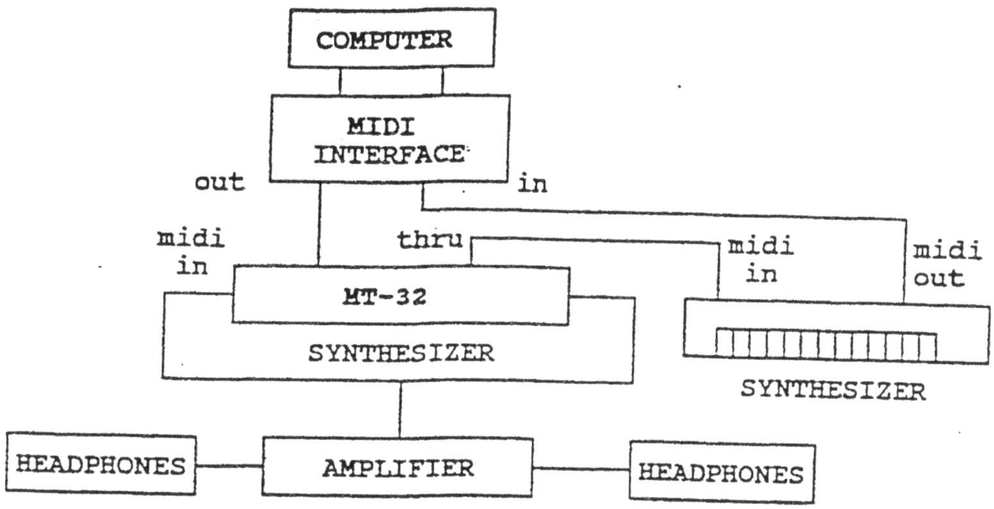

2. Connect one of the gray MIDI cables from the OUT of the MIDI interface on the computer to the IN of the MT-32 synthesizer. Run the short black MIDI cable from the THRU port of the MT-32 to the IN of the keyboard. The other MIDI cable is connected from the OUT port of the keyboard to the IN of the computer. (Remember for the black MIDI cables it is IN to IN and OUT to OUT.)

3. Connect the split wires to the back of the MT-32. Make sure that the "dead" plug without any wire lead is plugged into the headphone jack of the keyboard.

4. Boot-up ***Record Producer*** and switch to the Track/View window. You may want to set the tempo to a very low value (60) when entering the notes and then readjust it on playback. Click the Enable Record (R) column for Track 1.

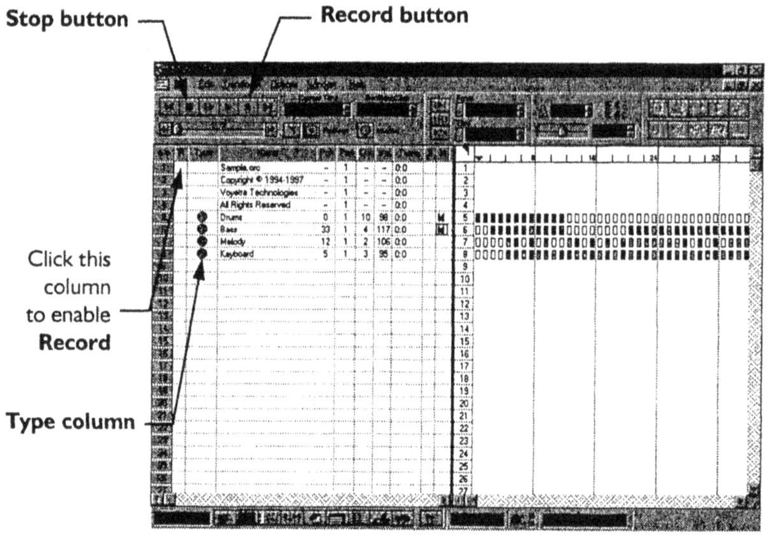

5. Click the Type column for Track 1. Make sure the Enable Record column is set to R. Check to see that the MIDI plug symbol is in the Type column. Select a channel number other than one. Select a Patch (instrument) number for the instrument you will be using to record.

TRANSPORT CONTROLS

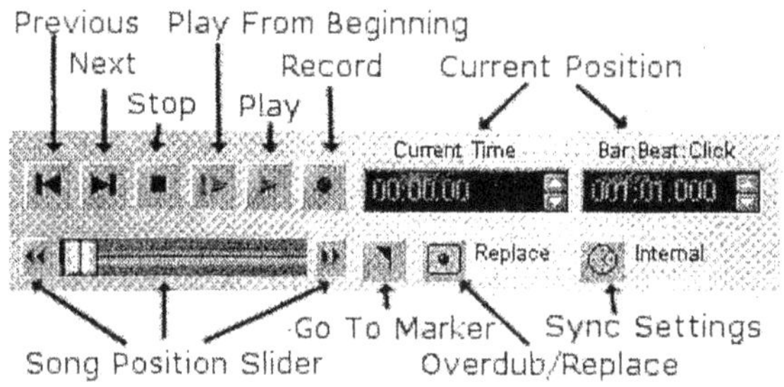

6. Click the Record button on the Transport Bar. After a four beat count-in, the Transport Status indicator will change from *Rec Ready* to *Recording* and recording actually begins.

7. Input the notes from the song provided using the piano keyboard. Depending on your keyboard skills you may want to input only four bars at a time and then stop by pressing the spacebar.

8. Move the Song Position pointer to the last bar recorded by clicking the right mouse button at the bar number just above the MIDI data boxes. This will move the Song Position pointer to that bar and then you may start the record procedure once more for the next four bars, and so on until the whole song is recorded. Play it.

9. Use the Piano Roll screen to edit and to adjust misplayed notes until the selection is perfect.

10. Orchestrate the musical piece for six instruments as was done in the previous laboratory activity. Use the Notepad to document your work and save it on your disk. Also record your work on Class Flash Drive provided by the instructor. Print a copy of the Track/View screen.

ACTIVITY 14: PROJECT PRESENTATION AND REPORT

PURPOSE

The purpose of this activity is to present the student with an opportunity to demonstrate the knowledge and skills in computer music they have gained in the laboratory over the past semester.

SUMMARY

In this activity each student or each group of students (up to four in a group) will have the opportunity to present an original project. The project may consist of a piece of music, original or not, entered and orchestrated by the student, the musical background for a video, etc. This is your chance to be creative.

PROCEDURE

In this laboratory activity the student or group will be required to document their work with a report that is as fully detailed as the previous laboratory activity reports and it must contain the names of all the student contributors, the nature of their contribution, and any formal musical training they may have had.

The grade for the project presentation will be computed by allotting 50% to the average grade determined by each member of the class casting a secret ballot scoring each presentation (other than their own) on a scale of zero to ten. The other 50% will be based on the written documentation presented. If it is a group presentation, each student in the group will receive the same grade for the project. The project must be ready to be presented during the time allotted for this laboratory activity. Preparations may be done during "open hours" in the lab.

The instructor must be informed of any special audio-visual equipment needed for the presentation at least a few days in advance.

m/pod-product-compliance
K Ltd.
IK11 3LW, UK

.5B/98

9 780757 56228